T0266302

Cryptography

RIVER PUBLISHERS SERIES IN SECURITY AND DIGITAL FORENSICS
INFORMATION SCIENCE AND TECHNOLOGY

Series Editors

K. C. CHEN
National Taiwan University
Taipei, Taiwan

SANDEEP SHUKLA
Virginia Tech, USA
and
Indian Institute of Technology Kanpur, India

Indexing: All books published in this series are submitted to Thomson Reuters Book Citation Index (BkCI), CrossRef and to Google Scholar.

The "River Publishers Series in Information Science and Technology" covers research which ushers the 21st Century into an Internet and multimedia era. Multimedia means the theory and application of filtering, coding, estimating, analyzing, detecting and recognizing, synthesizing, classifying, recording, and reproducing signals by digital and/or analog devices or techniques, while the scope of "signal" includes audio, video, speech, image, musical, multimedia, data/content, geophysical, sonar/radar, bio/medical, sensation, etc. Networking suggests transportation of such multimedia contents among nodes in communication and/or computer networks, to facilitate the ultimate Internet.

Theory, technologies, protocols and standards, applications/services, practice and implementation of wired/wireless networking are all within the scope of this series. Based on network and communication science, we further extend the scope for 21st Century life through the knowledge in robotics, machine learning, embedded systems, cognitive science, pattern recognition, quantum/biological/molecular computation and information processing, biology, ecology, social science and economics, user behaviors and interface, and applications to health and society advance.

Books published in the series include research monographs, edited volumes, handbooks and textbooks. The books provide professionals, researchers, educators, and advanced students in the field with an invaluable insight into the latest research and developments.

Topics covered in the series include, but are by no means restricted to the following:

- Communication/Computer Networking Technologies and Applications
- Queuing Theory
- Optimization
- Operation Research
- Stochastic Processes
- Information Theory
- Multimedia/Speech/Video Processing
- Computation and Information Processing
- Machine Intelligence
- Cognitive Science and Brian Science
- Embedded Systems
- Computer Architectures
- Reconfigurable Computing
- Cyber Security

For a list of other books in this series, www.riverpublishers.com

Cryptography

William J. Buchanan, OBE

Edinburgh Napier University, UK

River Publishers

Routledge
Taylor & Francis Group
LONDON AND NEW YORK

Published 2017 by River Publishers
River Publishers
Alsbjergvej 10, 9260 Gistrup, Denmark
www.riverpublishers.com

Distributed exclusively by Routledge
4 Park Square, Milton Park, Abingdon, Oxon OX14 4RN
605 Third Avenue, New York, NY 10017, USA

Cryptography / by William J. Buchanan.

Routledge is an imprint of the Taylor & Francis Group, an informa business

ISBN 978-87-93379-10-7 (print)

While every effort is made to provide dependable information, the publisher, authors, and editors cannot be held responsible for any errors or omissions.

Contents

Acknowledgement

I wish to thank my university for their continued support for our work, and to all the students who have prompted me to think in ways that I would have never thought possible.

I would especially like to also thank my family – Julie, Billy, Jamie, David, Sandra, Rebecca, Lynn, and my Mum – for their continued support and in listening to tales of Bob and Alice, and to my grandson – Conor – for making an already bright world, a whole lot brighter.

List of Figures

List of Tables

Introduction

The Internet we have created is often untrustworthy and full of risks. This is partly due to the lack of security built into the services and protocols that we use. In the 21st Century, for example, we often still cannot tell if the email we have received is actually from the person who says it is from, and that no-one has read our email on the way. In fact, we can't even tell if the email has been tampered with. The next generation of the Internet and its services must be built in a trustworthy way, and it is cryptography which provides the core techniques for us to keep things secret, to identify things, and to validate trustworthiness. Unfortunately, there is no one technique which will provide all of these things, and we often have to intertwine methods together in order to create trustworthy systems.

Thus we will see that we often use secret key encryption to protect the contents of a message, and public key encryption to prove the identity of the sender and the receiver. We then use hashing methods to provide the validity of messages, and message signatures to prove the identity of the sender and the contents of messages. Unfortunately, we are a long way of building a world where we use digital signatures in the same way that we sign our name, but with the ever increasing impact of data hacks and cyber crime, there can only be one way forward.

The book has been designed so that the web based material integrates exactly with the key topics covered, and should provide a way for the reader to investigate the methods. Along with this there are many online activities which investigate key areas, and where practical experience re-enforces a strong coverage of the principles of cryptography.

Within the book there are 12 chapters, and which each cover important areas, but each should re-enforce each other, and it is especially important in cryptography to plug gaps in knowledge. As much as possible the key fundamentals are covered, but, where possible, specific methods are outlined. The reader should thus aim to read each chapter carefully, and try and lay down a foundation which allows them to critically appraise current systems, but also be ready to design systems which use cryptography as a core element.

For many, cryptography opens up fundamental questions about the rights of our citizens to privacy and on how our societies protect themselves. The book thus aims to better inform those involved, and for us to understand the risks that citizens and our society face on the usage of cryptography. We need to increasingly design systems which put our citizens at the core of the design, and where their rights to privacy and security will become a key driver within the Information Age.

The book uses a range of programming languages, such as with C# and Python, as it is important to understand how the methods are coded in a range of programming languages and environments. The associated web site (http://asecuritysite.com/encryption) thus has a range of coding methods using C++, Java, C# and Python, and it is thus highly recommended that readers download the associated code and setup their own evaluation environment.

If you have any questions about the implementation of the code or any other cryptography-related question, please me email at w.buchanan@napier. ac.uk. The main Web site links can be found at:

http://asecuritysite.com/cryptobook

1

Ciphers and Fundamentals

1.1 Introduction

The future of the Internet, especially in expanding the range of applications, involves a much deeper understanding of privacy, integrity checking and authentication. Without this the Internet cannot properly expand and be trusted in its provision of services. One of the best ways to preserve privacy, check integrity and prove identity is data encryption, and which is known as the science of cryptographics.

Within encryption we often define the concept of Bob and Alice, who are involved with the communications, and Eve, who could listen or even modify their communications, or who could even pretend to be them. Bob and Alice thus communicate over a communication channel and which Eve is likely to have access to. In a secure environment Bob and Alice should be able to communicate freely, and identify themselves to each other, without Eve ever being able to reveal any of the messages involved, or being able to pretend to be them (Figure 1.1). The process typically involves taking some **plaintext**, and then converting it into **ciphertext**, which Eve should not be able to interpret, and then to convert it back into plaintext. Normally the conversion of plaintext to ciphertext is known as encryption, and the reverse is known as decryption.

In order to keep things secret, the two main methods that Bob and Alice can use are:

- A unique algorithm. This is an algorithm that both Bob and Alice know, but do not tell Eve. The algorithm for encoding and decoding is thus kept secret.
- Use a well-known algorithm. In this method Eve knows the algorithm, but where Bob and Alice use a special electronic key to uniquely define how the message is converted into cipertext, and then back again.

3

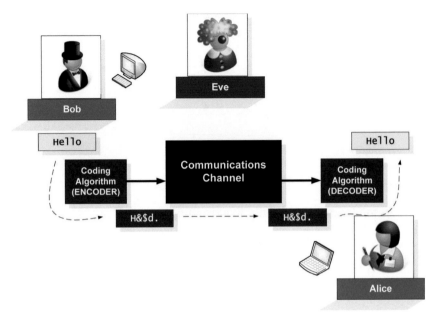

Figure 1.1 Bob, Alice and Eve.

A particular problem in any type of encryption is the passing of the information to define the secret, such as for the algorithm to be used or for an electronic key, as Eve may be listening to their communications.

This chapter looks at some of the basic principles of encryption, with the following chapters investigating the usage of secret-key (symmetric encryption) and public-key (asymmetric) methods. In secret-key encryption, we use a single electronic key to encrypt the plaintext, and the same key is then used to decrypt (normally involving a reversing of the encryption process). For public-key methods, we generate two electronic keys, and of which one is used encrypt the plaintext, and the other is used to decrypt it back to plaintext.

The concept of secret key encryption can be likened to Bob and Alice using a lockable box, of which only they have the key. Unfortunately, neither Bob nor Alice knows if Eve has taken a copy of their key. With public key encryption, Bob can create a number of identical padlocks, of which only he has the key to open them. Then if Alice wants to send him something, she will use one of his padlocks, and lock the box. Eve will thus not be able to open the box, as she will not have the required key. Bob must, obviously, keep the key to the padlock safe, so that Eve can't get access to it. The padlock can then be defined as his public key, and the other key as his private key.

As we will find, public and secret key methods often work together in perfect harmony, where secret key methods provide the actual core encryption, and public key methods provides ways to authenticate identities, and to pass encryption keys.

1.2 Simple Cipher Methods

One method of converting a message into cipher text is for Bob and Alice to agree on an algorithm which Bob will use to scramble his message, and then for Alice to do the opposite in order to unscramble the scrambled message. Thus, as long as Eve does not know the scrambling method, the cipher text will be secure. For example if Bob and Alice are sitting in a room where Eve is present, and then Bob taps on the table with a series of short (di) or long taps (dit), Bob can then pass a secret message to Alice, as long as they have agreed on what the codes identify. He might thus tap *di-di-di-dit*, and then pauses and taps *di-dit*, and where he passes the message in a standard Morse Code alphabet. In this way Alice decodes the message as "hi". Eve might eventually see that Bob is passing a message to Alice, but needs to know the type of encoding that they are using. In this way Bob and Alice agree on their method before they encode their messages, but where Eve may have heard them discussing the method that will be used. Eve, could also analyse their transmissions and then determine the codes by looking at common patterns.

With cipher methods we can use a mono-alphabetic code, where we create a single mapping from our alphabet to a cipher alphabet. This type of alphabet coding remains constant, whereas a polyalphabet can change its mapping depending on a variable keyword.

1.2.1 Morse Code

In a time when it was only possible to send sound pulses through a communications channel, Samuel F. B. Morse created a code mapping which sent pulses of electric current along wires with a silence in-between. Morse code is thus an encoding method, rather than a cipher, and works by translating characters into sequences of dots (.) and dashes (-). When transmitted as a sound pattern the dash lasts around three times longer than a dot, and with a longer delay between words as there is between letters.

The code was designed so that each of the characters varies in length approximately with the occurance of the letter in common English (Figure 1.2). For example there is a short code for an 'e' (dot), and a longer one for a less common letter, such as 'j' (dot dash dash dash). For many years Morse code was used by radio operators, and provided the standard sequence of a ship in distress: Dot Dot Dot ...Dash Dash Dash ...Dot Dot Dot (or SOS).

📖 **Web link (Morse code):** http://asecuritysite.com/coding/morse

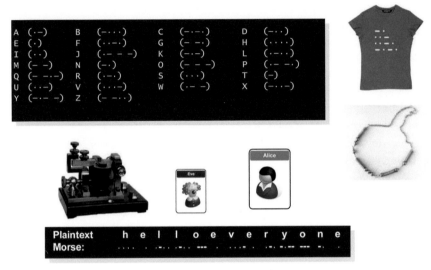

Figure 1.2   Morse code.

Web link (Morse code): http://asecuritysite.com/encryption/morse

As an extension, the Fractionated Morse Cipher uses a 26 character key mapping and converts a plaintext input to Morse code. It then converts this into fixed-length chunks of Morse code, which are then converted into ciphertext letters. In converting the plaintext to Morse code, it uses 'x's between characters and 'xx' between words. For example, "Hello World" is Morse Code is:

```
.... . .-.. .-.. --- /        .-- --- .-. .-.. -..
H   E   L   L   O  SPACE    W   O   R   L   D
```

We can then make this into a string with an 'x' between characters:

```
Plain text:  H   e  l   l   o   w   o   r   l   d
Morse string: ....x.x.-..x.-..x---xx.--x---x.-.x.-..x-..
```

We can now use three-character mappings to convert these back to text:

```
['...,'.x.','x.-','..x',.-.','.x-','---x','x.-','-x-',
'---x','.-.','x.-','..x,'-..']
```

and where the mapping are:

```
A B C D E F G H I J K L M N O P Q R S T U V W X Y Z
. . . . . . . . . - - - - - - - - - x x x x x x x x
. . . - - - x x x . . . - - - x x x . . . - - - x x
. - x . - x . - x . - x . - x . - x . - x . - x . -
```

We can then use this mapping (such as A is defined as '...', B as '..-' and C as '..x'. Next we can convert them back with:

```
AGTCDHOTQODTCJ
```

For "Peter piper picked" we get:

```
.--.x.x-x.x.-.xx.--.x..x.--.x.x.-.xx.--.x..x-.-.x-.-x.x-..xx
P    e t e r ' ' p    i   p e r ' ' p    i   c    k   e   d '
```

📖 **Web link (Fractionated Morse code):** http://asecuritysite.com/encryption/frac

1.2.2 Pigpen

Within ciphers, it is useful if Bob and Alice can create a cipher mapping that is easy to remember. One of the best methods is to use a graphical method, as the human eye often finds it easier to map graphical characters than to map alphabetic ones. The Pigpen cipher is a good example of this and uses a mono-alphabet substitution method.

For the Pigpen cipher, we initially created four grids in a square and a diagonal shape, with a dot placed in the second grid version (Figure 1.3). Next the alphabet characters are laid-out in sequence within the grids. Figure 1.4 outlines the mapping of the plaintext string of "biometric".

The problem with Pigpen is that once the mapping is known, it is difficult to keep the message secret. Bob could, though, embed it into a valid looking graphic, and send it to Alice. Eve, then, might not be able to see the embedded Pigpen symbols, but where Alice knows where to look.

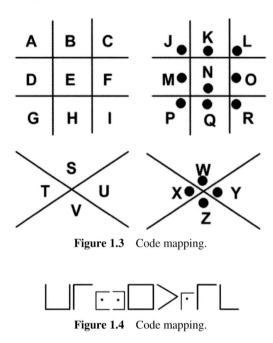

Figure 1.3 Code mapping.

Figure 1.4 Code mapping.

📖 **Web link (Pigpen code):** http://asecuritysite.com/challenges/pigpen

1.2.3 Rail Code

A useful method of hiding the cipher method is to scramble the plaintext letters in some way, and where it is not possible for the human eye to spot a pattern. Someone who knows the method will then be able to quickly decode. One method which scrambles in a defined pattern is the rail fence cipher. With this the message is written in a sequence across a number of rails. For example, if we use three rails, with a message of 'WE ARE DISCOVERED. FLEE AT ONCE', we get:

```
W . . . E . . . C . . . R . . . L . . . T . . . E
. E . R . D . S . O . E . E . F . E . A . O . C .
. . A . . . I . . . V . . . D . . . E . . . N . .
```

and where we then read across the rails to give a cipher code of "WECRL TEERD SOEEF EAOCA IVDEN". When we reverse of the process, we count the number of characters in the cipher, and map out with an 'X' for a position on the rail. The cipher is then written in sequence across the rails. So, for example, if we have cipher text of "AALHP", we write out for five missing characters:

```
X . . . X
. X . X .
. . X . .
```

and next we layout across each row:

```
A . . . A
. L . H .
. . P . .
```

which we can then read as "alpha".

📖 **Web link (Rail fence):** http://asecuritysite.com/challenges/rail

1.2.4 BIFID Cipher

The BIFID cipher uses a grid and was invented by Felix Delastelle in 1901. In its simplest form it creates a grid and which maps the letters into numeric values. In creating the grid, we scramble the alphabetic characters, such as:

```
  1 2 3 4 5
1 B G W K Z
2 Q P N D S
3 I O A X E
4 F C L U M
5 T H Y V R
```

Next we look up the grid, and then arrange the two-character value into two rows. For example is we have a plaintext of "maryland", then "m" is "4" and "5", so we place "4" in the first row, and "5" in the second row, and continue to do this for all the letters:

```
maryland
43554322
53533334
```

Next we read along the rows and merge, to give:

```
43 55 43 22 53 53 33 34
```

And finally we convert them back to letters from the grid:

```
L R L P Y Y A X
```

Let's try the reverse, with DXETE, and when looking at the grid we get:

24 34 35 51 35

We can then put then into rows to give:

2 4 3 4 3
5 5 1 3 5

This gives us 25 (s) 45 (m), 31 (i), 43 (l) and 35 (e) – which is "smile".

📖 **Web link (Bifid cipher):** http://asecuritysite.com/coding/Bifid

We can make the grids more complex, such as with the four-square cipher. This method uses four 5 × 5 matrices arranged in a square, are where each matrix contains 25 letters. The upper-left and lower-right matrices are the "plaintext squares" and each contains a standard alphabet. The upper-right and lower-left squares are the "ciphertext squares" and have a mixture of characters.

First we break the message into bigrams, such as with "ATTACK AT DAWN" which gives:

```
AT TA CK AT DA WN
```

We now use the four squares and locate the bigram to cipher in the plain alphabet squares. With 'AT', we take the first letter from the top left square, the second letter from the bottom right square:

```
a b c d e    Z G P T F
f g h i k    O I H M U
l m n o p    W D R C N
q r s t u    Y K E Q A
v w x y z    X V S B L
```

```
M F N B D     a b c d e
C R H S A     f g h i k
X Y O G V     l m n o p
I T U E W     q r s t u
L Q Z K P     v w x y z
```

Now, we determine the characters in the ciphertext around the corners of the rectangle for 'AT':

```
AT TA CK AT DA WN
IT

a b c d e     Z G P T F
f g h i k     O I H M U
l m n o p     W D R C N
q r s t u     Y K E Q A
v w x y z     X V S B L

M F N B D     a b c d e
C R H S A     f g h i k
X Y O G V     l m n o p
I T U E W     q r s t u
L Q Z K P     v w x y z
```

And so we pick off 'TI'. The result becomes:

```
ATTACKATDAWN
TIYBFHTIZBSY
```

📖 **Web link (Four square cipher):** http://asecuritysite.com/challenges/four

1.2.5 Playfair

The Playfair cipher was created by Charles Wheatstone, but was made famous by Lord Playfair. Initally a grid is created with a secret phrase, such as:

```
napierrun
```

Next we write out the 5 × 5 matrix, but do not repeat characters (and get rid of 'J'):

```
N  A  P  I  E
R  U  B  C  D
F  G  H  K  L
M  O  Q  S  T
V  W  X  Y  Z
```

If we use the phrase of "GREATS", we split into sequences of two character sets:

```
GR EA TS
```

The rules are then:

[1] If they are in different columns, take from the rectangle defined between them and pick off the opposite ends.

[2] If they are in the same column, select the letter one below (and wrap-round if necessary).

[3] If they are in the same row, select the letter one along (and wrap-round if necessary).

The cipher is then created with:

- 'G','R' are bounded by 'FU' (Rule 1).
- 'E', 'A' are in the same row so we select one letter along (Rule 3) to give 'NP'.
- 'T','S' are in the same row so we select one letter along (Rule 3) to give 'MT'.

The cipher is thus "FUNPMY" (Figure 1.5).

📖 **Web link (Playfair):** http://asecuritysite.com/coding/playfair

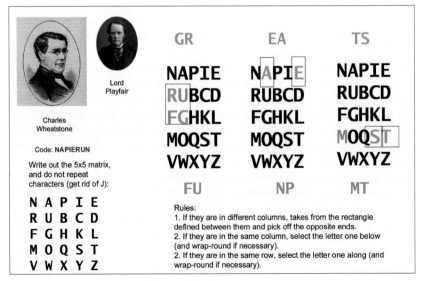

Figure 1.5   Code mapping.

1.2.6 Homophonic Substitution Code

Cipher codes can often be analysed using the probability of the letters/symbols in the ciphertext. A homophonic substitution code aims to overcomes this problem, as it varies the number of codes assigned to each character, and relates this to the probability of the characters. For example the character 'e' might have 12 codes assigned to it, but 'z' would only have one. An example code is given in Table 1.1. The homophonic substitution code is monoalphabet, even though it as uses one translation for the code mappings, as several codes can be used for a single plaintext letter.

With this, each of the codes is randomly assigned, with the number of codes assigned relating to the probability of their occurrence. Thus, using the code table in Table 1.1, the code mapping for "helloeveryone" would be:

```
Plaintext    h   e   l   l   o   e   v   e   r   y   o   n   e
Ciphertext:  19  25  42  81  16  26  22  28  04  55  30  00  32
```

In this case there are four occurrences of the letter 'e', and each one has a different code. As the number of codes depends on the number of occurrences of the letter, where each code will roughly have the same probability. It will thus be difficult to determine the code mapping from the probabilities of codes. Unfortunately the code is not perfect as the English language still contains certain relationships which can be traced. For example the letter 'q' would be represented by a single code, but there is a high probability that the next character will be a 'u'. Thus, using Table 1.1, there would only be three codes which would follow the value for a 'q'. If a given ciphertext contains a code followed by one of three codes, then it is likely that the plaintext is a 'q' and a 'u'.

📖 **Web link (Homophonic cipher):** http://asecuritysite.com/coding/ho

Table 1.1 Example homophonic substitution

a	b	c	d	e	f	g	h	i	j	k	l	m	n	o	p	q	r	s	t	u	v	w	x	y	z
07	11	17	10	25	08	44	19	02	18	41	42	40	00	16	01	15	04	06	05	13	22	45	12	55	47
31	64	33	27	26	09	83	20	03			81	52	43	30	62		24	34	23	14		46			93
50		49	51	28			21	29			86		80	61			39	56	35	36					
63		76	32				54	53			95		88	65			58	57	37						
66			48				70	68					89	91			71	59	38						
77			67				87	73					94				00	90	60						
84			69										96						74						
			72																78						
			75																92						
			79																						
			82																						
			85																						

1.2.7 Caesar Coding and Scrambled Alphabet

The problem we have with encoding methods is that they can often be cracked with a simple lookup between the plaintext value and the equivalent cipher code. To make it more difficult we need to create a cipher which has a shared secret, and which only Bob and Alice know. The cipher text then changes with respect to this secret. An example of this is with the Caesar cipher, and which was created by Julius Caesar (using a 3-letter shift). Bob and Alice then simply agree to the number of shifts that the alphabet needs to be moved by.

In the example in Figure 1.6 the letters for the cipher have been moved forward by two positions, where a 'c' becomes an 'A'. Thus 'the' will be coded as 'RFC'. There are, though, several problems with this type of coding. The main one is that it is not secure as there are only 25 unique codings, and will thus be fairly easy for someone to find the mapping.

An improvement is to scramble the mapping using a code mapping (Figure 1.7), and where a random mapping is used to determine the cipher mappings. For the first character to be mapped ('a'), we would have 26 possible mapping. If we then move to the next character mapping ('b') we would have 25 remaining possible mappings. We can then continue on and would end up with:

26! mappings which gives approximately 4.03×10^{26} mappings

As we now have many more possible mappings, the cipher becomes more secure, as it is likely that Eve will have to search through many mappings until she finds the right one. This type of approach is know as a **brute force** method, as Eve tries all the possible code mappings, until she finds a solution. The worst attempt will see her search through all the possible mappings, but she might also find it on her first attempt. On average, though, she will search through half the possible mappings to find the right solution.

To work out how long she will take, we can assume that, on average, she will search through half of the code mappings. So if she takes one second to check each mapping, the time taken, on average, will be:

$T_{average} = (4.03 \times 10^{26})/2$ seconds

which is around 6.4×10^{18} years (over 6,400,000,000,000,000,000 years).

The scrambled alphabet code thus looks secure from a brute force viewpoint. Unfortunately it can be cracked fairly quickly by using frequency analysis. For the code in Figure 1.7, we can see 'A' in the cipher appears most often, and since 'e' is the most popular English letter, it is likely that it maps to a plaintext 'e'. Next we can see that 'Q' appears four times, thus it is most likely to be mapped to a 't', which is the next most probable letter in the English alphabet.

A more formal analysis of the probabilities is given in Table 1.2 and where we can see that the letter 'e' is the most probable, followed by 't', and then 'o', and so on. Along with analysing single letter occurrences, it is also possible to look at two-letter occurrences (*digrams*), or even three-letter occurrences (*trigrams*). We could also analyse the occurrences of words (which are separated by spaces), and where 'the' is the most common word.

A scrambled alphabet cipher scheme is easy to implement, but, unfortunately, once it has been 'cracked', it is easy to decrypt the ciphered data. Normally, to improve the cipher process, the cipher has extra parameters which change the mapping. This might include changing the mapping over time, such as for the time-of-day or the date. In this way, Bob and Alice would know the mappings of the code for a given time and/or date, such as having different mapping for each day of the week. The Enigma machine was used in the War where operators re-configured the machines every day with a code book (or key sheet). Each key sheet contained the daily Enigma settings over the period of a month, and where the machine was reconfigured each day.

Figure 1.6   Caesar code.

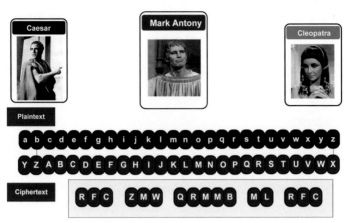

Figure 1.7   Code mapping.

Table 1.2   Probability of occurrences

Letters (%)		Digrams (%)		Trigrams (%)		Words (%)	
E	13.05	TH	3.16	THE	4.72	THE	6.42
T	9.02	IN	1.54	ING	1.42	OF	4.02
O	8.21	ER	1.33	AND	1.13	AND	3.15
A	7.81	RE	1.30	ION	1.00	TO	2.36
N	7.28	AN	1.08	ENT	0.98	A	2.09
I	6.77	HE	1.08	FOR	0.76	IN	1.77
R	6.64	AR	1.02	TIO	0.75	THAT	1.25
S	6.46	EN	1.02	ERE	0.69	IS	1.03
H	5.85	TI	1.02	HER	0.68	I	0.94
D	4.11	TE	0.98	ATE	0.66	IT	0.93
L	3.60	AT	0.88	VER	0.63	FOR	0.77
C	2.93	ON	0.84	TER	0.62	AS	0.76
F	2.88	HA	0.84	THA	0.62	WITH	0.76
U	2.77	OU	0.72	ATI	0.59	WAS	0.72
M	2.62	IT	0.71	HAT	0.55	HIS	0.71
P	2.15	ES	0.69	ERS	0.54	HE	0.71
Y	1.51	ST	0.68	HIS	0.52	BE	0.63
W	1.49	OR	0.68	RES	0.50	NOT	0.61
G	1.39	NT	0.67	ILL	0.47	BY	0.57
B	1.28	HI	0.66	ARE	0.46	BUT	0.56
V	1.00	EA	0.64	CON	0.45	HAVE	0.55
K	0.42	VE	0.64	NCE	0.43	YOU	0.55
X	0.30	CO	0.59	ALL	0.44	WHICH	0.53
J	0.23	DE	0.55	EVE	0.44	ARE	0.50
Q	0.14	RA	0.55	ITH	0.44	ON	0.47
Z	0.09	RO	0.55	TED	0.44	OR	0.45

Refer to the following pages:

📖 **Web link (Caeser code):** http://asecuritysite.com/coding/caeser
   Web link (Scrambled code): http://asecuritysite.com/coding/scramble
   Web link (Scrambled code challenge): http://asecuritysite.com/
challenges/scramb

1.2.8 Vigenère Cipher

An improved code over the scrambled alphabet approach was developed by
Vigenère, where a different mapping, based on a keyword, is used for each
character of the cipher. This is known as a *polyalphabetic* cipher as it uses
a number of cipher alphabets. The way that the cipher mapping changes
is agreed by Bob and Alice. One of the most popular methods is to use a
code word which they agree on, and then move the mapping based on the
characters in the keyword.

For example, if we use the mapping of Table 1.3, and if the code word
is "GREEN", then the rows used are: Row 6 (G), Row 17 (R), Row 4 (E),
Row 4 (E), Row 13 (N), Row 6 (G) and Row 17 (R). The message of
"hellohowareyou" is thus converted as:

Keyword	GREENGREENGREE
Plaintext	hellohowareyou
Ciphertext	NVPPBNFAEEKPSY

The great advantage of this type of cipher is that the same plaintext character
is likely to be coded to different mappings, depending on the position of the
keyword. For example, for a keyword of GREEN, 'e' can be coded as 'K' (for
G), 'V' (for R), 'I' (for E) and 'R' (for N). The method, though, was cracked
by Major Friedrich Wilhelm Kasiski, a German infantry officer. He was the
first to propose a method of attacking polyalphabetic substitution ciphers,
and, in 1863, published a 95-page book on cryptography:

Die Geheimschriften und die Dechiffrir-Kunst "Secret writing and the Art
of Deciphering"

Its main focus was on the Vigenère cipher and where he developed a method
known as Kasiski examination. In it he analysed the gaps between repeated
ciphertext fragments, so that he could gain a hint on the key length. In this,
we take the cipher message and analyse for repeated patterns, and which

gives a hint towards the key size. For example, if we have a message of "theywillnotkeeptheburningdeck" and then with a key of "abc", we get:

```
theywillnotkeeptheburningdeck
abcabcabcabcabcabcabcabcabcab
TIGYXKLMPOUMEFRTIGBVTNJPGEGCL
```

We can see that the "the" word has aligned to the key:

```
the   ywillnotkeep   the   burningdeck
abc   abcabcabcabc   abc   abcabcabcab
TIG   YXKLMPOUMEFR   TIG   BVTNJPGEGCL
```

So we could reason that we might have a key size of three. Normally, though, we need a considerable amount of cipher text to accurately guess the key size. We can then use a frequency analysis method to get a shortlist for the possible key values.

📖 **Web link (Kasiski analysis):** http://asecuritysite.com/encryption/kasiski

Web link (Vigenère analysis): http://asecuritysite.com/encryption/vig_crack

To improve security, the greater the size of the code word, the more the rows that can be included in the cipher process. It is also safe from analysis of common two- and three-letter occurrences, if the keysize is relatively long. For example 'ee' could be encrypted with 'KV' (for GR), 'VI' (for RE), 'II' (for EE), 'IR' (for EN) and 'RK' (for NG).

1.2.9 One-Time Pad (OTP)

The problem with the ciphers previously defined is that once Eve knows the method, she can normally crack all the codes created. Also if Bob and Alice use a keyword, Eve could try lots of different keywords to see if one works. In this way most ciphers can be broken, and where it is just a matter of time before it is cracked. If the time relevance of the message is greater than the average time to crack, the provenance of message can be preserved. For example if an army sends a message of "ATTACK" to their troops, the message just has been be secret until the time that they attack. After they have attacked then it would not matter if their cipher was cracked.

If we want an uncrackable cipher, we must use a one-time pad, and which is a cipher code mapping that is used only once. The one-time mapping is

Table 1.3 Coding

Row	a	b	c	d	e	f	g	h	i	j	k	l	m	n	o	p	q	r	s	t	u	v	w	x	y	z
1	B	C	D	E	F	G	H	I	J	K	L	M	N	O	P	Q	R	S	T	U	V	W	X	Y	Z	A
2	C	D	E	F	G	H	I	J	K	L	M	N	O	P	Q	R	S	T	U	V	W	X	Y	Z	A	B
3	D	E	F	G	H	I	J	K	L	M	N	O	P	Q	R	S	T	U	V	W	X	Y	Z	A	B	C
4	E	F	G	H	I	J	K	L	M	N	O	P	Q	R	S	T	U	V	W	X	Y	Z	A	B	C	D
5	F	G	H	I	J	K	L	M	N	O	P	Q	R	S	T	U	V	W	X	Y	Z	A	B	C	D	E
6	G	H	I	J	K	L	M	N	O	P	Q	R	S	T	U	V	W	X	Y	Z	A	B	C	D	E	F
7	H	I	J	K	L	M	N	O	P	Q	R	S	T	U	V	W	X	Y	Z	A	B	C	D	E	F	G
8	I	J	K	L	M	N	O	P	Q	R	S	T	U	V	W	X	Y	Z	A	B	C	D	E	F	G	H
9	J	K	L	M	N	O	P	Q	R	S	T	U	V	W	X	Y	Z	A	B	C	D	E	F	G	H	I
10	K	L	M	N	O	P	Q	R	S	T	U	V	W	X	Y	Z	A	B	C	D	E	F	G	H	I	J
11	L	M	N	O	P	Q	R	S	T	U	V	W	X	Y	Z	A	B	C	D	E	F	G	H	I	J	K
12	M	N	O	P	Q	R	S	T	U	V	W	X	Y	Z	A	B	C	D	E	F	G	H	I	J	K	L
13	N	O	P	Q	R	S	T	U	V	W	X	Y	Z	A	B	C	D	E	F	G	H	I	J	K	L	M
14	O	P	Q	R	S	T	U	V	W	X	Y	Z	A	B	C	D	E	F	G	H	I	J	K	L	M	N
15	P	Q	R	S	T	U	V	W	X	Y	Z	A	B	C	D	E	F	G	H	I	J	K	L	M	N	O
16	Q	R	S	T	U	V	W	X	Y	Z	A	B	C	D	E	F	G	H	I	J	K	L	M	N	O	P
17	R	S	T	U	V	W	X	Y	Z	A	B	C	D	E	F	G	H	I	J	K	L	M	N	O	P	Q
18	S	T	U	V	W	X	Y	Z	A	B	C	D	E	F	G	H	I	J	K	L	M	N	O	P	Q	R
19	T	U	V	W	X	Y	Z	A	B	C	D	E	F	G	H	I	J	K	L	M	N	O	P	Q	R	S
20	U	V	W	X	Y	Z	A	B	C	D	E	F	G	H	I	J	K	L	M	N	O	P	Q	R	S	T
21	V	W	X	Y	Z	A	B	C	D	E	F	G	H	I	J	K	L	M	N	O	P	Q	R	S	T	U
22	W	X	Y	Z	A	B	C	D	E	F	G	H	I	J	K	L	M	N	O	P	Q	R	S	T	U	V
23	X	Y	Z	A	B	C	D	E	F	G	H	I	J	K	L	M	N	O	P	Q	R	S	T	U	V	W
24	Y	Z	A	B	C	D	E	F	G	H	I	J	K	L	M	N	O	P	Q	R	S	T	U	V	W	X
25	Z	A	B	C	D	E	F	G	H	I	J	K	L	M	N	O	P	Q	R	S	T	U	V	W	X	Y

then shared between Bob and Alice, and is used once to send a message, and where another pad is created for another message (Figure 1.8).

For example, we can first create a code book, which only Bob and Alice know:

```
yehq  medlg  yaqif  xygfs  vlznx
llyyk ikbsy  tvoon  nvtuq  qzvvn
ucyio nftsj  bffbx  ozxkl  ckrsf
asfxg mqdlp  gltek  obvfm  hqrxc
rbljl jlgcn  vzwlw  kctlq  cftzx
bpmgy kaiup  lftaf  ufqrp  ofjib
fwfgz lilmk  uzaed  urbwl  eitgw
xpbji wfees  oubvd  dthpk  vfmnv
wdnww xczkb  wgcdo  pvvlp  zpfti
ladva scool  sshhv  lvrtg  wrebv
```

In this case there are 25 characters on each line of the one-time pad, and we thus go from [0] through to [24] on the first row, then from [25] to [49] on the second row, and so on. Next if Bob wants to send a message he will select a key based on the positions of the letters:

[5][92][4][232][203][70][225][195]

If we look up the positions for this, the key becomes:

```
mvvowclv
```

Next we take our secret word, such as "newhampshire", and shift each letter depending on the position of our key. In this case we translate a "n" to row "m" and get a "z" (which is 12 character shifts ... n[opqrstuvwxy]z:).

```
    a b c d e f g h i j k l m n o p q r s t u v w x y z
. . .
10 k l m n o p q r s t u v w x y z a b c d e f g h i j
11 l m n o p q r s t u v w x y z a b c d e f g h i j k
12 m n o p q r s t u v w x y z a b c d e f g h i j k l
```

The cipher then becomes: "zzrvwoantdms", which Bob will send to Alice, and Alice does the reverse of Bob's operation, based on the shared secret key. Unfortunately the OTP cipher suffers from having to regenerate the pad each time, or, at least, to regenerate a new key.

📖 **Web link (OTP):** http://asecuritysite.com/coding/otp

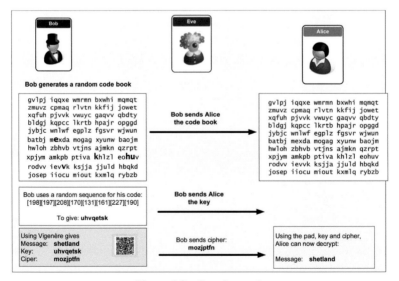

Figure 1.8   One-time pad.

1.3 Encoding Methods

On a computer system, code and data are represented as binary, but humans find it difficult to deal with binary formats, so other formats are used to represent binary values. Two typical formats used to represent characters are ASCII and UTF-16. With ASCII we have 8-bit values and it can thus supports up to 256 different characters (2^8). UTF-16 extends the characters to 16-bit values, and thus gives a total of 65,536 characters (2^{16}). Within ASCII coding, we map printable characters, such as 'a', and 'b', to decimal, binary and hexadecimal values:

ASCII	Binary	Hex	Decimal
'e'	0110 0101	0x65	101
'E'	0100 0101	0x45	69
' '	0010 0000	0x20	32

We also have other 'non-printing' characters which typically have a certain control function. These include CR (Carriage Return), LF (Line Feed), and Horizontal Tab (HT):

ASCII	Binary	Hex	Decimal	Character representation
CR	0110 0101	0x0D	13	\r
LF	0100 0101	0x0A	10	\n
HT	0000 0111	0x07	7	\t

📖 **Web link (ASCII):** http://asecuritysite.com/coding/ascii

Within text files we are likely to have line breaks, and which are created by the CR and LF characters. In Microsoft Windows-type systems, we use CR and LF at the end of a line (\n\r), while a Linux/Mac-type system only uses CR for a new line (\r)

Normally when we encrypt into ciphertext it produces a bit stream which contains non-printing characters, and we thus need to represent the cipher in a printable way. We may also be required to represent our encryption keys in a printable and/or distributable format. For this we often use a hexadecimal or Base-64 format as these allow us to represent the cipher into a printable format (Figure 1.9).

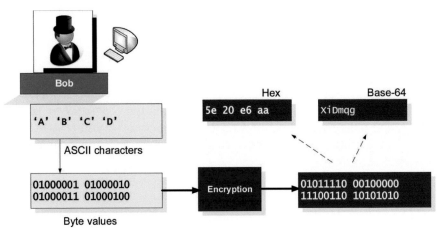

Figure 1.9   Conversion from binary into hexadecimal or Base-64.

The most common format for representing standard English characters is ASCII. In its standard form it uses a 7-bit binary code to represent characters (letters, giving a range of 0 to 127), but it is rather limited in its scope as it does not support symbols such as Greek letters. To increase the number of symbols which can be represented, extended ASCII is used which has a 16-bit code. Appendix A shows the standard ASCII character set (in binary, decimal, hexadecimal and also as a character).

Some important non-printable ASCII characters are: New line (0x13); Carriage Return (0x10); Tab (0x07); and Backspace (0x08), while a Space is represented by 0x20. The representations are for 'A' and 'B' are:

Char	Decimal	UTC-16	ASCII	Hex	Oct	HTML
A	65	00000000 01000001	01000001	41	101	A
B	66	00000000 01000010	01000010	42	102	B

📖 **Web link (ASCII table):** http://asecuritysite.com/coding/asc
   Web link (UTF-16 table): http://asecuritysite.com/coding/asc2
   Web link (ASCII conversion): http://asecuritysite.com/coding/ascii

1.3.1 Hexadecimal and Base-64

The conversation to a hexadecimal format involves splitting the bit stream into groups of four bits (Figure 1.10) and for Base-64 into groups of six bits (Figure 1.11). With a hexademical format, we have values from 0 to 15, and which are represented by four-bit values from 0000 to 1111. For Base-64, we take six bits at a time. For example, if we take an example of "fred", then we get:

```
ASCII     f          r          e          d
Binary    01100110  01110010  01100101  01100100
```

To convert to Base-64, we group in 6-bits:

```
Binary  011001  100111  001001  100101  011001  00
```

And then map these to a Base-64 table:

```
Binary    011001    100111    001001    100101    011001  00
Decimal   25        39        9         37        25      0
Base-64   Z         n         J         l         Z       A
```

The result is ZnJlZA

With Base-64, we create groups of four Base-64 characters, and we pad with zeros to fill-up the six-bit values, and then use the "=" character to pad to create groups of four Base-64 characters:

```
test -> 01110100 01100101 01110011 01110100
test -> 011101 000110 010101 110011 011101 00[0000] = =
test -> d      G      V      z      d      A        = =


help -> 01101000 01100101 01101100 01110000
help -> 011101 000110 010101 110011 011101 00[0000] = =
help -> a      G      V      s      c      A        = =
```

Unfortunately some of the characters look similar when they are printed, such as whether we have a zero ('0') or an 'O'. To avoid this we can convert to a Base-64 format, but there are similar-looking letters: 0 (zero), O (capital o), I (capital i) and l (lower case L), and non-alphanumeric characters of + (plus) and / (slash). The solution is Base-58, used in Bitcoin applications, and where we remove the characters which are similar looking.

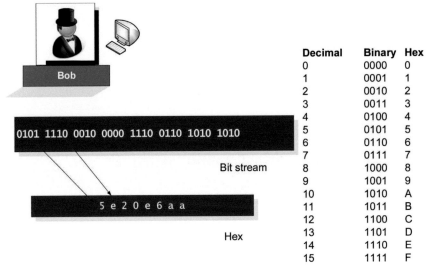

Decimal	Binary	Hex
0	0000	0
1	0001	1
2	0010	2
3	0011	3
4	0100	4
5	0101	5
6	0110	6
7	0111	7
8	1000	8
9	1001	9
10	1010	A
11	1011	B
12	1100	C
13	1101	D
14	1110	E
15	1111	F

Figure 1.10   Conversion to hex.

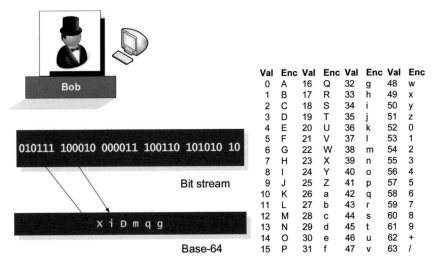

Val	Enc	Val	Enc	Val	Enc	Val	Enc
0	A	16	Q	32	g	48	w
1	B	17	R	33	h	49	x
2	C	18	S	34	i	50	y
3	D	19	T	35	j	51	z
4	E	20	U	36	k	52	0
5	F	21	V	37	l	53	1
6	G	22	W	38	m	54	2
7	H	23	X	39	n	55	3
8	I	24	Y	40	o	56	4
9	J	25	Z	41	p	57	5
10	K	26	a	42	q	58	6
11	L	27	b	43	r	59	7
12	M	28	c	44	s	60	8
13	N	29	d	45	t	61	9
14	O	30	e	46	u	62	+
15	P	31	f	47	v	63	/

Figure 1.11   Conversion to Base-64.

For Base-58, we convert the ASCII characters into binary, and then keep dividing by 58 and convert the remainder to a Base58 character. The alphabet becomes:

'123456789ABCDEFGHJKLMNPQRSTUVWXYZabcdefghijkmnopqrstuvwxyz'

It we take an example of 'e', where e have a decimal value of 101, so we divide by 58 to get:

1 remainder 43

and next we divide 1 by 58 and we get:

0 remainder 1

We then take character at position 1 and at position 43, to give:

123456789ABCDEFGHJKLMNPQRSTUVWXYZabcdefghijkmnopqrstuvwxyz

and then get:

2k

If we now take 'ef', we get 958 ($102 + 101 \times 256$), where we move each character up one byte. Basically we take the binary value of the string and then divide by 58 and take the remainder. So 'ef' is '01100101 01100110'.

📖 **Web link (Base-58 conversion):** http://asecuritysite.com/encryption/ base58

1.4 Huffman Coding and Lempel-Viz Welsh (LZW)

Along with encoding methods, we often try to compress our data by either looking at patterns within the binary digits or within the metadata contained in an object. One of the most widely used methods is Huffman Coding which uses a variable length code for each of the elements within the data. This normally involves analyzing the data to determine the probability of its elements, and where the most probable elements are coded with a few bits, and the least probable elements coded with a greater number of bits. This could be done on a character-by-character basis within text-based data, or on a byte-by-byte basis on other binary data (such as for graphics files).

The following example relates to characters. First, the textual data is scanned to determine the number of occurrences of a given letter. For example:

Letter	'b'	'c'	'e'	'i'	'o'	'p'
No. of occurrences:	12	3	57	51	33	20

Next the characters are arranged in order of their number of occurrences, such as:

'e'	'i'	'o'	'p'	'b'	'c'
57	51	33	20	12	3

After this the two least probable characters are assigned either a 0 or a 1. Figure 1.12 shows that the least probable ('c') has been assigned a 0 and the next least probable ('b') has been assigned a 1. The addition of the number of occurrences for these is then taken into the next column and the occurrence values are again arranged in descending order (that is, 57, 51, 33, 20 and 15). As with the first column, the least probable occurrence is assigned a 0 and the next least probable occurrence is assigned a 1. This continues until the last column. When complete, the Huffman-coded values are read from left to right and the bits are listed from right to left.

The final coding will be:
'e' 11
'i' 10
'o' 00
'p' 011
'b' 0101
'c' 0100

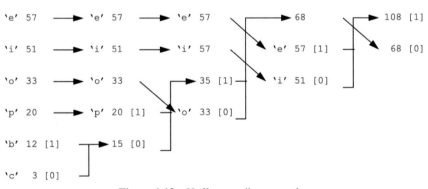

Figure 1.12   Huffman coding example.

📖 **Web link (Huffman):** http://asecuritysite.com/coding/huff

Around 1977, Abraham Lempel and Jacob Ziv developed the Lempel–Ziv class of adaptive dictionary data compression techniques (also known as LZ-77 coding), and which is now the basis of many popular compression methods. The LZ coding scheme is especially suited to data which has a high degree of repetition, and then makes back-references to these repeated parts. Typically a special flag is used to identify coded and unencoded parts, where the flag creates a back reference to the repeated sequence. An example piece of text could be:

```
'The receiver requires a receipt for it. This is automatically
sent when it is received.'
```

This text has several repeated sequences, such as 'is', 'it', 'en', 're' and 'recei'. For example, we could identify the repetitive sequence of 'recei' (as shown by the underlined highlighted text). If we use an encoded sequence for a flag sequence of *#m#n* then *m* can represents the number of characters to trace back to find the character sequence and *n* is the number of replaced characters. The encoded message would become:

```
'The receiver#9#3quires a#20#5pt for it. This is
automatically sent wh#6#2 it #30#2#47#5ved.'
```

Normally, a long sequence of text has many repeated words and phrases, such as 'and', 'there', and so on. Note that in some cases, this could lead to longer conversions if short sequences were replaced with codes that were longer than the actual sequence itself.

The Lempel–Ziv–Welsh (LZW) algorithm (also known LZ-78) extends LZ-77 by building a dictionary of frequently used groups of characters (or 8-bit binary values), and then rather than storing the actual value, a reference is added to it in a table. Then, before the conversion is decoded, we must read the dictionary. In Figure 1.13 we store the words in a table, and then refer to this in the stored data.

A typical method to then apply to the data is RLE (Run Length Encoding) which takes long sequences of a repeated value and then refers to them in the stored data. For example a sequence of number of:

6,5,5,5,5,5,5,5,5,5,5,10

could become:

6,5 [10],10

where [10] represents ten repeated values.

If we have an input phrase of:

Cows graze in groves on grass which grows in grooves in groves

Then the compressed version could become:

['C', 'o', 'w', 's', ' ', 'g', 'r', 'a', 'z', 'e', ' ', 'i', 'n', 260, 'r', 'o', 'v', 'e',
259, 'o', 268, 261, 'a', 's', 259, 'w', 'h', 'i', 'c', 'h', 269, 257, 259, 267, 286,
271, 273, 266, 276, 270, 272, 's']

The table would then contain:

Adding: [256] Co
Adding: [257] ow
Adding: [258] ws
Adding: [259] s
Adding: [260] g
Adding: [261] gr
Adding: [262] ra
Adding: [263] az
Adding: [264] ze
Adding: [265] e
Adding: [266] i

📖 **Web link (LZW):** https://asecuritysite.com/comms/lz

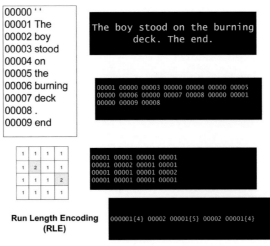

Figure 1.13   LZW and RLE.

1.5 Data Integrity (CRC-32)

Along with keeping things secret, and in proving the identity of an entity, we also need to integrate **integrity**, where we can prove that something has not been changed. A simple method of doing this is to add a checksum, in order to detect errors in the binary digits. CRC (Cyclic Redundancy Check) is one of the most reliable error detection schemes and can detect up to 95.5% of all errors. The most commonly used code is CRC-32 and provides a 32-bit CRC signature (eight hex characters), and which is normally appended onto the data. When data is read, the system checks the CRC-32 value, and if it is different from the expected value there is likely to be an error in the data.

The basic idea of a CRC can be illustrated using an example. Suppose that the Bob and Alice both agree that the numerical value that Bob sends to Alice will always be divisible by 9. Then if Alice receives a value which is not divisible by 9 she knows that the data has an error or has been modified. If the value that Bob is sending is 32, he could multiply the value by 10 to give 320, and then add a value for the least significant digit that would make it divisible by 9. In this case Bob would add 4, making a transmitted value of 324. If this transmitted value were to be corrupted in transmission, there would only be a 10% chance that an error would not be detected. When received without an error, Alice would ignore the least significant digit.

A standard test vector for CRC-32 is "The quick brown fox jumps over the lazy dog" and which generates a CRC-32 value of: 414fa339 (0100 0001 0100 1111 1010 0011 0011 1001).

 📖 **Web link (CRC-32):** http://asecuritysite.com/encryption/crc32

CRC-32 is fairly strong in detecting an error in the transmission, but cannot actually detect the bits what are in error (error correction). More complex schemes exist which can not only detect errors but correct them. One of the most popular schemes is Reed-Solomon:

Web link (CRC-32): http://asecuritysite.com/encryption/reed

1.6 Little Endian or Big Endian

Memories store data in bytes, and where each byte has a unique memory location. The order that the byte values are stored depends on the computer architecture type. Most PC systems using Intel processors use a **Little Endian** format, where the least significant byte is stored in the lowest memory address. Thus if we have an unsigned 32-bit value of 0x01020304 (16,909,060), the value is stored at:

Location (100h): 01 (Most significant byte)
Location (101h): 02
Location (102h): 03
Location (103h): 04 (Least significant byte – at the end)

Most processors now use the Little Endian format. The Big Endian format has been used in IBM z/Architecture mainframes, where the most significant byte is stored in the lowest memory address. It is also used in network packets such as with TCP and IP headers.

1.7 Introduction to Probability and Number Theory

Encryption requires a background in some basic maths principles, including for the usage of integers and in some basic operations.

1.7.1 Combinations and Permutations

Often we have calculations that involve a number of combinations or permutations. With combinations we do not care the order of the selections. For example if we have four countries: UK, France, Germany and Ireland, there are four combinations of three countries:

[{UK, France, Germany}, {UK, France, Ireland}, {France, Germany, Ireland} and {UK, Germany, Ireland}]

The formula for this is:

$$nC_k = \frac{n!}{k!(n-k)!}$$

Where ! is the factorial operator, n is the total to choose from, and k is the number of options to choose. For example 5! is $5 \times 4 \times 3 \times 2 \times 1$. So for our example we get:

$$_4C_3 = \frac{4!}{3!(4-3)!} = \frac{4 \times 3 \times 2 \times 1}{3 \times 2 \times 1} = 4$$

With permutations we look at all the options including their sequence. For example, the number of permutations for three countries with our example is:

[{UK, France, Germany}, {UK, France, Ireland}, {UK, Germany, France}, {UK, Germany, Ireland}, {UK, Ireland, France}, {UK, Ireland, Germany}, ... {France, Germany, Ireland}]

The formula for this is:

$$_nP_k = \frac{n!}{(n-k)!}$$

So for our example we get:

$$_4P_3 = \frac{4!}{(4-3)!} = \frac{4 \times 3 \times 2 \times 1}{1} = 24$$

1.7.2 Probability Theory

With probability theory we determine the likelihood of an event happening, typically by understanding the chances of how each of the elements involved in an event interact, and the likelihood of them happening. For a dice, each of the numbers of a dice are equally likely, thus the probability of us rolling a specific value of n is:

$$P(n) = \frac{1}{6} \approx 0.167$$

If we have two events (A and B) that are **independent**, the probability of both occurring is:

$$P(A \text{ and } B) = P(A).P(B)$$

If the events are **mutuality exclusive**, such as, if we toss a coin, and if it is heads, it cannot also be tails. So for mutuality exclusive events:

$$P(A \text{ and } B) = 0$$

If we want to determine the probability of one of two events (A or B) happening:

$$P(A \text{ or } B) = P(A) + P(B)$$

For example, the probability of rolling a two or a three on dice, the probability will be:

$$P(2 \text{ or } 3) = 1/6 + 1/6 = 1/3$$

If we throw twice dice, each of the die are independent, so the chances of two 1's being thrown is 1/6 multiplied by 1/6, which equals 1/36.

If the events are **dependent**, we need to understand the dependence. For example what is the probability of drawing two Aces from a pack of 52 cards? The first pick has a probability of 4/52 (1/13), but if we selected an Ace first, then there will only be 3 Aces left out of 51 cards, so the chances of drawing another Ace will be 3/51. Overall the probably of drawing two Aces will thus be:

$$P(A \text{ and } A) = \frac{1}{13}\frac{3}{51} = \frac{3}{663} = 0.0045$$

In general, for two dependent events (A and B), we have:

$$P(A \text{ and } B) = P(A) \cdot P(B|A)$$

Where P(B|A) is the probability of B happening given that A happened.

1.7.3 Set Theory

With set theory we define a range objects that make up a set. If we create two sets named Players and Spectators:

Players — {mike, fred, bert}
Spectators — {ian, michael, mike}

The main symbols that we use are:

Symbol	Symbol Name	Description
|	such that	so that
A∩B	intersection	objects belong to set A **and** set B
A∪B	union	objects belong to set A **or** set B
A⊆B	subset	subset has fewer elements or equal to the set
∈	belongs to	when an object is within a set
∉	does not belong to	when an object is not in a set

Thus A∩B — {mike} and A∪B —{mike,fred,bert,ian,michael}.
Then 'mike' ∈ Players, and 'ian' ∉ Players.

1.7.4 Number Representations

Many of the important concepts in cryptography are based on number theory around the study of integers, with a special focus on divisibility. The main classifications for numbers are: integers; rational numbers; real numbers; and complex numbers. In maths we define these as:

- Integers can be positive or negative numbers and have no fractional part. They are represented with the $\mathbb{Z}$ symbol $\{\ldots-2, -1, 0, +1, +2,\ldots\}$.
- Rational numbers are fractions ($\mathbb{Q}$).
- Real numbers ($\Re$) include both integers and rational numbers, and any other number that can be used in a comparison.
- Prime numbers ($\mathbb{P}$) represent the integers which can only be divisible by itself and unity.
- Natural numbers ($\mathbb{N}$) represent positive numbers which are integers $\{1,2\ldots\}$.

1.7.5 Logarithms

There are some methods in cryptography which base themselves on logarithms. They were discovered by John Napier, and who first proposed that we can multiply two numbers together (a, b), by finding the finding the log of a and adding it to the log of b. We can then take the inverse log to determine the result. This changed the face of calculations, where we could multiply large numbers together, just by looking up a table for the log value, and adding the results, and again apply the reverse through a look-up table. To multiply we get:

$$a \times b = \text{Inverse Log } (\text{Log } (a) + \text{Log } (b))$$

The base of the log is important for the calculation. For our decimal system we use a base of 10 ($\log_{10}(x)$ and 10^x), but for many mathematical operations we use a natural log base ($\text{Log}_e(x)$ or e^x, where e has a value of approximately 2.718). The base of e is used in many naturally occurring changes, such as within electrical circuits. The rules are thus:

$g = a.b$
$\log(g) = \log(a)+\log(b)$
$g = \text{Inverse Log } (\log(a)+\log(b))$

$g = a/b$
$\log(g) = \log(a)-\log(b)$
$g = \text{Inverse Log } (\log(a)-\log(b))$

$g = a^x$
$\log(g) = x.\log(a)$
$g = \text{Inverse Log } (x.\log(a))$

For example:

$$g = 10^3$$
$$\log_{10}(g) = 3.\log_{10}(10)$$
$$g = 10^{(3 \times 1)} = 1,000$$

1.8 Prime Numbers

A prime number is a value which only has factors of 1 and itself, and are used in areas such as key exchange and in public key encryption. Their core protection is that it is a significant challenge for a computer to factorise the result of the multiplication of two prime numbers. The simplest test for a prime number is to divide the value from all the integers from 2 to the value divided by 2. If any of the results leaves no remainder, the value is not prime, otherwise it is composite. We can obviously improve on this by getting rid of even numbers which are greater than 2, and also that the highest value to be tested is the square root of the value.

So if $n = 37$, then our maximum value will be $\sqrt{n}$, which, when rounded down is 6. So we can try: 2, 3, and 5, of which of none of these divide exactly into 37, so it is a prime number. Now let us try 55, where we will then try 2, 3, 5 and 7. In this case 5 does divide exactly into 55, so the value is not prime.

Another improvement we can make is that prime numbers (apart from 2 and 3) fit into the equation of:

$$6k \pm 1$$

where $k=0$ gives 0 and 1, $k=1$ gives 5 and 7, $k=2$ gives 11 and 13, $k=3$ gives 17 and 19, and so on. Thus we can test if we can divide by 2 and then by 3, and then check all the numbers of $6k \pm 1$ up to $\sqrt{n}$.

📖 **Web link (Prime Numbers):** http://asecuritysite.com/encryption/isprime

1.9 Encryption Operators (mod, EX-OR and shift)

It is important that the operators used in encryption do not lose any information in the encryption process, and that the operations must be reversible in some way. Along with this, the encryption process is often fairly processor-intensive, so the operators must be fairly simple in their approach in order to be fast for Bob and Alice, but which involves extensive processing for Eve. The main operators which match best to this profile are: bit rotate ($<<$ or $>>$); eXclusive-OR (**X-OR** - ⊕); and **mod** operations. These can typically be achieved in a single operation, and can thus be used for fast encryption and decryption. Along with this the rotate and X-OR functions are fairly easy to reverse.

1.9.1 Mod Operator

The **mod** operator provides the remainder of an integer divide. For example for 31 divided by 8 gives the result of 3 remainder 7. Thus 31 (**mod** 8) equals 7. Often in cryptography the mod operation uses a prime number, such as:

$$\text{Result} = \text{value}^x \; \textbf{mod} \; (\text{prime number})$$

For example, if we have a prime number of 269, and a value of 8 with an x value of 5, the result of this operation will be:

$$\text{Result} = 8^5 \, (\textbf{mod} \; 269) = 32,768 \, (\textbf{mod} \; 269) = 219$$

With prime numbers, if we know the result, it is difficult to find the value of x that has been used, even though we have the other values, as there can be many values of x that can produce the same result. It is this feature which makes it difficult to determine a secret value (in this case the secret is x). In Python, Java and C#, the mod operator is "%".

1.9.2 Shift-Operators

The bit-shift operators can either be left- or right-shift (or more precisely rotate left, or rotate right operators), where the shifting process normally takes the bits which exit from one end, and put them into the other end. This is normally defined as a rotation – where we can have a rotate left or a rotate right. An encryption process might thus operate by taking one byte at a time and rotate them left by four bit positions:

```
Input    1010 1000 1111 0000 0101 1100 0000 0001
Output   1000 1010 0000 1111 1100 0101 0001 0000
```

Thus the decryption process would merely rotate each of the bits of the bytes by four places to the right.

1.9.3 Integers and Big Integers

In computer systems we represent integers with a number of bits. Normally in cryptography we use unsigned integers in order to apply simple operations on the values. A typical integer representations are:

C# data type Representation Range

byte	byte	uses 8 bits and ranges from 0 to 255
ushort	unsigned short	uses 16 bits and ranges from 0 to 65,535
uint	unsigned int	uses 32 bits and ranges from 0 to 42,949,67,295
ulong	unsigned long	uses 64 bits and ranges from 0 to 18,446,744,073,709,551,615

Thus when we use the shift operator, the variables are automatically cast against their variable types. In C the "<<" operator shifts left, and the ">>" operator shifts right. Unfortunately in most software development languages there is no rotate operator, so the bits which move off the end are required to be pushed back onto the other end. In C a function to produce a rotate right for a variable (v) by n bits is:

```
var ror(var v,unsigned int b) {
      return (v>>n)|(v<<(8*sizeof(var)-n));
   }
```

The number of bits used to define an integer is often defined by the size of the registers which are used in the processor. In most cases the maximum size is 64 bits – and which is represented by an unsigned long value (ulong). In cryptography we often have values which are much greater than this and where our integer values can have 2,048 bits or more. Thus the normal data integer types will not support these operators, and there can be an overflow within the operations. We thus use Big Integers to perform the operations, and which store their values as string entities, and not as numeric values. This can thus support almost any number that we need to generate. When the values are operated on, the strings are converted into a numerical format, and the operations performed, and the result placed back into a string format.

A popular implement of Big Integers is the Bouncy Castle library, where, in C#, the following calculates 2 to the power of a given number (*i*):

```
BigInteger b = new BigInteger(''2'');
BigInteger c = b.Pow(i);
```

📖 **Web link (Big Integers):** http://asecuritysite.com/encryption/keys3

The values are then declared as Big Integers objects and can be displayed by converting to a string. For example, if we want to calculate:

$$A = g^x \bmod (n)$$
$$B = g^y \bmod (n)$$

$$k_1 = B^x \bmod (n)$$
$$k_2 = A^y \bmod (n)$$

we can implement the following (where *x* and *y* are random values between 0 and 90, and *g* and *n* are constant values):

```
int x = Global.random(90);
int y = Global.random(90);

BigInteger g = new BigInteger("153d5d6172adb4cb9a428cc", 16);
BigInteger n = new BigInteger("9494fec095f3b8ca98cdf3b", 16);

BigInteger A = g.Pow(x).Mod(n);
BigInteger B = g.Pow(y).Mod(n);

BigInteger k1=B.Pow(x).Mod(n);
BigInteger k2=A.Pow(y).Mod(n);

String k1value = g.ToString();
String k2value = n.ToString();
```

Web link (Example): http://asecuritysite.com/encryption/diffie2

The following defines the maximum value that can be represented for various integer bit sizes:

Int size	Number of values
16	**65,536**
32	**4,294,967,296**
48	281,474,976,710,656
64	**18,446,744,073,709,551,616**
80	1,208,925,819,614,629,174,706,176
96	79,228,162,514,264,337,593,543,950,336
112	5,192,296,858,534,827,628,530,496,329,220,096
128	340,282,366,920,938,463,463,374,607,431,768,211,456
144	22,300,745,198,530,623,141,535,718,272,648,361,505,980,416
160	1,461,501,637,330,902,918,203,684,832,716,283,019,655,932,542,97 6
176	95,780,971,304,118,053,647,396,689,196,894,323,976,171,195,136,4 75,136
192	6,277,101,735,386,680,763,835,789,423,207,666,416,102,355,444,46 4,034,512,896
208	411,376,139,330,301,510,538,742,295,639,337,626,245,683,966,408, 394,965,837,152,256
224	26,959,946,667,150,639,794,667,015,087,019,630,673,637,144,422,5 40,572,481,103,610,249,216
240	1,766,847,064,778,384,329,583,297,500,742,918,515,827,483,896,87 5,618,958,121,606,201,292, 619,776

1.9.4 X-OR

Along with the shift operators another important operator is the bitwise X-OR operation ($\oplus$). Its basic function is:

Bit1	Bit2	Output
0	0	0
1	0	1
0	1	1
1	1	0

An example of an operation with an X-OR of 0101 0101 for each byte is:

Input	1010 1000	1111 0000	0101 1100	0000 0001
X-OR	0101 0101	0101 0101	0101 0101	0101 0101
Output	**1111 1101**	**1010 0101**	**0100 1001**	**0101 0100**

The great advantage of the X-OR bitwise operation is that, like the bit rotate operators, it preserves the information in the processed output, and can be undone by merely operating on the output with the same value that was used to generate the result. For example:

Output	1111 1101	1010 0101	0100 1001	0101 0100
X-OR	0101 0101	0101 0101	0101 0101	0101 0101
Input	**1010 1000**	**1111 0000**	**0101 1100**	**0000 0001**

Same value

The following shows an example conversion, where we have a string ("Test") and apply a key, with an resulting encoded format of "IBEHAA==":

	ASCII	Hex	Hex (Base-64)	Hex (Binary)
	(Result)	Hex		
Input	Test	54657374	VGVzdA==	01010100 01100101 01110011 01110100
Key				01110100 01110100 01110100 01110100
Encoded		20110700	IBEHAA==	00100000 00010001 00000111 00000000

A simple encryption process might be:

- Take 32 bits at a time.
- Shift bits by four spaces to the left.
- X-OR the value by 1010 1000.
- Shift bits by two spaces to the right.
- X-OR the value by 1010 1000.

Then, the decryption process would be (reading 32 bits at a time):

- X-OR the value by 1010 1000
- Shift bits by two spaces to the left.
- X-OR the value by 1010 1000.
- Shift bits by four spaces to the right.

1.9.5 Modulo-2 Operations

In cryptography we try and avoid complex mathematical operations which involve carry-overs for bits. This type of operation is known as Modulo-2, or GF(2) – which is a Galois field of two elements – and is used in many areas including with checksums and ciphers. The multiplication function involves multiplying the binary values and ignoring the remainder from each carry forward. This type of operation simplifies the implementation and is fast in its operation. It basically involves some bit shifts and an EX-OR function, and which makes it fast in computing the multiplication.

The basic operations are:

$$0+0=0 \qquad 1+1=0$$
$$0+1=1 \qquad 1+0=1$$

It performs the equivalent operation to an exclusive-OR (XOR) function. For modulo-2 arithmetic, subtraction is the same operation as addition:

0–0=0	1–1=0
0–1=1	1–0=1

Multiplication is performed with the following:

$0\times0=0$	$0\times1=0$
$1\times0=0$	$1\times1=1$

which is an equivalent operation to a logical AND operation.

Binary digit representation, such as 101110, is often difficult to use when multiplying and dividing, so a typical representation is to manipulate the binary value as a polynomial of bit powers. This technique represents each bit as an x to the power of the bit position and then adds each of the bits, such as:

10111	x^4+x^2+x+1
1000 0001	x^7+1
1111 1111 1111 1111	$x^{11}+x^{10}+x^9+x^8+x^7+x^6+x^5+x^4+x^3+x^2+x+1$
10101010	$x^7+x^5+x^3+x$

For example:	101×110
is represented as:	$(x^2+1)\times(x^2+x)$
which equates to:	$x^4+x^3+x^2+x$
which is thus:	11110

📖 **Web link (Example):** http://asecuritysite.com/calculators/mod2

The addition of the bits is treated as a Modulo-2 addition, where any two variables which have the same powers are equal to zero (1+1=0). For example:

$$x^4 + x^4 + x^2 + 1 + 1$$

is equal to x^2 as x^4+x^4 equates to zero and 1+1 equates to 0 (in modulo-2). An example which shows this is the multiplication of 10101 by 01100.

Thus:	10101×01110
is represented as:	$(x^4+x^2+1)\times(x^3+x^2+x)$
which equates to:	$x^7+x^6+x^5+x^5+x^4+x^3+x^3+x^2+x$
which equates to:	$x^7+x^6+x^4+x^2+x$
which is thus:	11010110

The division process uses an exclusive-OR operation instead of subtraction and can be implemented with a shift register and a few XOR gates. For example, 101101 divided by 101 is implemented as follows:

```
          1011
100 | 101101
       100
       ----
        110
        100
        ----
         101
         100
         ----
           1
```

Thus, the modulo-2 division of 101101 by 100 is thus 1011 remainder 1. As with multiplication, this modulo-2 division can also be represented with polynomial values.

📖 **Web link (Example):** http://asecuritysite.com/comms/mod_div

1.10 GCD

GCD is known as the greatest common divisor, or greatest common factor (gcf), and is the largest positive integer that divides into two numbers without a remainder. For example, the GCD of 9 and 15 is 3. It is an operation that is used many encryption algorithms, and example of some code to calculate the GCD for two values (a and b) is:

```
static int GCD(int a, int b)
{
    int Remainder;
    while( b != 0 )
    {
        Remainder = a % b;
        a = b;
        b = Remainder;
    }
    return a;
}
```

If we run with a value of 54 and 8, we get:

a:54, b:8, Remainder:6
a:8, b:6, Remainder:2
a:6, b:2, Remainder:0
Return value:2

   Web link (Example): http://asecuritysite.com/encryption/gcd

1.11 Random Number Generators

Within cryptography random numbers are used to generate things like encryption keys. If the generation of these keys can be predicted in some way, it may be possible to guess them. The two main types of random number generators are:

- **Pseudo-Random Number Generators** (PRNGs). This method repeats the random numbers after a given time (periodic). They are fast and are also deterministic, and are useful in producing a repeatable set of random numbers.
- **True Random Number Generators** (TRNGs). This method generates a true random number, and uses some form of random process. One approach is to monitor the movements of a mouse pointer on a screen or from the pauses between keystrokes. Overall the method is generally slow, especially if it involves human interaction, but is non-deterministic and aperiodic.

Normally simulation and modelling applications use PRNG, so that the values generated can be repeated for different runs, while cryptography, lotteries, gambling and games use TRNG, as each value should not repeat or be predictable. If the generation of key was deterministic, Eve could possibly guess the key created. So, in the generation of encryption keys for public key encryption, users are often asked to generate some random activity, and where a random number is then generated based on this activity. This random number is then used to generate the encryption keys.

Computer programs, though, often struggle to generate truly random numbers, so hardware generators are often used within highly secure applications. One method is to generate a random number based on low-level, statistically random *noise* signals. This includes things like thermal noise and from the photoelectric effect.

📖 **Web link (Random number):** http://asecuritysite.com/encryption/random

1.11.1 Linear Congruential Random Numbers

One method of creating a simple random number generator is to use a sequence generator of the form:

$$X_{i+1} \leftarrow (a \times X_i + c) \bmod m$$

Where *a*, *c* and *m* are integers, and where X_0 is the seed value of the series. If we take the values of *a*=21, X_0=35, *c*=31 and *m*=100 we get a series of:

(21×35+31) mod 100 gives 66
(21×66+31) mod 100 gives 17
(21×17+31) mod 100 gives 88
and so on.

```
66 17 88 79 90 21 72 43 34 45 76 27 98 89 0 31 82 53
```

📖 **Web link (Linear congruential):** http://asecuritysite.com/encryption/linear

Within cryptography, it is important that we are generating values which are as near random as possible, so that Eve cannot guess the random numbers that Bob and Alice have used. With randomness we cannot determine how random the values are, by just taking a few samples. For this we need a large number of samples, and take an estimation of the overall randomness.

There are various tests for randomness. For example, we could define an average value which is half way between the number range, and then determine the ratio of the values above and below the half way point. This will work, but will not show us if the values are well distributed. Along with this we could determine the arithmetic mean of the values, and match it to the centre value within the range of numbers.

An improved method to test for the distribution of values is the Monte Carlo value for Pi test. With this method, we take our random numbers and scale them between –1.0 (scaled from the minimum value) and 1.0 (scaled from the maximum value). Next we take two values at a time and calculate:

$$\sqrt{x^2 + y^2}$$

If this value is less than or equal to one, we place in the circle (with a radius of 1), otherwise it is out of the circle. The estimation of PI is then four times the number of points in the circle (M) divided by the total number of points (N). In Figure 1.14, the blue points are outside the circle and the yellow ones are inside.

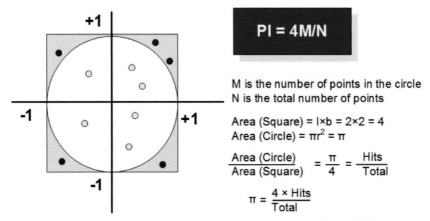

Figure 1.14   Analysis of cipher text compared with normal probabilities.

📖 **Web link (Monte Carlo):** http://asecuritysite.com/encryption/mc

Another method for determining randomness is to measure the entropy of the data. Entropy was defined by Claude E. Shannon in his 1948 paper, and where the maximum entropy occurs when there is an equal distribution of all bytes across the data. Normally we define these in terms of bytes. The method we use is to take the frequencies of the byte values and calculate how many bits are used:

$$E_n = \sum_{n=1}^{n=255} f_n \log_2(f_n)$$

where f_n relates to the frequency of the byte values. A maximum entropy is 8 bits (for a byte value). For values from 0 to 255, we would expect a result around 8 bits if the values are random.

   Web link (Entropy): http://asecuritysite.com/encryption/ent

1.12 Frequency Analysis

Finally, we will do a little bit of frequency analysis, as it is often used in cipher cracking, especially to spot variations in the probably of codes. It is best illustrated with an example. If our cipher text is:

```
LQ A EAOONM WC A CNI UNHAUNZ OKN IWMRU KAZ HKAQDNU CMWE AQ
LQUSZOMLAR ADN LQOW AQ LQCWMEAOLWQ ADN. LO LZ WQN IKLHK,
SQRLPN NAMRLNM ADNZ, NQHATZSRAONZ FLMOSARRB OKN IKWRN IWMRU.
LO LZ ARZW WQN IKLHK ARRWIZ OKN QNI LQUSZOMLNZ OW XN XAZNU
LQ AQB RWHAOLWQ ILOKWSO MNVSLMLQD AQB QAOSMAR MNZWSMHNZ,
WM OW XN LQ AQB AHOSAR TKBZLHAR RWHAOLWQZ. OBTLHARRB ARR OKAO
LZ MNVSLMNU LZ A MNRLAXRN QNOIWMP HWQQNHOLWQ. WSM IWMRU LZ
HKAQDLQD XB OKN UAB, AZ OMAULOLWQAR CMWEZ WC XSZLQNZZ AMN
XNLQD MNTRAHNU, LQ EAQB HAZNZ, XB EWMN MNRLAXRN AQU CAZONM
IABZ WC WTNMAOLQD. WSM TWZOAR ZBZONE, IKLRN ZOLRR SZNU
CWM EAQB SZNCSR ATTRLHAOLWQZ, KAZ XNNQ RAMDNRB MNTRAHNU XB
NRNHOMWQLH EALR. ILOK FWOLQD, OKN ZRWI AQU HSEXNMZWEN OAZP WC
EAMPLQD FWOLQD TA-TNMZ ILOK OKN TMNCNMMNU HAQULUAON, LZ
QWI XNLQD MNTRAHNU XB NRNHOMWQLH FWOLQD. OKN OMAULOLWQAR
ZBZONEZ,OKWSDK, KAFN XNNQ AMWSQU CWM KSQUMNUZ LC QWO
OKWSZAQUZ WC BNAMZ, AQU OBTLHARRB SZN INRR OMLNU-AQU-ONZONU
ENHKAQLZEZ. CWM OKN EWZO TAMO, CWM NJAETRN, IN OMSZO
A TATNM-XAZNU FWOLQD ZBZONE, NFNQ OKWSDK LO LZ INRR PQWIQ
OKAO A HWSQO WC OKN FWONZ ILOKLQ AQ NRNHOLWQ ILRR WCONQ
TMWUSHN ULCCNMNQO MNZSROZ NAHK OLEN OKAO OKN FWON LZ HWSQONU,
AQU OKNQ MNHWSQONU. AQ NRNHOMWQLH ENOKWU ILRR, WQ OKN
WOKNM KAQU, EWZO RLPNRB KAFN A ZSHHNZZ MAON WC 100%.
```

We can now determine the frequency of the characters:

a	b	c	d	e	f	g	h	i	j	k	l	m
90	23	22	18	21	10	0	37	24	1	43	75	61
[8.5%]	[2.2%]	[2.1%]	[1.7%]	[2.0%]	[0.9%]	[0.0%]	[3.5%]	[2.3%]	[0.1%]	[4.1%]	[7.1%]	[5.8%]

n	o	p	q	r	s	t	u	v	w	x	y	z
119	90	6	76	59	31	19	38	2	73	16	0	67
[11.2%]	[8.5%]	[0.6%]	[7.2%]	[5.6%]	[2.9%]	[1.8%]	[3.6%]	[0.2%]	[6.9%]	[1.5%]	[0.0%]	[6.3%]

And we can plot the occurrence against how the characters relate to standard English character probabilities (Figure 1.15).

The following table shows how the text matches the normal probability to the text (where 'E' has the highest level of occurrence and 'Z' has the least). The rows with lower case show what would be expected for the order, and the upper case ones shows what the cipher text gives for the order:

e	t	a	o	i	n	s	h	r	d	l	c	u
N	O	A	Q	L	W	Z	M	R	K	U	H	S
m	w	f	g	y	p	b	v	k	x	j	q	z
I	B	C	E	T	D	X	F	P	V	J	G	Y

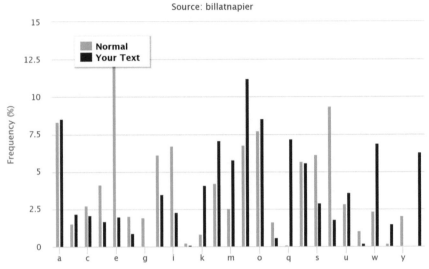

Figure 1.15 Analysis of cipher text compared with normal probabilities.

From this we predict:

• That N in the cipher text maps to e in plaintext.

The next best guess is to analyse the one, two and three-letter characters:

One letter (Most pop: a, I)	Two letter (Most pop: of, to, in, it, is, be, as, at, so, we, he, by, or, on, do, if, me, my, up, an, go, no, us, am)	Three letter (Most Pop: the, and, for, are, but, not, you, all, any, can, had, her, was, one, our, out, day, get, has, him, his, how, man, new, now, old, see, two, way, who, boy, did, its, let, put, say, she, too, use)
a [90]	lq [20] wc [8] aq [20] lo [9] lz [9] ow [3] xn [7] wm [12] xb [4] az [8] lc [2] in [3] wq [15]	cni [1] okn [13] kaz [2] adn [3] wqn [2] qni [1] aqb [5] arr [5] wsm [3] amn [1] aqu [8] cwm [6] qwi [2] qwo [1] szn [3]

Probably the best guess is that an 'A' is an 'a', and we can locate "the" with "OKN", which is the most popular for three letter words. Thus we have 'O' mapped to 't', 'K' to 'h'. If we have mapped 'T' to 'a', then "AQU" looks like it is "and", and thus gives us 'Q' maps to 'n' and 'U' maps to 'd'. This gives:

```
Ln a EatteM WC a CeI deHadeZ the IWMRd haZ HhanDed CMWE an
LndSZtMLaR aDe LntW an LnCWMEatLWn aDe. Lt LZ Wne IhLHh,
SnRLPe eaMRLeM aDeZ, enHaTZSRateZ FLMtSaRRB the IhWRe
IWMRd. Lt LZ aRZW Wne IhLHh aRRWIZ the neI LndSZtMLeZ tW Xe
XaZed Ln anB RWHatLWn ILthWSt MeVSLMLnD anB natSMaR MeZWSMHeZ,
WM tW Xe Ln anB aHtSaR ThBZLHaR RWHatLWnZ. tBTLHaRRB aRR that
LZ MeVSLMed LZ a MeRLaXRe netIWMP HWnneHtLWn. WSM IWMRd LZ
HhanDLnD XB the daB, aZ tMadLtLWnaR CWMEZ WC XSZLneZZ aMe
XeLnD MeTRaHed, Ln EanB HaZeZ, XB EWMe MeRLaXRe and CaZteM
IaBZ WC WTeMatLnD. WSM TWZtaR ZBZteE, IhLRe ZtLRR SZed CWM
EanB SZeCSR aTTRLHatLWnZ, haZ Xeen RaMDeRB MeTRaHed XB
eReHtMWnLH EaLR. ILth FWtLnD, the ZRWI and HSEXeMZWEe taZP
WC EaMPLnD FWtLnD Ta-TeMZ ILth the TMeCeMMed HandLdate, LZ
nWI XeLnD MeTRaHed XB eReHtMWnLH FWtLnD. the tMadLtLWnaR
ZBZteEZ, thWSDh, haFe Xeen aMWSnd CWM hSndMedZ LC nWt
thWSZandZ WC BeaMZ, and tBTLHaRRB SZe IeRR tMLed-and-teZted
EeHhanLZEZ. CWM the EWZt TaMt, CWM eJaETRe, Ie tMSZt a
TaTeM-XaZed FWtLnD ZBZteE, eFen thWSDh Lt LZ IeRR PnWIn
that a HWSnt WC the FWteZ ILthLn an eReHtLWn ILRR WCten
TMWdSHe dLCCeMent MeZSRtZ eaHh tLEe that the FWte LZ
HWSnted, and then MeHWSnted. an eReHtMWnLH EethWd ILRR,
Wn the WtheM hand, EWZt RLPeRB haFe a ZSHHeZZ Mate WC 100%.
```

It now becomes easy by scanning an eye over it, where "nWt" looks like 'W' is an 'o', and "Xe" looking like "be", so 'X' becomes a 'b', "Ln" looks like it is "In", which makes a 'L' mapping to 'i'. If we just look at the first line we get:

```
in a EatteM oC a CeI deHadeZ the IoMRd haZ HhanDed CMoE an
indSZtMiaR aDe into an inCoMEation aDe. it iZ one IhiHh,
SnRiPe eaMRieM aDeZ, enHaTZSRateZ FiMtSaRRB the IhoRe IoMRd.
it iZ aRZo one IhiHh aRRoIZ the neI indSZtMieZ to be baZed
in anB RoHation IitloSt MeVSiMinD anB natSMaR MeZoSMHeZ,
oM to be in anB aHtSaR ThBZiHaR RoHationZ.
```

And we can spot ... "EatteM" maps to "matter" and "oC" to "of", "inCoMEation" maps to "information":

```
in a matter of a feI deHadeZ the IorRd haZ HhanDed from an
indSZtriaR aDe into an information aDe. it iZ one IhiHh,
SnRiPe earRier aDeZ, enHaTZSRateZ FirtSaRRB the IhoRe IorRd.
it iZ aRZo one IhiHh aRRoIZ the neI indSZtrieZ to be baZed
in anB RoHation IitloSt reVSirinD anB natSraR reZoSrHeZ,
or to be in anB aHtSaR ThBZiHaR RoHationZ.
```

"feI" then maps to "few", "worRd" to "world", "indSZtriaR" to "industrial" to give:

```
in a matter of a few deHades the world has HhanDed from an
industrial aDe into an information aDe. it is one whiHh,
unliPe earlier aDes, enHaTsulates FirtuallB the whole world.
```

And it doesn't take too many guesses to end up with:

```
in a matter of a few decades the World has changed from
an industrial age into an information age. it is one which,
unlike earlier ages, encapsulates virtually the whole World.
```

 📖 **Web link (Frequency Analysis):** http://asecuritysite.com/coding/freq
 📖 **Web link (Challenge):** http://asecuritysite.com/challenges/scramb

1.13 Lab/tutorial

The lab and tutorial related to this chapter is available on-line at:

http://asecuritysite.com/crypto01

A few cipher challenge is available at:

http://asecuritysite.com/challenges

2

Secret Key Encryption

2.1 Introduction

Cryptography has been a fundamental element of keeping messages secret, and is often used within military operations. The ciphers covered in the previous chapter often used a secret algorithm that was known by Bob and Alice, but Eve could discover the secret algorithm, and thus crack their communications. Normally, though, the ciphering process uses a special electronic key which makes it easy for Bob and Alice to cipher and decipher, but extremely difficult if the key is not known. Eve can then either try to find a loop-hole in the algorithm, or resort to searching for the key (brute force). Obviously Eve could also trick Bob or Alice to reveal their key, thus we need to make sure that the key is kept secret. So, if she observed it, she could make a copy of it.

2.1.1 Early Days

In the days before transistors, ciphers were often generated through mechanical systems, such as for the Enigma rotor cipher machine. Enigma used a polyalphabetic substitution cipher, which did not repeat within a reasonable time period, along with a secret key. For the cracking of the Enigma cipher, the challenge was thus to determine both the algorithm used and the key. Enigma's main weakness, though, was that none of the plain text letters could be ciphered as itself. This made the challenge easier, as the crackers could dismiss any of the codes which had a mapping to the same letter. But this still left many code translations to be analysed and which would have been impossible for a human to crack. So the cipher crackers in the UK used a commercial version of the Enigma machine, and where Dilly Kox discovered the rotor wiring for secret radio messages. The cipher crackers still struggled to crack the codes produced by the military-strength Enigma machine, which used a 3-rotor machine, with the addition of a plugboard, and which further scrambled the cipher – adding "salt" to it.

The breakthrough came in 1932 when a Polish mathematician Marian Rejewski, who had joined the Polish Cipher Bureau, broke the military Enigma ciphering system. For this the Bureau obtained ciphered documents and two pages of the daily keys for Enigma. He then used a number of permutations to decode Enigma's scrambler wiring. An important inspiration was that the setup of the keyboard mapping to the entry ring was alphabetic (ABC...) rather than mapping to the order of the keys on a German keyboard.

From September 1938, Alan Turing dedicated himself to cracking the Enigma code with Dilly Knox, and, armed with the information from the Polish Cipher Bureau, they set about defining a crib-based encryption approach. With this the most important break-through occurred on 4 September 1939 (the day after war broke out). His two break-through papers on the work were actually kept secret until April 2012, as the information was so sensitive to military operations. Turing received an OBE for his work in 1945, but much of his work was kept private.

2.1.2 Encryption

There has been a long history of defence agencies blocking the development of high-grade cryptography. In the days before powerful computer hardware, the Clipper chip was used, and where a company would have register to use it, and then be given a chip to use. Government agencies then created an escrow key, which was a copy of the key being used.

A major step change came in 1973 when the National Bureau of Standards (which later became the National Institute of Standards and Technology – NIST) of the U.S. Department of Commerce called for a new method of creating a cipher with an encryption algorithm and a key. By 1976, this became the Data Encryption Standard (DES) cipher, and which quickly became a world-wide standard. The core strength of the method was a 64-bit key (but where only 56-bits were actually used for the key, as the other eight bits are used as parity bits).

The weakness of DES was that both Bob and Alice had to share a secret key, and had the problem of passing it to each other, as Eve could be listening. This problem was investigated in 1976 by Whitfield Diffie, Martin Hellman, and Ralph Merkle who proposed a method of using a publicly known key – the public key – which could be used to encrypt the data, and only an associated private key could be used to decrypt it. The system would then be equivalent to distributing a padlock to anyone who wanted to secure something, and using a secret key to open the padlocks. This system is defined as **asymmetric encryption** and uses with two keys, whereas **symmetric encryption**, as used with DES, only has one key.

Diffie, Hellman and Merkle then created a method for key exchange using a one-way function, and which was named the **Diffie-Hellman** method. With this Bob and Alice share two values: G (a generator); and a prime number p. Bob then creates a random number (x) and Alice creates a random number (y). Next Bob calculates A, and Alice calculates B:

$$A = G^x \bmod (p)$$
$$B = G^y \bmod (p)$$

They exchange these values, and then Bob calculates the shared key as:

$$\text{Shared Key} = B^x \bmod (p)$$

and Alice calculates the same value:

$$\text{Shared Key} = A^y \bmod (p)$$

The method could thus be used with a symmetric key method, and where either side creates random values, and then compute values which they could share openly. At the end of this they will end up with the same encryption key. This key could be permanent, or re-negotiated after a given time, or even recreated for each new session.

While Whitfield Diffie proposed the possibilities of a public key method, a major breakthrough came when Ronald L. Rivest, Adi Shamir, and Leonard M. Adlemen created the RSA method. Martin Gardner, in his Mathematical Games column in Scientific American, was so impressed by the method that he published an RSA challenge for which readers could send a stamped address envelope for the full details.

They thus defined a method (covered in more detail in a later chapter) which involves the difficulty in factorizating values for their prime number factors. For example if we have a value of 133, then the prime number factors are 7 and 19, as:

$$133 = 7 \times 19$$

Ron Rivest went on to create hashing methods (MD2, MD4, MD5 and MD6), along with several secret key methods (RC2, RC3 and RC5 – where RC stands for Ron's Cipher).

Phil Zimmerman was one of the first to face up to defence agencies with his PGP (Pretty Good Privacy) software, which, when published in 1991, allowed users to send encrypted and authenticated emails. For this the United States Customs Service filed a criminal investigation related to a violation in the Arms Export Control Act, and where cryptographic software was seen as a munition. Eventually the charges were dropped.

2.1.3 Secure Communications

The application of encryption on network communications traces its roots by to Netscape, and who created one of the original Web browsers. They used a handshaking method and a digital certificate to create a shared key and which was used by either side of the communication. The first definition of SSL (Secure Socket Layer) Version 1.0 was in 1993, and eventually, in 1996, they released a standard which is still widely used: SSL 3.0 (RFC 6101 – written by Netscape engineers Phil Karlton and Alan Freier). Since then TLS (Transport Layer Sockets) have been used to replace SSL, but has suffered from many security problems, such as with FREAK ("Factoring RSA Export Keys"), and which exposed a vulnerability which was introduced to comply with US Cryptography Export Regulations. This related to the keys being used for exportable software being limited to 512-bits or less (and were defined as RSA EXPORT keys). This initially allowed the NSA to break the encryption, as they had powerful computers, but many computers can now crack 512-bit public keys within reasonable time limits.

2.1.4 Modern Methods

With DES struggling to keep up with its 56-bit equivalent keys, the industry looked for improved methods, and included a short-term fixed named 3-DES. With 3-DES we use two keys, of which one key is used to encrypt, the next key then decrypts, and then the first key then encrypts again. This gives an equivalent key size of 112 bits, but it has the overhead of three encryption rounds.

3-DES was a short-term fix, so, in 1997, NIST opened a competition for a new encryption method, with the main contenders of: CAST-256; RC6; Rijndael; SAFER+; Serpent; and Twofish. A final test proved that the Rijndael was the fastest, and which was free of security flaws. It was developed by two Belgian cryptographers: Joan Daemen and Vincent Rijmen, and is now standardised as AES (Advanced Encryption Standard).

By the mid-1990s, the standard hashing methods – which allowed for a thumbprint to be created for data – were showing their age, with flaws being discovered in MD5. NIST then defined a new method named SHA (Secure Hash Algorithm) and which was standardized as SHA-1 (a 160-bit hash) and SHA-256 (a 256-bit hash).

2.2 Key-based Cryptography

The main objective of cryptography is to provide a mechanism for two (or more) entities to communicate without any other entity being able to read or change their messages. Along with this it can provide other functions, such as:

- **Integrity check**. This makes sure that the message has not been tampered with by non-legitimate source.
- **Providing authentication**. This verifies the senders identity. Unfortunately most of the current Internet infrastructure has been build on a fairly open system, and where users and devices can be easily spoofed, thus authentication is now a major factor in verifying users and devices.

One of the main problems with using a secret algorithm for encryption is that it is difficult to determine if Eve has discovered the algorithm used, thus most encryption methods use a key-based approach where an electronic key is applied to a well-known algorithm. Another problem with using different algorithms for the encryption is that it is often difficult to keep devising new algorithms and to tell the receiving party that the data is being encrypted with a new algorithm. Thus, using electronic keys, there are no problems with everyone knowing the encryption/decryption algorithm, because, without the key, it should be computationally difficult to decrypt the ciphertext (Figure 2.1).

The three main methods of encryption are (Figure 2.2):

- **Symmetric key-based encryption**. This involves the same key being applied to the encrypted data, in order that the original data is recovered. Typical methods are DES, 3-DES, RC2, RC4, and AES.
- **Asymmetric key-based encryption**. This involves using a different key to decrypt the encrypted data, in order that the original data is recovered. Typical method is RSA, DSA and ElGamal.
- **One-way hash functions**. With a one-way hash function it should not be mathematically possible to reverse the derived cipher back to the original data. Unfortunately it can typically be broken by knowing the mapping of the data to the hash value, or by performing a brute force analysis on the stored hash value. The one-way hash function is typically used in authentication applications, such as generating a hash value for a message, and also to store ciphered versions of passwords. The method of knowing the mapping between the hashed values and the original data is a **rainbow table attack**, while a brute force analysis is known as a **dictionary-type attack**.

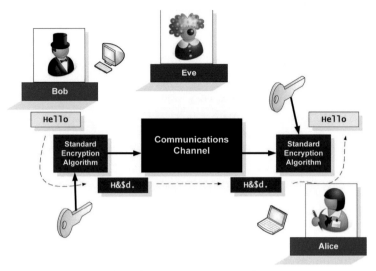

Figure 2.1   Key-based encryption.

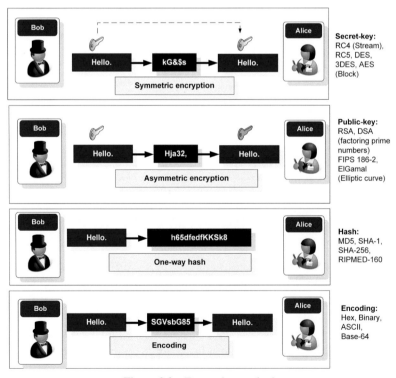

Figure 2.2   Encryption methods.

The major advantage that secret-key encryption has over public-key is that it is typically much faster to decrypt, and can thus be used where a fast conversion is required, such as in real-time encryption.

2.2.1 Computation Difficulty

Every cipher code is crackable and a measure of the security of a code is the amount of time that Eve will take to crack it. For this Eve could continually try different encryption keys, and then determine the one that matches the key which has been used to encrypt the data. On average it will take a search of half the key space, when there is a total of 2^n keys for encryption keys with n bits. Thus a 1-bit key would only have two keys; a 2-bit code would have four keys; and so on.

Table 2.1 shows the number of possible keys as a function of the number of bits in the key. For example it can be seen that a 64-bit key has 18 400 000 000 000 000 000 different keys. If every key is tested in 10 µs then it would take a total time of 1.84×10^{14} seconds (5.11×10^{10} hours or 2.13×10^8 days or 58,346,02 years) to try all of the possible keys.

So, for example, if it takes 1 million years for a person to crack the code, it can be considered safe. Unfortunately, from the point of security of an encrypted message, the performance of computer systems increases by the year. Then, if a computer takes one million years to crack a code, and assuming an increase in computing power of a factor of two per year, it would take 500,000 years in the next year to crack. Then, as shown in Table 2.2, after almost 20 years, it would take only one year to decrypt the same message. There is thus no guarantee that although a cipher is secure at the present time, and that it will be secure in the future.

The increasing power of computers is one factor in reducing the processing time, but another is the increasing usage of parallel processing. With this key-based cipher cracking is well suited to parallel processing as each processing element can be assigned a number of encryption keys to check. Each of these can then work independently of the other[1].

Table 2.3 gives a timing example, and assumes a doubling of processing power each year, and for processor arrays of 1, 2, 4...4,096 elements. It can thus be seen that with an array of 4,096 processing elements it takes only seven years before the code is decrypted within two years. Thus an organization which is serious about deciphering messages is likely to have the resources to invest in large arrays of processors, or networked computers.

---

[1]This differs from many applications in parallel processing which suffer from interprocess (or) communication.

Table 2.1   Number of keys related to the number of bits in the key

Code Size	Number of Keys	Code Size	Number of Keys	Code Size	Number of Keys
1	2	12	4096	52	4.5×10^{15}
2	4	16	65 536	56	7.21×10^{16}
3	8	20	1048576	60	1.15×10^{18}
4	16	24	16777216	64	1.84×10^{19}
5	32	28	2.68×10^{8}	68	2.95×10^{20}
6	64	32	4.29×10^{9}	72	4.72×10^{21}
7	128	36	6.87×10^{10}	76	7.56×10^{22}
8	256	40	1.1×10^{12}	80	1.21×10^{24}
9	512	44	1.76×10^{13}	84	1.93×10^{25}
10	1024	48	2.81×10^{14}	88	3.09×10^{26}

Table 2.2   Time to decrypt a message assuming an increase in computing power

Year	Time to Decrypt (Years)	Year	Time to Decrypt (Years)
0	1 million	10	977
1	500000	11	489
2	250000	12	245
3	125000	13	123
4	62500	14	62
5	31250	15	31
6	15625	16	16
7	7813	17	8
8	3907	18	4
9	1954	19	2

Table 2.3   Time to decrypt a message with increasing power and parallel processing

Processors	Year 0	Year 1	Year 2	Year 3	Year 4	Year 5	Year 6	Year 7
1	1000000	500000	250000	125000	62500	31250	15625	7813
2	500000	250000	125000	62500	31250	15625	7813	3907
4	250000	125000	62500	31250	15625	7813	3907	1954
8	125000	62500	31250	15625	7813	3907	1954	977
16	62500	31250	15625	7813	3907	1954	977	489
32	31250	15625	7813	3907	1954	977	489	245
64	15625	7813	3907	1954	977	489	245	123
128	7813	3907	1954	977	489	245	123	62
256	3906	1953	977	489	245	123	62	31
512	1953	977	489	245	123	62	31	16
1024	977	489	245	123	62	31	16	8
2048	488	244	122	61	31	16	8	4
4096	244	122	61	31	16	8	4	2

2.2.2 Stream Encryption and Block Encryption

The encryption method can either be applied by selecting blocks of a data, and then encrypting them, or it can operate on the data stream, where one bit at a time is encrypted (Figure 2.3). Typical block sizes are 128, 192 or 256 bits. Overall stream encryption is often much faster, and can typically be applied in real-time applications. The main methods are:

- **Stream encryption:** RC4 (one of the fast streaming methods around) and ChaCha.
- **Block encryption:** RC2 (40-bit key size), RC5 (variable block size), IDEA, DES, 3-DES, AES (Rijndael), Blowfish and Twofish.

Stream-based encryption have, in the past, been used with wireless systems, where an infinitely long key is created from the secret key. This is then exclusive-OR-ed with the data stream to produce the cipher stream.

The most widely used secret-key encryption (symmetric) methods are:

- RC2 (40-bit key size, 64-bit blocks).
- RC4 (stream cipher) – used in SSL and WEP.
- RC5 (variable key size, 32, 64 or 128-bit block sizes).
- AES (128, 192 or 256-bit key size, 128-bit block size).
- DES (56 bit key size, 64 bit block size).
- 3-DES (112 bit key size, 64 bit block size).

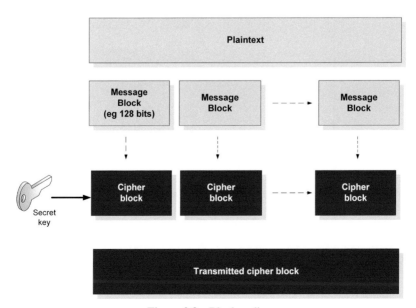

Figure 2.3   Block coding.

An example of a stream conversion is:

Data stream:	0101110101010111
Pseduo-infinite key:	1001100000111010
Result:	1100010101101101

where the receiver will then generate the same infinite key, and simply X-OR it with the received stream to recover the original data stream (Figure 2.4). A weakness of the system is obviously in the way that the pseudo-infinite key is created, and which is typically generated from a pass phrase (which limits the actual range of keys). To overcome the same pseudo-infinite key being used for different communications, an initialization vector (IV) is normally used (the random seed). This can then be incremented for each data frame sent, and will thus result in a different key for each transmission. Unfortunately the IV value has a limited range, and will eventually roll-over to the same value, after which an intruder can use a statistical analysis technique to crack the cipher stream.

📖 **Web link (RC4):** http://asecuritysite.com/encryption/rc4_wep

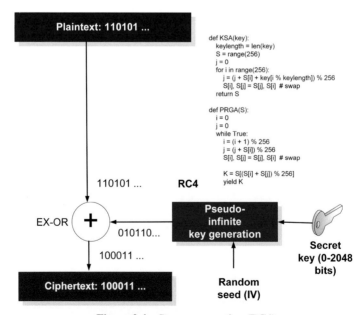

Figure 2.4   Stream encryption (RC4).

2.2.3 Padding

Block ciphers work by creating blocks of given sizes, such as for 128-bit blocks. Each of the blocks are then encrypted and converted to cipherblocks. The data will often not fit perfectly into all of the blocks, with the one at the end missing some data. The problem with this is that intruders could analyse the end blocks and look for patterns. Overall the simplest method is to just pad it with NULL characters (ASCII zero values) so that it fills the last block. Unfortunately a C string also contains a NULL character at the end of the string, so the NULL character could not be misinterpreted. Bruce Schneier recommends that the stuffed values have a 0x80 value followed by zero bytes, or with *n* bytes containing a value of *n*. Typical block sizes are:

- DES, CAST5, and Blowfish: 64-bit block size.
- AES: 128-bit block size.
- RC4, RC5, RC6 can have variable block sizes.

We can then define a number of padding methods:

- **CMS** (Cryptographic Message Syntax). This pads with the same value as the number of padding bytes. Defined in RFC 5652, PKCS#5, PKCS#7 and RFC 1423 PEM.
- **Bits**. This pads with 0x80 (10000000) followed by zero (null) bytes. Defined in ANSI X.923 and ISO/IEC 9797-1.
- **ZeroLength**. This pads with zeros except for the last byte which is equal to the number (length) of padding bytes.
- **Null**. This pads will NULL bytes and is only used with ASCII text.
- **Space**. This pads with spaces and is only used with ASCII text.
- **Random**. This pads with random bytes with the last byte defined by the number of padding bytes.

📖 **Web link (Padding):** http://asecuritysite.com/encryption/padding
    Web link (Padding): http://asecuritysite.com/encryption/padding_des

If we use "hello" (0x68 – 'h', 0x65 – 'e, and so on), for AES, we must pad to 16 bytes, this means there are 11 padding bytes (0xB) to give:

```
After padding (CMS): 68656c6c6f0b0b0b0b0b0b0b0b0b0b0b
Cipher (ECB): 0a7ec77951291795bac6690c9e7f4c0d
After padding (Bit): 68656c6c6f8000000000000000000000
Cipher (ECB): 731abffc2e3b2c2b5caa9ca2339344f9
After padding (ZeroLen): 68656c6c6f0000000000000000000000a
Cipher (ECB): d28e2f7e8e44e068732b292bde444245
After padding (Null): 68656c6c6f0000000000000000000000
Cipher (ECB): 444797422460453d95856eb2a1520ece
After padding (Space): 68656c6c6f0000000000000000000000
Cipher (ECB): 444797422460453d95856eb2a1520ece
After padding (Random): 68656c6c6fffc6ecfd884a38798d62a0a
Cipher (ECB): c2c88b4364d2c2dc6f2cac9ab73c995d
```

For random padding we see we have a padded hexacademical block of "68656c6c6fffc6ecfd884a38798d62a0a" where "68656c6c6" is the message ("hello"), "ffc6ecfd884a38798d62a" is the random collection of bytes, and "0a" identifies that there are 10 bytes used for the random data padding.

For CMS, if we use "hello123", for AES, we must pad to eight bytes, this means there are eight padding bytes (0x8) to give:

```
After padding (CMS): 68656c6c6f3132330808080808080808
Cipher (ECB): a20bd93e1af5c0433b68e537ddc70d9a
decrypt: hello123
```

PKCS (Public-Key Cryptography Standards) was designed and published, in the 1990s, by RSA Security Inc, and has now been standardised in the form of RFCs. PKCS #5 (RFC 2859) is a standard used for password-based encryption, and PKCS #7 (RFC 2815) is used to sign and/or encrypt messages for PKI.

Stream cipher methods do not require any byte stuffing as they work on bit streams, so that the output cipher is the same size as the input data stream. Extra characters may thus be added, if required, in order to hide the length of the message.

2.3 Brute-Force Analysis

It is important to understand how well the cipher text will cope against a brute force attack, and where an intruder tries all the possible keys. As an example, let's try a 64-bit encryption key which gives us: 1.84×10^{19} combinations (2^{64}). If we now assume that we have a fast processor that tries one key every billionth of second (1GHz clock), then the average[2] time to crack the code will be:

$$T_{average} = 1.84 \times 10^{19} \times 1 \times 10^{-9} \div 2 \approx 9,000,000,000 \text{ seconds}[3]$$

It will thus take approximately 2.5 million hours (150 million minutes or 285 years) to crack the code, which is likely to be strong enough in most cases. Unfortunately, as we have seen, computing power often increases by the year, so if we assume a doubling of computing power, then:

Date	Hours	Days	Years
0	2,500,000	104,167	285
+1	1,250,000	52,083	143
+2	625,000	26,042	71
+3	312,500	13,021	36
+4	156,250	6,510	18
+5	78,125	3,255	9
+6	39,063	1,628	4
+7	19,532	814	2
+8	9,766	407	1
+9	4,883	203	1
+10	2,442	102	0.3
+11	1,221	51	0.1
+12	611	25	0.1
+13	306	13	0
+14	153	6	0
+15	77	3	0
+16	39	2	0
+17	20	1	0

and we can see that it now only takes 17 years to crack the code in a **single day**! If we then apply parallel processing, the time to crack reduces again. In the following an array of 2×2 (4 processing elements), 4×4 (16 processing elements), and so on, are used to determine the average time taken to crack the code. If, thus, it currently takes 2,500,000 minutes to crack the code, it can be seen that by Year 6, it takes less than one minute to crack the code, with a 256×256 processing matrix.

---

[2]The average time will be half of the maximum time.
[3]9,223,372,036 seconds to be more precise.

Processing Elements	Year 0 (minutes)	Year 1 (min)	Year 2 (min)	Year 3 (min)	Year 4 (min)	Year 5 (min)	Year 6 (min)	Year 7 (min)
1	2500000	1250000	625000	312500	156250	78125	39062.5	19531.3
4	625000	312500	156250	78125	39062.5	19531.3	9765.7	4882.9
16	156250	78125	39062.5	19531.3	9765.7	4882.9	2441.5	1220.8
64	39063	19531.5	9765.8	4882.9	2441.5	1220.8	610.4	305.2
256	9766	4883	2441.5	1220.8	610.4	305.2	152.6	76.3
1024	2441	1220.5	610.3	305.2	152.6	76.3	38.2	19.1
4096	610	305	152.5	76.3	38.2	19.1	9.6	4.8
16384	153	76.5	38.3	19.2	9.6	4.8	2.4	1.2
65536	38	19	9.5	4.8	2.4	1.2	0.6	0.3

The use of parallel processing is now well-known in the industry, and the Electronic Frontier Foundation (EFF) set out to prove that DES was weak. For this they created a 56-bit DES cracking machine which had an array of 29 circuits of 64 chips (1856 elements), and processed 90,000,000 keys per seconds. It, in 1998, eventually cracked the code within 2.5 days. A more recent machine is the COPACOBANA (Cost-Optimized Parallel COde Breaker) which cost less than $10,000, and cracked a 64-bit DES code in less than nine days.

The ultimate in distributed applications is to use unused processor cycles of machines connected to the Internet. For this, applications such as **distributed.net** allow for the analysis of a key space when the screen saver is on (Figure 2.5). It has since been used the method to crack a number of challenges, such as in 1997 with a 56-bit RC5 Encryption Challenge. It was cracked in 250 days, and has since moved on, in 2002, to crack 64-bit RC5 Encryption Challenge in 1,757 days (with 83% of the key space tested). The current challenge involves a 72-bit key.

Along with the increasing power of computers, and parallel processing, another method of improving the performance of brute force analysis is to use supercomputers. Three of the most powerful machines in the world are:

- **Tianhe-2 (MilkyWay-2):** National Super Computer Center in Guangzhou, 3,120,000 cores, 1PB memory. Intel Xeon processors. 54,902.4 TFlop/s. Manufacturer: NUDT.
- **Titan:** DOE/SC/Oak Ridge National Laboratory. 560,640 cores, 710,144 GB, 27,112.5 TFlop/s. Opteron 6274 processors. Manufacturer: Cray Inc.
- **BlueGene/Q:** DOE/NNSA/LLNL, IBM Department of Energy's (DOE) National Nuclear Security Administration's (NNSA), 1 PB memory, 20,132.7 TFlop/s, 1,572,864 cores using Power BQC 16C processors. Manufacturer: IBM.

An encryption algorithm which is cracked in a million minutes on a standard PC, could take BlueGene less than a second to crack.

  📖 **Web link (Key cracking):** http://asecuritysite.com/encryption/key
      Web link (Key cracking): http://asecuritysite.com/encryption/keys
      Web link (Key cracking): http://asecuritysite.com/encryption/keys2

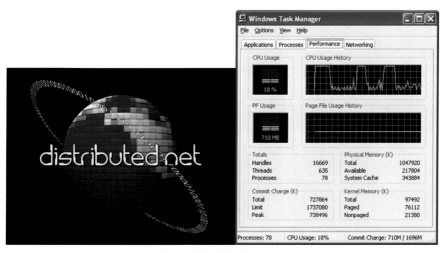

Figure 2.5   Distributed.net

2.4 Adding Salt

A major problem in encryption is that the ciphertext can be played back where an intruder can copy an encrypted message and play it back, as the same plaintext will always give the same ciphertext. The solution to this is to add **salt** to the ciphering process, so that it changes its operation from block-to-block (for block encryption) or data frame-to-data frame (for stream encryption). The most basic method, and which does not use salt, is Electronic Code Book (ECB). With this the cipher text is processed with the key for each block, thus the same ciphertext will be the same for the same plain text message:

Hello -> 5ghd%43f=
Hello -> 5ghd%43f=

If the intruder knew that the plaintext was "Hello", they would be able to play back this message. Salt is added with an IV (Initialisation Vector) which must be the same on both sides. In WEP, the IV is incremented for each data frame, so that the cipher text changes. As can be seen in Figure 2.6, blocks of the same data for ECB will be encrypted to give the same output.

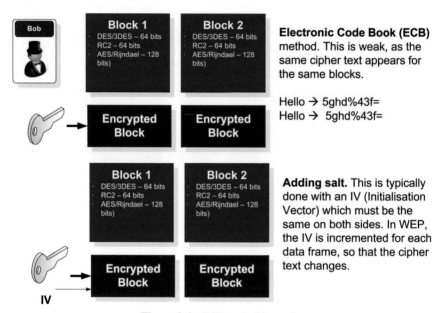

Figure 2.6   ECB and adding salt.

If we take "eeeeeeeeeeeeeeeeeeeeeeeeeeeeeeeeeeeeeeeeeeeeeeeeeeeeeeee" and encrypt with 3-DES and a key of "bill12345" we get (Figure 2.7):

1122900B30BF1183   1122900B30BF1183   1122900B30BF11831   122900B30BF1183
1122900B30BF1183   1122900B30BF1183   7591F6A1D8B4FC8A

where we can see that the "e..e" values are always coded with the same cipher text. As 3-DES has message blocks of 64-bits, then 8 'e' values will fill each block.

[eeeeeeee] [eeeeeeee] [eeeeeeee]
[eeeeeeee] [eeeeeeee] [eeeeeeee]
[eeeeee <PADDING>]

Thus we can say that "eeeeeeee" maps to the cipher text of 1122900B30BF1183.

📖 **Web link (Key cracking):** http://asecuritysite.com/encryption/threedes? word= eeeeeeeeeeeeeeeeeeeeeeeeeeeeeeeeeee eeeeeeeeeeeeeeeeeeeeeeee&key=bill12345

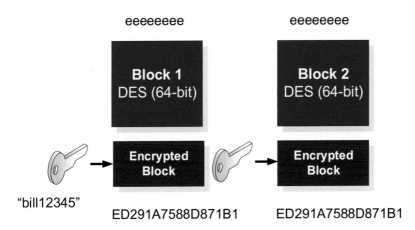

eeeeeeeeeeeeeeeeeeeeeeeeeeeeeeeeeee
eeeeeeeeeeeeeeeeeeeeeeeeeeeeee

[eeeeeeee] [eeeeeeee] [eeeeeeeee][eeeeeeee] [eeeeeeee] [eeeeeeee][eeeeee <PADDING>]

eeeeeeee   eeeeeeee

Block 1
DES (64-bit)

Block 2
DES (64-bit)

Encrypted Block   Encrypted Block

"bill12345"

ED291A7588D871B1   ED291A7588D871B1

ED291A7588D871B1ED291A7588D871B1ED291A7588D
871B1ED291A7588D871B1ED291A7588D871B1ED291A
7588D871B18D6DF6795DDEDACD

Figure 2.7   ECB.

2.4.1 Cipher Block Chaining CBC

An improvement over ECB is to use **Cipher Block Chaining** (CBC). This method uses the IV for the first block, and then the results from the previous block is used to encrypt the current block (Figure 2.8). As defined in Table 2.4 (where C represents the cipher blocks and P represents the data blocks), the IV value is used in the first block, and must be passed from the sender to the receiver or it will not be possible to decrypt the first and, thus, the subsequent blocks. The IV value must be sent with the cipher text, in order for it to be deciphered.

Table 2.4　CBC

Method	First Block	Successive Blocks
Cipher Block Chaining (CBC) Encryption	$C_0 = Encrypt(P_0 \oplus IV)$	$C_i = Encrypt(P_i \oplus C_{i-1})$
Cipher Block Chaining (CBC) Decryption	$P_0 = IV \oplus Decrypt(C_0)$	$P_i = C_{i-1} \oplus Encrypt(C_i)$

If we encrypt two files with the same contents with 256-bit AES CBC, and with a password for "123456" for a Base-64 output, we see that the output changes:

```
> openssl enc -e -aes-256-cbc -in test.txt -pass pass:123456 -base64
U2FsdGVkX199kL+Z/toFkPeG8GXjO/90el40HDqE4nY=
> openssl enc -e -aes-256-cbc -in test.txt -pass pass:123456 -base64
U2FsdGVkX1/Jegqlude2pERWnpargPI/4kdDlIltfY8=
```

The reason the starting part is the same, is that there is a signature of "Salted_" at the start, and which is seen when we do not use Base-64:

```
> openssl enc -e -aes-256-cbc -in test.txt -pass pass:123456
Salted___\ᴸT3ᵣ∞U*π=ÿΩ∩!Uñ├)⊥│çu!
> openssl enc -e -aes-256-cbc -in test.txt -pass pass:123456
Salted__åÜ▓!‼◢╟°┘z ᴸ]Év♀Ékß9S≡çΓ
```

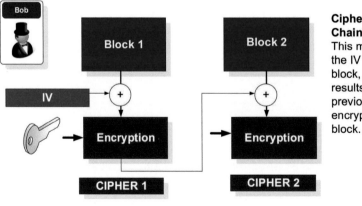

Cipher Block Chaining (CBC). This method uses the IV for the first block, and then the results from the previous block to encrypt the current block.

Figure 2.8　CBC.

2.4.2 Cipher Feedback (CFB)

The Cipher Feedback (CFB) mode is similar to CBC, but it makes the block cipher into a self-synchronising **stream cipher**. With a non-synchronising cipher, if we lose any part of the ciphertext, we could not rebuild the rest of the cipher stream as the current cipher block is based on a previous one, and these build together as a chain, so that subsequent blocks cannot be decrypted (Table 2.5 and Figure 2.9). In order to protect against the loss of a single bit or byte, CBC is converted into a stream cipher and where one bit is processed at a time. The X-OR process typically works on a stream rather than for a block.

In a block cipher we must wait for the whole block to be filled before we can decrypt, but with feedback ciphers we have an output generated by single-bit X-OR operators, which can also be done within the receiver. Figure 2.10 shows the first stage of the encryption process, where the IV value is encrypted, and then EX-OR'ed with the data stream, one bit at a time. This converts the cipher text into a bit stream. On the other side the decryption process will take the IV value and encrypt it, and then take the received bit steam one bit at a time, and X-OR it with the received cipher stream. In this way we have full synchronisation, one bit at a time.

Table 2.5   CFB

Method	First Block	Successive Blocks
Cipher Feedback (CFB) Encryption	$C_0 = P_0 \oplus Encrypt(IV)$	$C_i = P_i \oplus Encrypt(C_{i-1})$
Cipher Feedback (CFB) Decryption	$P_0 = C_0 \oplus Encrypt(IV)$	$P_i = C_i \oplus Encrypt(C_{i-1})$

2.4.3 Output Feedback (OFB)

In the Output Feedback (OFB) method, the first stage takes the data blocks and X-OR's with the encrypted version of the IV value. The output of the first stage encryption is then feed into the next stage, and encrypted, with the output being X-OR'ed with the second block. With CFB the output of the X-OR stage is used instead.

As with CFB, Output Feedback (OFB) creates a synchronous stream output, but takes the output from the cipher stage, rather than from the output of the X-OR process.

Table 2.6   OFB

Method	First Block	Successive Blocks
Cipher Feedback (OFB) Encryption	$O_0 = Encrypt(IV)$ $C_0 = P_0 \oplus O_0$	$I_i = O_{i-1}$ $O_i = Encrypt(I_i)$ $C_i = P_i \oplus O_i$
Cipher Feedback (OFB) Decryption	$O_0 = Encrypt(IV)$ $C_0 = P_0 \oplus O_0$	$I_i = O_{i-1}$ $O_i = Encrypt(I_i)$ $C_i = P_i \oplus O_i$

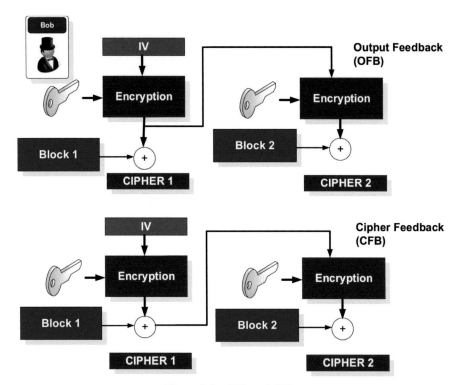

Figure 2.9   OFB and CFB.

2.4.4 Counter Mode

The Counter (CTR) mode, as shown in Figure 2.11, converts the block cipher into a stream cipher. With this it generates a counter value and a nonce, and encrypts this, in order to EX-OR with the plain text block. The counter can

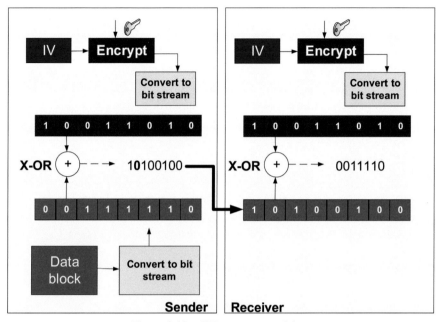

Figure 2.10 CFB as a stream cipher.

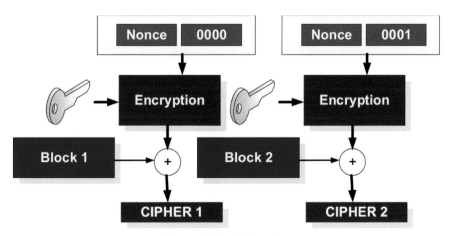

Figure 2.11 CTR mode.

increment by one each time, or can be generated by a given algorithm (known only be the sender and trusted receiver). The advantage of this method is that the processing of each block is independent to the others, so we can apply parallel processing to each. In the previous methods we required the feedback from the other stages to feed into the current one, so we cannot apply parallel processing.

A nonce is a random number and is only used once, and is generated by one party and sent using a secure handshaking process. Often the nonce is used for the generation of an IV, so that the counter does not always start on the same value.

2.4.5 CBC Example

So let's look at applying CBC with Blowfish. Let's start with a message of "fred", and a key of "bert", and use and IV of 1:

http://www.asecuritysite.com/encryption/blowfishcbc?word=fred,bert,1

which gives: 1AC9C54C951E180E0000000000000000

Next we'll change to an IV of 2:

http://www.asecuritysite.com/encryption/blowfishcbc?word=fred,bert,2

which gives: D27FA68C6AC79420000000000000000000

Next we will apply it to 3-DES, which uses a 112-bit key, and an IV value which is 8 bytes. Let's take an example with a message of:

The quick brown fox jumped over the lazy dog

and a key of:

1234567890123456ABCDEFGH

and if we use an IV of "12345678" we get:

E6B6345F1015380284481BBCFFB9052A227FC14F73072E8D5
007AC01DFEDCC2BCBCE1EB14A95ED60BA1A44700F4E18AE

but if we use an IV of "23456789" we get:

5BF29657E6064EB99E52ACC8E3A6808A761A86A7EE85C25C
327022C30D939D3A8A41A9CD42689AA4481FF20155816A8C

So, we can see that it changes of different IV values.

2.5 AES

With AES we have blocks of 16 bytes (128 bits) and with key sizes of 16 (128-bit), 24 (192-bit), and 32 bytes (256-bit). The input takes 16 bytes for each block as a 4×4 matrix, such as:

01	02	03	04
05	06	06	07
08	09	0A	0B
0C	0D	0E	0F

For different key sizes we go through a number of rounds (N):

128-bit (16 bytes) key -> N=10 rounds
192-bit (24 bytes) key -> N=12 rounds
256-bit (32 bytes) key -> N=14 rounds

Figure 2.14 outlines the process for 128-bit encryption, and where we have 10 rounds. The key is expanded to 44 words with 33 bits each, and where each round uses four words (128 bits) as an input. With Round 0, the initial transformation consists of an add round key. The following rounds (apart from the final round) consists of:

- Substitute Bytes.
- Shift Row.
- Mix Column.
- Add round key.

and final round consists of:

- Substitute Bytes.
- Shift Row.
- Add round key.

2.5.1 Substitution Bytes (S-box)

This process transforms the inputs to a new value as an output using an S-box table. In this case the S-Box table is a 16×16 matrix which takes each input value, and where the first 4 bits is used to define row of the table, and the next four bits defines the column (Figure 2.12).

	x0	x1	x2	x3	x4	x5	x6	x7	x8	x9	xa	xb	xc	xd	xe	xf
0x	63	7c	77	7b	f2	6b	6f	c5	30	01	67	2b	fe	d7	ab	76
1x	ca	82	c9	7d	fa	59	47	f0	ad	d4	a2	af	9c	a4	72	c0
2x	b7	fd	93	26	36	3f	f7	cc	34	a5	e5	f1	71	d8	31	15
3x	04	c7	23	c3	18	96	05	9a	07	12	80	e2	eb	27	b2	75
4x	09	83	2c	1a	1b	6e	5a	a0	52	3b	d6	b3	29	e3	2f	84
5x	53	d1	00	ed	20	fc	b1	5b	6a	cb	be	39	4a	4c	58	cf
6x	d0	ef	aa	fb	43	4d	33	85	45	f9	02	7f	50	3c	9f	a8
7x	51	a3	40	8f	92	9d	38	f5	bc	b6	da	21	10	ff	f3	d2
8x	cd	0c	13	ec	5f	97	44	17	c4	a7	7e	3d	64	5d	19	73
9x	60	81	4f	dc	22	2a	90	88	46	ee	b8	14	de	5e	0b	db
ax	e0	32	3a	0a	49	06	24	5c	c2	d3	ac	62	91	95	e4	79
bx	e7	c8	37	6d	8d	d5	4e	a9	6c	56	f4	ea	65	7a	ae	08
cx	ba	78	25	2e	1c	a6	b4	c6	e8	dd	74	1f	4b	bd	8b	8a
dx	70	3e	b5	66	48	03	f6	0e	61	35	57	b9	86	c1	1d	9e
ex	e1	f8	98	11	69	d9	8e	94	9b	1e	87	e9	ce	55	28	df
fx	8c	a1	89	0d	bf	e6	42	68	41	99	2d	0f	b0	54	bb	16

Figure 2.12   S-box.

For example if the input byte is CF, then the output will be 8A. The inverse S-box does the reverse of the S-box process, so the 8A maps back to CF (Figure 2.13).

	x0	x1	x2	x3	x4	x5	x6	x7	x8	x9	xa	xb	xc	xd	xe	xf
0x	52	09	6a	d5	30	36	a5	38	bf	40	a3	9e	81	f3	d7	fb
1x	7c	e3	39	82	9b	2f	ff	87	34	8e	43	44	c4	de	e9	cb
2x	54	7b	94	32	a6	c2	23	3d	ee	4c	95	0b	42	fa	c3	4e
3x	08	2e	a1	66	28	d9	24	b2	76	5b	a2	49	6d	8b	d1	25
4x	72	f8	f6	64	86	68	98	16	d4	a4	5c	cc	5d	65	b6	92
5x	6c	70	48	50	fd	ed	b9	da	5e	15	46	57	a7	8d	9d	84
6x	90	d8	ab	00	8c	bc	d3	0a	f7	e4	58	05	b8	b3	45	06
7x	d0	2c	1e	8f	ca	3f	0f	02	c1	af	bd	03	01	13	8a	6b
8x	3a	91	11	41	4f	67	dc	ea	97	f2	cf	ce	f0	b4	e6	73
9x	96	ac	74	22	e7	ad	35	85	e2	f9	37	e8	1c	75	df	6e
ax	47	f1	1a	71	1d	29	c5	89	6f	b7	62	0e	aa	18	be	1b
bx	fc	56	3e	4b	c6	d2	79	20	9a	db	c0	fe	78	cd	5a	f4
cx	1f	dd	a8	33	88	07	c7	31	b1	12	10	59	27	80	ec	5f
dx	60	51	7f	a9	19	b5	4a	0d	2d	e5	7a	9f	93	c9	9c	ef
ex	a0	e0	3b	4d	ae	2a	f5	b0	c8	eb	bb	3c	83	53	99	61
fx	17	2b	04	7e	ba	77	d6	26	e1	69	14	63	55	21	0c	7d

Figure 2.13   Inverse S-box.

The following is some sample code:

```
sbox =     [0x63, 0x7c, 0x77, 0x7b, 0xf2, 0x6b, 0x6f, 0xc5, 0x30, 0x01,
            0x67, 0x2b, 0xfe, 0xd7, 0xab, 0x76, 0xca, 0x82, 0xc9, 0x7d,
            0xfa, 0x59, 0x47, 0xf0, 0xad, 0xd4, 0xa2, 0xaf, 0x9c, 0xa4,
            0x72, 0xc0, 0xb7, 0xfd, 0x93, 0x26, 0x36, 0x3f, 0xf7, 0xcc,
            0x34, 0xa5, 0xe5, 0xf1, 0x71, 0xd8, 0x31, 0x15, 0x04, 0xc7,
            0x23, 0xc3, 0x18, 0x96, 0x05, 0x9a, 0x07, 0x12, 0x80, 0xe2,
            0xeb, 0x27, 0xb2, 0x75, 0x09, 0x83, 0x2c, 0x1a, 0x1b, 0x6e,
            0x5a, 0xa0, 0x52, 0x3b, 0xd6, 0xb3, 0x29, 0xe3, 0x2f, 0x84,
            0x53, 0xd1, 0x00, 0xed, 0x20, 0xfc, 0xb1, 0x5b, 0x6a, 0xcb,
            0xbe, 0x39, 0x4a, 0x4c, 0x58, 0xcf, 0x51, 0xa3, 0x40, 0x8f,
            0x92, 0x9d, 0x38, 0xf5, 0xbc, 0xb6, 0xda, 0x21, 0x10, 0xff,
            0xf3, 0xd2, 0xcd, 0x0c, 0x13, 0xec, 0x5f, 0x97, 0x44, 0x17,
            0xc4, 0xa7, 0x7e, 0x3d, 0x64, 0x5d, 0x19, 0x73, 0x60, 0x81,
            0x4f, 0xdc, 0x22, 0x2a, 0x90, 0x88, 0x46, 0xee, 0xb8, 0x14,
            0xde, 0x5e, 0x0b, 0xdb, 0xe0, 0x32, 0x3a, 0x0a, 0x49, 0x06,
            0x24, 0x5c, 0xc2, 0xd3, 0xac, 0x62, 0x91, 0x95, 0xe4, 0x79,
            0xe7, 0xc8, 0x37, 0x6d, 0x8d, 0xd5, 0x4e, 0xa9, 0x6c, 0x56,
            0xf4, 0xea, 0x65, 0x7a, 0xae, 0x08, 0xba, 0x78, 0x25, 0x2e,
            0x1c, 0xa6, 0xb4, 0xc6, 0xe8, 0xdd, 0x74, 0x1f, 0x4b, 0xbd,
            0x8b, 0x8a, 0x70, 0x3e, 0xb5, 0x66, 0x48, 0x03, 0xf6, 0x0e,
            0x61, 0x35, 0x57, 0xb9, 0x86, 0xc1, 0x1d, 0x9e, 0xe1, 0xf8,
            0x98, 0x11, 0x69, 0xd9, 0x8e, 0x94, 0x9b, 0x1e, 0x87, 0xe9,
            0xce, 0x55, 0x28, 0xdf, 0x8c, 0xa1, 0x89, 0x0d, 0xbf, 0xe6,
            0x42, 0x68, 0x41, 0x99, 0x2d, 0x0f, 0xb0, 0x54, 0xbb, 0x16]

sboxInv = [0x52, 0x09, 0x6a, 0xd5, 0x30, 0x36, 0xa5, 0x38, 0xbf, 0x40,
            0xa3, 0x9e, 0x81, 0xf3, 0xd7, 0xfb, 0x7c, 0xe3, 0x39, 0x82,
            0x9b, 0x2f, 0xff, 0x87, 0x34, 0x8e, 0x43, 0x44, 0xc4, 0xde,
            0xe9, 0xcb, 0x54, 0x7b, 0x94, 0x32, 0xa6, 0xc2, 0x23, 0x3d,
            0xee, 0x4c, 0x95, 0x0b, 0x42, 0xfa, 0xc3, 0x4e, 0x08, 0x2e,
            0xa1, 0x66, 0x28, 0xd9, 0x24, 0xb2, 0x76, 0x5b, 0xa2, 0x49,
            0x6d, 0x8b, 0xd1, 0x25, 0x72, 0xf8, 0xf6, 0x64, 0x86, 0x68,
            0x98, 0x16, 0xd4, 0xa4, 0x5c, 0xcc, 0x5d, 0x65, 0xb6, 0x92,
            0x6c, 0x70, 0x48, 0x50, 0xfd, 0xed, 0xb9, 0xda, 0x5e, 0x15,
            0x46, 0x57, 0xa7, 0x8d, 0x9d, 0x84, 0x90, 0xd8, 0xab, 0x00,
            0x8c, 0xbc, 0xd3, 0x0a, 0xf7, 0xe4, 0x58, 0x05, 0xb8, 0xb3,
            0x45, 0x06, 0xd0, 0x2c, 0x1e, 0x8f, 0xca, 0x3f, 0x0f, 0x02,
            0xc1, 0xaf, 0xbd, 0x03, 0x01, 0x13, 0x8a, 0x6b, 0x3a, 0x91,
            0x11, 0x41, 0x4f, 0x67, 0xdc, 0xea, 0x97, 0xf2, 0xcf, 0xce,
            0xf0, 0xb4, 0xe6, 0x73, 0x96, 0xac, 0x74, 0x22, 0xe7, 0xad,
            0x35, 0x85, 0xe2, 0xf9, 0x37, 0xe8, 0x1c, 0x75, 0xdf, 0x6e,
            0x47, 0xf1, 0x1a, 0x71, 0x1d, 0x29, 0xc5, 0x89, 0x6f, 0xb7,
            0x62, 0x0e, 0xaa, 0x18, 0xbe, 0x1b, 0xfc, 0x56, 0x3e, 0x4b,
            0xc6, 0xd2, 0x79, 0x20, 0x9a, 0xdb, 0xc0, 0xfe, 0x78, 0xcd,
            0x5a, 0xf4, 0x1f, 0xdd, 0xa8, 0x33, 0x88, 0x07, 0xc7, 0x31,
            0xb1, 0x12, 0x10, 0x59, 0x27, 0x80, 0xec, 0x5f, 0x60, 0x51,
            0x7f, 0xa9, 0x19, 0xb5, 0x4a, 0x0d, 0x2d, 0xe5, 0x7a, 0x9f,
            0x93, 0xc9, 0x9c, 0xef, 0xa0, 0xe0, 0x3b, 0x4d, 0xae, 0x2a,
            0xf5, 0xb0, 0xc8, 0xeb, 0xbb, 0x3c, 0x83, 0x53, 0x99, 0x61,
```

```
            0x17, 0x2b, 0x04, 0x7e, 0xba, 0x77, 0xd6, 0x26, 0xe1, 0x69,
            0x14, 0x63, 0x55, 0x21, 0x0c, 0x7d]


def subBytes(state):
    for i in range(len(state)):
        #print "state[i]:", state[i]
        #print "sbox[state[i]]:", sbox[state[i]]
  state[i] = sbox[state[i]]

def subBytesInv(state):
    for i in range(len(state)):
    state[i] = sboxInv[state[i]]

 state=[1,2,3,4,5,6,7,8,9,10,11,12,13,14,15,16]
 subBytes(state)
 print state

 subBytesInv(state)
 print state
```

If we run we get:

```
Output from S-box: [124, 119, 123, 242, 107, 111, 197, 48, 1, 103,
43, 254, 215, 171, 118, 202]
Inverse S-box: [1, 2, 3, 4, 5, 6, 7, 8, 9, 10, 11, 12, 13, 14, 15,
16]
```

2.5.2 Shift Row Transformation

With this process, the following transformation is applied:

1- First row remains unchanged.
2- Second row has a one-byte circular left shift.
3- Third row has a two-byte circular left shift.
4- Fourth row has a three-byte circular left shift.

For example:

54	33	AB	C1
32	15	8D	BB
5A	73	D5	52
31	91	CC	98

->

54	33	AB	C1
15	8D	BB	32
D5	52	5A	73
98	31	91	CC

For the reverse process, a right shift will be used.

2.5.3 Mix Column Transformation

Within this transformation, each column is taken one at a time and each byte within the column is transformed to a new value based on all four bytes in the column. For each column (a_0, a_1, a_2 and a_3) we have (where we use Galois Multiplication):

$$\begin{bmatrix} a'_0 \\ a'_1 \\ a'_2 \\ a'_3 \end{bmatrix} = \begin{bmatrix} 2\ 3\ 1\ 1 \\ 1\ 2\ 3\ 1 \\ 1\ 1\ 2\ 3 \\ 3\ 1\ 1\ 2 \end{bmatrix} \begin{bmatrix} a_0 \\ a_1 \\ a_2 \\ a_3 \end{bmatrix}$$

The inverse is given by:

$$\begin{bmatrix} a'_0 \\ a'_1 \\ a'_2 \\ a'_3 \end{bmatrix} = \begin{bmatrix} 14 & 11 & 13 & 9 \\ 9 & 14 & 11 & 13 \\ 13 & 9 & 14 & 11 \\ 11 & 13 & 9 & 14 \end{bmatrix} \begin{bmatrix} a_0 \\ a_1 \\ a_2 \\ a_3 \end{bmatrix}$$

The Python code for the mix column transformation for a single column is:

```
from copy import copy

def galoisMult(a, b):
    p = 0
    hiBitSet = 0
    for i in range(8):
        if b & 1 == 1:
            p ^= a
        hiBitSet = a & 0x80
        a <<= 1
        if hiBitSet == 0x80:
            a ^= 0x1b
        b >>= 1
    return p % 256

def mixColumn(column):
    temp = copy(column)
    column[0] = galoisMult(temp[0],2) ^ galoisMult(temp[3],1) ^ \
                galoisMult(temp[2],1) ^ galoisMult(temp[1],3)
    column[1] = galoisMult(temp[1],2) ^ galoisMult(temp[0],1) ^ \
                galoisMult(temp[3],1) ^ galoisMult(temp[2],3)
    column[2] = galoisMult(temp[2],2) ^ galoisMult(temp[1],1) ^ \
                galoisMult(temp[0],1) ^ galoisMult(temp[3],3)
```

```
        column[3] = galoisMult(temp[3],2) ∧ galoisMult(temp[2],1) ∧ \
                    galoisMult(temp[1],1) ∧ galoisMult(temp[0],3)

def mixColumnInv(column):
    temp = copy(column)
    column[0] = galoisMult(temp[0],14) ∧ galoisMult(temp[3],9) ∧ \
                galoisMult(temp[2],13) ∧ galoisMult(temp[1],11)
    column[1] = galoisMult(temp[1],14) ∧ galoisMult(temp[0],9) ∧ \
                galoisMult(temp[3],13) ∧ galoisMult(temp[2],11)
    column[2] = galoisMult(temp[2],14) ∧ galoisMult(temp[1],9) ∧ \
                galoisMult(temp[0],13) ∧ galoisMult(temp[3],11)
    column[3] = galoisMult(temp[3],14) ∧ galoisMult(temp[2],9) ∧ \
                galoisMult(temp[1],13) ∧ galoisMult(temp[0],11)
g = [1,2,3,4]

mixColumn(g)
print 'Mixed: ',g
mixColumnInv(g)
print 'Inverse mixed', g
```

The result gives:

```
Mixed: [3, 4, 9, 10]
Inverse mixed: [1, 2, 3, 4]
```

2.5.4 Add Round Key Transformation

With this transformation, we implement an XOR operation between the round key and the input bits.

```
def addRoundKey(state, roundKey):
    for i in range(len(state)):
        state[i] = state[i] ^ roundKey[i]

state=[1,2,3,4,5,6,7,8,9,10,11,12,13,14,15,16]
roundkey=[2,3,4,5,6,7,8,9,10,11,12,13,14,15,16,1]

addRoundKey(state,roundkey)
print state
addRoundKey(state,roundkey)
print state
```

A sample run gives:
[3, 1, 7, 1, 3, 1, 15, 1, 3, 1, 7, 1, 3, 1, 31, 17]
[1, 2, 3, 4, 5, 6, 7, 8, 9, 10, 11, 12, 13, 14, 15, 16]

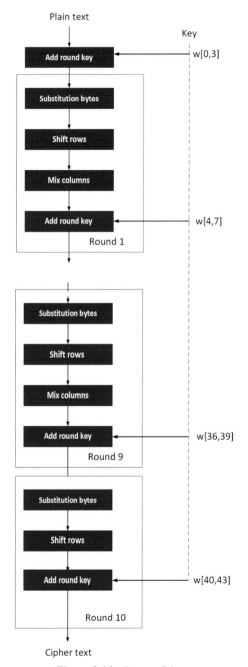

Figure 2.14   Inverse S-box.

2.6 Secret-Key Encryption

Private-key (or secret-key) encryption techniques use a secret key which is only known by the two communicating parties, as illustrated in Figure 2.15. This key can generated by a pass-phrase, or can be passed between the two parties over a secure communications link.

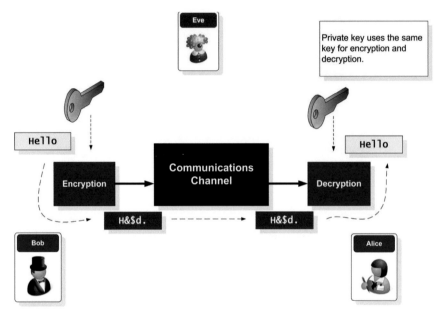

Figure 2.15   Secret key encryption/decryption process.

2.6.1 DES/3-DES

DES (Data Encryption Standard) is a block cipher scheme which operates on 64-bit block sizes. The secret key has only **56 useful bits,** as eight of its bits are used for parity (which gives 2^{56} or 10^{17} possible keys). DES uses a complex series of permutations and substitutions, and the result of these operations is XOR'ed with the input. This is then repeated 16 times using a different order of the key bits each time. DES is a strong code and has never been broken, although several high-powered computers are now available which use brute force to crack the cipher. A possible solution is **3-DES** (or triple DES) which uses DES three times in a row. First to encrypt, next to decrypt and finally to encrypt. This system allows a key-length of around 112 bits. The technique uses two keys and three executions of the

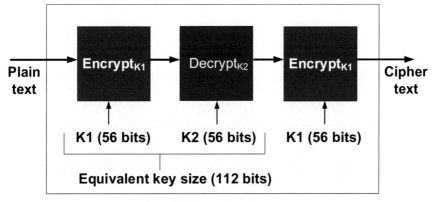

Figure 2.16 Triple DES process.

DES algorithm. A key, K_1, is used in the first execution, then K_2 in the second phase and in the final phase K_1 is used again. These two keys thus give an effective key length of 112 bits, that is 2×64 key bits minus 16 parity bits. The Triple DES process is illustrated in Figure 2.16. With 3-DES it is possible to implement DES, if we use the same keys for K_1 and K_2. The general form is:

Cipher = $E_{K1}(D_{K2}(E_{K1}(P)))$

If we use the same key, we get:

Cipher = $E_{K1}(D_{K1}(E_{K1}(P)))$

which is equivalent to:

Cipher = $E_{K1}(P))$ as $D_{K1}(E_{K1}(P)) = P$

2.6.2 RC4

RC4 is a **stream** cipher designed by RSA Data Security, Inc and was a secret until information on it appeared on the Internet. The secure socket layer (SSL) protocol and wireless communications (IEEE 802.11a/b/g) uses RC4. It uses a pseudo random number generator, and where the output of the generator is XOR'ed with the plaintext. It is a fast algorithm and can use a wide range of key lengths. Unfortunately the same key should not be used twice. Recently a 40-bit key version was broken in eight days without any special computing power.

2.6.3 AES/Rijndael

AES (Advanced Encryption Standard) is a new standard for encryption, and uses 128, 192 or 256 bits. It was selected by NIST in 2001 (after a five year standardisation process). The name Rijndael comes from its Belgium creators: Joan Daemen and Vincent Rijmen. The future of wireless systems (WPA-2) is likely to be based around AES (while WPA uses TKIP which is a session key method based on stream encryption using RC4).

2.6.4 IDEA

IDEA (International Data Encryption Algorithm) is similar to DES. It operates on 64-bit blocks of plaintext, using a 128-bit key, and has over 17 rounds with a complicated mangler function. During decryption this function does not have to be reversed and can simply be applied in the same way as during encryption (this also occurs with DES). IDEA uses a different key expansion for encryption and decryption, but every other part of the process is identical. The same keys are used in DES decryption, but in the reverse order. The key is devised in eight 16-bit blocks; the first six are used in the first round of encryption the last two are used in the second run. It is free for use in non-commercial application and appears to be a strong cipher.

2.6.5 RC5

RC5 is a fast block cipher designed by Ron Rivest for RSA Data Security. It has a parameterized algorithm with a variable block size (32, 64 or 128 bits), a variable key size (0 to 2048 bits) and a variable number of rounds (0 to 255). It has a heavy use of data dependent rotations, and the mixture of different operations.

2.6.6 Skipjack

Skipjack is a secret key encryption algorithm, designed by the NSA, and was used with the Clipper chip. It has an 80-bit key and operates on 64-bit data blocks. The key was thus to be held in escrow using Law Enforcement Access Field (LEAF), and the chip itself was tamperproof, where the key could not be determined. It was a secret but eventually declassified in 1998, and its method was also discovered through observations. NIST no longer certify Skipjack as a recommended method for encryption.

2.6.7 Blowfish

In 1993, Bruce Schneier created Blowfish as a general-purpose private key encryption algorithm, using either a 128-, 192- or a 256-bit encryption key. Unlikely many other encryption methods, it was unpatented, and could be freely used by anyone. It included the concept of a S-box.

2.6.8 Twofish

Bruce Schneier created Twofish as a general-purpose private key encryption algorithm, using either a 128-, 192- or a 256-bit encryption key.

2.6.9 Camellia

Camellia is a block cipher created by Mitsubishi and NTT.

2.6.10 XTEA

XTEA (eXtended TEA) is a block cipher which uses a 64-bit block size and a 128-bit key. It was designed by David Wheeler and Roger Needham at the Cambridge Computer Laboratory, and part of an unpublished technical report in 1997. The following provide some practical examples of symmetric encryption:

📖 **Web link (DES):**   http://asecuritysite.com/encryption/des
   Web link (3-DES):   http://asecuritysite.com/encryption/threedes
   Web link (RC2):   http://asecuritysite.com/encryption/rc2
   Web link (AES):   http://asecuritysite.com/encryption/aes
   Web link (RC4):   http://asecuritysite.com/encryption/rc4
   Web link (Skipjack): http://asecuritysite.com/encryption/skipjack
   Web link (Blowfish): http://asecuritysite.com/encryption/blowfish
   Web link (Camellia): http://asecuritysite.com/encryption/camellia
   Web link (XTEA):   http://asecuritysite.com/encryption/xtea

2.7 Key Entropy

The encryption key length is only one of the factors that can give a pointer to the security of the encryption process. Unfortunately most encryption processes do not use the full range of keys, as the encryption key itself is typically generated using an ASCII pass-phrase, which is often hashed. For example wireless systems typically use a pass phrase to generate the encryption key. Thus for 64-bit encryption, only five alphanumeric characters (40-bits) are used and 13 alphanumeric characters (104 bits) are used for 128-bits encryption[4]. These characters are often created from well-known words and phrases such as:

Nap1

Whereas 128-bit encryption could use:

NapierStaff1

Thus, this approach typically reduces the number of useable keys, as the keys themselves will be generated from dictionaries, such as:

About
Apple
Aardvark

where keys generated from obvious pass phases. On the other hand:

xyRg54d
io2Fddse

which will be less common (but could be checked if the standard dictionary pass phases did not yield a result).

Entropy measures the amount of unpredictability, and in encryption it relates to the degree of uncertainty of the encryption process. If all the keys in a 128-bit key were equally likely, then the entropy of the keys would be 128 bits. Unfortunately, due to the problems of generating keys through pass phrases, the entropy of standard English can be less than 1.3 bits per character, and typical passwords at less than 4 bits per character. Thus for a 128-bit encryption key in wireless, and using standard English, gives a maximum entropy of only 16.9 bits (1.3 times 13), which equates to a small size size of around 17 bits. So rather than having 20,282,409, 603,651,670,423,947,251,286,016 (2^{104}) possible keys, there are only 131,072 (2^{17}) keys.

---

[4]In wireless, a 64-bit encryption key is actually only a 40 bit key, as 24 bits is used as an initialisation vector. The same goes for a 128-bit key, where the actual key is only 104 bits.

As an example, let's say an organisation uses a 40-bit encryption key, and that it has the following possible phases:

Napier, napier, napier1, Napier1, napierstaff, Napierstaff, napierSoc, napierSoC, SoC, Computing, DCS, dcs, NapierAir, napierAir, napierair, Aironet, MyAironet, SOCAironet, NapierUniversity, napieruniversity

and which gives 20 different phases. The the entropy (in bits) is then equal to:

$$Entropy(bits) = log_2(N)$$
$$= log_2(20)$$
$$= \frac{log_{10}(20)}{log_{10}(2)}$$
$$= 4.3$$

So, the entropy of the 40-bit encryption key is only 4.3 bits.

Unfortunately many password systems and operating systems base their encryption keys on **pass-phases**, where the secret key is protected by a password. This is a major problem, as a strong encryption key can be used, but the password which protects it is open to a dictionary attack, and that the overall entropy will be low.

With key entropy we can thus measure the equivalent number of bits in a key when taking into account the number of keys (or passwords) that are actually used. For example, let's say that we generate our keys from a password which is 8 characters long, using the characters from a to z ([a-z]). We then have:

$$8^{26} \text{ different passwords which is } 3.02 \times 10^{23}$$

Now, we can use key entropy which will determine the equivalent key size for the limited range of key.

$$Key\ Entropy = log_2(Phrases) = \frac{log_{10}(Phrases)}{log_{10}(2)}$$

For our 8 character passwords, we can determine the equivalent key size of:

$$Key\ Entropy = log_2(26^8) = \frac{log_{10}(26^8)}{log_{10}(2)} = 37.6\ bits$$

Thus if we are using a 128-bit encryption key, we are not using an equivalent of 50 bits, which considerably reduces the strength of the encryption process. Table 2.7 outlines the equivalent entropy for differing pass phrase definitions.

Table 2.7   Key entropy for an 8 character password

Password Definition	Number of Possible Characters	Total Number of Passwords	Entropy (Bits)
[0-9]	10	100000000	26.6
[a-z]	26	2.08827×10^{11}	37.6
[a-zA-Z]	52	5.34597×10^{13}	45.6
[a-zA-Z0-9]	62	2.1834×10^{14}	47.6
[a-zA-Z0-9$%!@+=]	68	4.57163×10^{14}	48.7

📖 **Web link (Entropy):** http://asecuritysite.com/Encryption/en

2.8 OpenSSL

The OpenSSL library is used by many applications to implement cryptography. It started with **Eric A Young** and Tim Hudson, in Dec 1998, who created the first version of OpenSSL (SSLeay – SSL **E**ric **A Y**oung), and which then became Version 0.9.1. Eric finished his research and was involved with Cryptsoft (www.cryptsoft.com) where he gained the honour of having a Distinguished Engineer role. After Eric left, it was then left to Steve Marequess (from the US) and Stephen Henson (from the UK) to continue its development through the OpenSSL Software Foundation (OSF).

Web link (OpenSSL): http://asecuritysite.com/encryption/opensslp

The lab at the end of this chapter will outline the usage of OpenSSL.

2.9 Pohlog-Hellman

In the Pohlog-Hellman method we have two secret values, where one is used to encrypt and the other to decrypt. We first select a prime number, such as:

$p = 5$

Now we determine a value for the encryption key (*e*) and the decryption key (*d*) so that:

$e \times d \ (\text{mod } (p\text{-}1)) = 1$

For example:

$$e \times d \pmod{4} = 1$$

If we select $(e.d) = 21$, then we can use $e=3$, $d=7$:

$$3 \times 7 \pmod{4} = 1$$

To cipher a message (*Message*) we raise the message to the power of e and then take the modulus of p:

Cipher = Message^e mod p

To decrypt we take the cipher (*Cipher*) and raise to the power of d, and take the modulus of p:

Plaintext = Cipher^d (mod p)

For example if we have a *Message* of 2 we get:

$$Cipher = 2^3 \pmod{5} = 3$$

and:

$$Message = 3^7 \pmod{5} = 2$$

This matches the original message.

📖 **Web link (Pohlog-Hellman):** http://asecuritysite.com/encryption/poh

We can solve $(e \times d)$ mod (N) using the Euclidean method:

📖 **Web link (Euclidean):** http://asecuritysite.com/encryption/inve

2.10 Lab/tutorial

The lab and tutorial related to this chapter is available on-line at:

http://asecuritysite.com/crypto02

3

Hashing

3.1 Introduction

The key focus of cryptography is to provide privacy, prove identity and show integrity. We have seen that a secret key can be used to define secrecy, and that public key can be used to pass this key, but we need a method of proving an identity and also to check the integrity of a message. Hashing is normally used to either hide the original contents of a message (such as hiding a password), or to check the integrity of data.

In the past a simple checksum was often used to check the integrity of data, such as adding a checksum value to a list of numbers so that the total would be a multiple of 9. This would work for many errors, but there is a chance that the errors would produce a valid checksum. For example if 4, 5 and 13 were to be sent, the checksum would be 5 (as the total becomes 27 which is a multiple of 9). An error in one of the values such as for 5, 5 and 13 with a checksum of 9 would give an error, and the data would be rejected. Unfortunately if the values were changed to 13, 5 and 13, the checksum of 5 would be valid (36) and the receiver would think the values were correct. Thus in using a hashing method to check for integrity we aim to provide a fairly unique value, and where it is almost impossible to change the bits in the data to produce the same hash signature.

3.2 Hashing Methods

The fingerprinting of data was aided by Ron Rivest, in 1991, with the MD5 algorithm (Figure 3.1). It uses a message hash which is a simple technique to mix up the bits within a message, and uses exclusive-OR operations, bit-shifts, and/or character substitutions. These are typically used to either provide: some form of conversion between binary and text; support for the storage of passwords; or in authentication techniques to create a unique signature for a given sequence of data. The main techniques are:

- **Base-64 encoding**. This is used in electronic mail, and is typically used to change binary data into a standard 7-bit ASCII form. It takes 6-bit characters, at a time, and converts them to a printable character.
- **UNIX password hashing**. This is used in the **passwd** file which contains hashed version of passwords. It is a one-way function, so that it is typically not possible to guess the password from the hashed code. But if the hashed code for the given word is known, Eve can determine the password. Weak passwords can obviously be broken with a dictionary attack, where an off-line program can be used to search through a known dictionary of common words and which matches the hashed codes against the one in the password file. These problems have been partially overcome with a shadow password file (/etc/shadow) and which can only be viewed by the administrator.
- **NT password hashing**. In many versions of Microsoft Windows, there was is no password file, as in UNIX. These passwords are stored as hashes in the system registry. It is thus open to a dictionary attack in the same way that UNIX is exposed to it. Along with this, it has several other weaknesses which reduce the strength of the password. This includes converting the password into upper case between hashes, and in splitting it into two parts.
- **MD5**. This is used in several encryption and authentication methods, and is standardized in RFC1321. It produces a 32 hexadecimal character output (128-bits), and which can also be converted into a text format, as shown in Figure 3.1.
- **SHA** (Secure Hash Algorithm). This is an enhanced message hash, which produces a 40 hexadecimal character output (160-bits). It will thus produces a 40 hexadecimal character signature for any message from 1 to 2,305,843,009,213,693,952 characters. At present it is computationally difficult to produce two messages which produce the same hashed result, as illustrated in Figure 3.2. For SHA-2, it is possible to generate 256-, 384- or 512-bit signatures.

📖 **Web link:** http://asecuritysite.com/encryption/md5

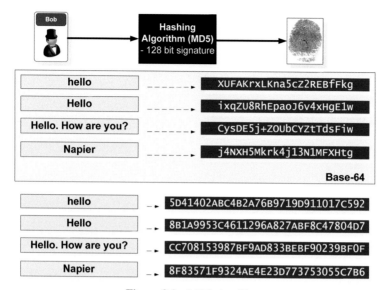

Figure 3.1 MD5 algorithm.

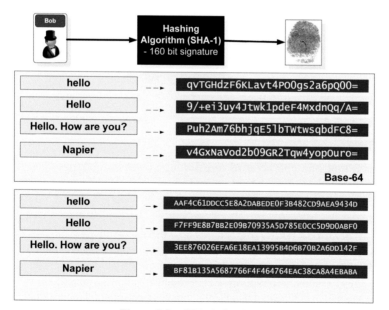

Figure 3.2 SHA-1 algorithm.

For example, if a message was:

```
Hello, how are you?
Are you feeling well?

Fred.
```

then the MD5 hash for this is:

```
518bb66a80cf187a20e1b07cd6cef585
```

For example, the text:

```
Security and mobility are two of the mОst important issues on the
Internet, as they will allow users to secure their data transmissions,
and also break their link their physical connections.
```

we get an MD5 hash of:

```
91E2AB34D0B2DE28700A0E94071BCC46
```

where as:

```
Security and mobility are two of the mаst important issues on the
Internet, as they will allow users to secure their data transmissions,
and also break their link their physical connections.
```

gives:

```
C0DA7FCC869C1E94687BF1CABAAB780B
```

It can be seen that one character of a difference complete changes the hash value. We can do the same for system and binary files, such as determining the hash code for the DLL's in Windows\system32 folder:

```
455D04D3EBDE98FB5AB92B7363DFF33D c:\windows\system32\6to4svc.dll
12B4C8208B5146C8D17F3F502E00A540 c:\windows\system32\aaaamon.dll
441086F355F0DEA94621984C9A3BE765 c:\windows\system32\acctres.dll
A9517EC6F843959566692570390C457F c:\windows\system32\acledit.dll
E92003F404A889BBADF70E8743E498B9 c:\windows\system32\aclui.dll
...
```

This allows us to check that files have not been changed. The hash function is thus useful in creating a one-way function which cannot be reversed. It has a wide scope of applications, from authenticating: users and devices; applications; and DLLs, to fingering data, files and even the complete contents of disk drives.

3.3 Problems with Hashes

One-way hashes are used for digital fingerprints and for secure password storage. Typical methods are NT hash, MD4, MD5, and SHA-1, and are used to convert plaintext into a hash value (Figure 3.3). It has applications in storing passwords, such as in Unix/Windows and on Cisco devices (Figure 3.4). A weakness of one-way hashing is that the same piece of plaintext will result in the same ciphertext (unless some salt is applied). Thus it is possible for an intruder to generate a list of hash values for a standard dictionary (Figure 3.5), and possibly determine the plaintext which makes the one-way hash. Important factors with hash signatures are:

- **Collision**. This is where another match is found, no matter the similarity of the original message. This can be defined as a Collision attack.
- **Similar context**. This is where part of the message has some significance to the original, and generates the same hash signature. This is defined as a Pre-image attack.
- **Full context**. This is where an alternative message is created with the same hash signature, and has a direct relation to the original message. This is an extension to a Pre-image attack.

In 2006, for example, it was shown that MD5 can produce a collision within one minute, whereas it was 18 hours for SHA-1.

📖 **Web link (MD5):** http://asecuritysite.com/encryption/md5

A collision occurs when there are two different values that produce the same hash signature. In the following example of MD5 we use a hex string to define the data element (as the characters would be non-printing). We then have different values which create the same hash signature (the red characters identify the changes in the input data):

```
d131dd02c5e6eec4693d9a0698aff95c   2fcab58712467eab4004583eb8fb7f89
55ad340609f4b30283e488832571415a   085125e8f7cdc99fd91dbdf280373c5b
d8823e3156348f5bae6dacd436c919c6   dd53e2b487da03fd02396306d248cda0
e99f33420f577ee8ce54b67080a80d1e   c69821bcb6a8839396f9652b6ff72a70
```

```
d131dd02c5e6eec4693d9a0698aff95c    2fcab50712467eab4004583eb8fb7f89
55ad340609f4b30283e4888325f1415a    085125e8f7cdc99fd91dbd7280373c5b
d8823e3156348f5bae6dacd436c919c6    dd53e23487da03fd02396306d248cda0
e99f33420f577ee8ce54b67080280d1e    c69821bcb6a8839396f965ab6ff72a70
```

which should both give an MD4 hash of:
79054025255FB1A26E4BC422AEF54EB4

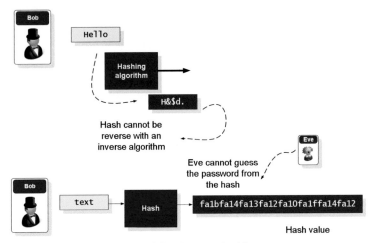

Figure 3.3   One-way hashing.

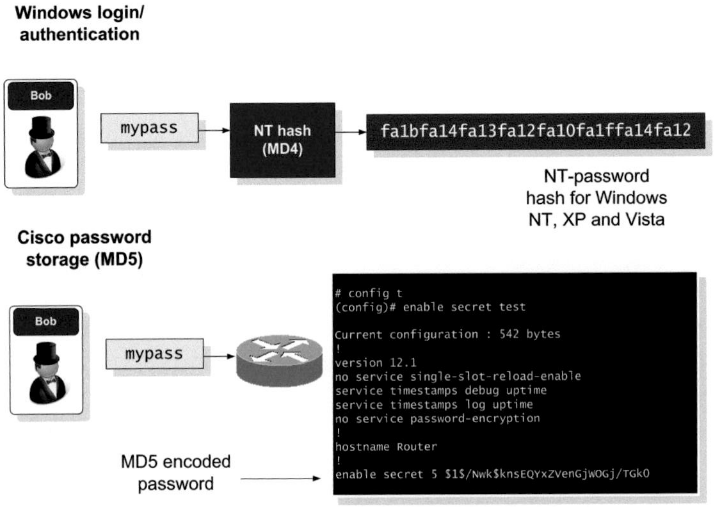

Figure 3.4   Application of one-way hashing.

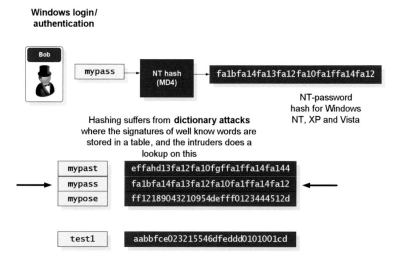

Figure 3.5 Application of one-way hashing.

3.3.1 Hash Cracking

Moore's Law predicted that computing power doubles every 18 months or so, so if we have a code which takes 100 years to crack, within 18 months, with the equivalent cost of a system, it will only take 50 years. To simplify things we can predict that computing power doubles every year, so we find that a code which takes 100 years to crack, will, after 10 years, only takes a matter of weeks to crack (7 weeks). But the trend of improving hardware is now being overtaken by the Cloud, and many of the standard cryptography methods we have been using for years is now being defined as crackable.

The first to feel the heat is MD5 and has been a standard method for creating a digital fingerprint of data. It is used extensively in checking that data has not been changed and in providing identity. In the past it has been used to store hashed values of passwords, but its application in this area is reducing fast, as many of the common hashed MD5 values for words have been resolved. One of the key things that is important for MD5 is that the different data does not produce a collision – where different data, especially in the same type of context, does not produce the same hash signature. Mat McHugh showed that he could produce the same hash signature for different images, using hashclash, and for just 65 cents on the Amazon GPU Cloud. For 10 hours of computing on the Amazon GPU Cloud, Mat created these two images which generate the same hash signature (Figure 3.6). If we check the hash signatures we get:

```
C:\openssl>openssl md5 hash01.jpg
MD5(hash01.jpg)= e06723d4961a0a3f950e7786f3766338

C:\openssl>openssl md5 hash02.jpg
MD5(hash02.jpg)= e06723d4961a0a3f950e7786f3766338
```

Figure 3.6   Application of one-way hashing.

Mat used the birthday attack which is one of the methods used for a brute-force attack, and is based on the birthday problem in probability theory. It defines that if we take a set of *n* randomly chosen people, and then there will be a certain percentage who will have the same birthday. A group size of only 70 people results in a 99.9% chance of two people sharing the same birthday. Using this method, if we take an *m*-bit output there are 2^m messages, and the same hash value would only require $2^{m/2}$ random messages. To understand the birthday attack, let us start with the probability that one person will not have the same birthday as themselves:

$$P(no\_match) = \frac{365}{365}$$

For two people, we have 364 days to choose from, so the probability that they will not have the sample birthday is:

$$P(no\_match) = \frac{365}{365}\frac{364}{365}$$

Then for three people we only have 363 days left after checking the first two people to give:

$$P(no\_match) = \frac{365}{365}\frac{364}{365}\frac{363}{365}$$

If we have n people, we can then write this as:

$$P(match) = 1 - \frac{365!}{365^n(365-n)!}$$

If we take 10 people in a room, the probablity that there will be a match in birthdays is:

$$P(match) = 1 - \frac{365!}{365^{10}(355)!}$$
$$= 1 - \frac{365 \times 364 \times 363 \times 362 \times 361 \times 360 \times 359 \times 358 \times 357 \times 356}{366^{10}}$$
$$= 0.117$$

So for 10 people in a room, there an approximate 12% chance that they will have the same birthday. Then using the equation given below, we can calculate:

- Same birthday with 20 people gives 41.14%.
- Same birthday with 30 people gives 70.63%.
- Same birthday with 60 people gives 99.41%.

📖 **Web link (Birthday):** http://asecuritysite.com/encryption/birthday
Web link (Birthday – Big Int): http://asecuritysite.com/encryption/birthday2

$$P(match) = 1 - \frac{M!}{M^n(M-n)!}$$

Computer programs too can be modified to give the same hash value. For example these files (goodbye.exe and hello.exe), produce different outputs but have the same hash value:

```
C:\openssl>openssl md5 erase.exe
MD5(erase.exe)= cdc47d670159eef60916ca03a9d4a007

C:\openssl>openssl md5 hello.exe
MD5(hello.exe)= cdc47d670159eef60916ca03a9d4a007

C:\openssl>erase.exe
This program is evil!!!
Erasing hard drive...1Gb...2Gb... just kidding!
Nothing was erased.

C:\openssl>hello.exe
Hello, world!
```

📖 **Web link:** http://asecuritysite.com/files01.zip

3.4 Salting the Hash Value

All of the methods previous covered allow for the easy reverse of the encryption or encoding process if the key is known, apart from one-way hashing where it should be almost impossible to reverse back the hashed value to the original data. Unfortunately, technology has moved on since the creation of hashing methods, and they can often be cracked using brute force (where an intruder keeps trying hashing values until the output matches the hash), using a dictionary of common passwords, or use a rainbow table (which has a pre-compiled list of hash values).

These days MD5 and SHA-1 are seen as weak for many reasons, so we often start with SHA-256 (which produces a 256-bit hashed value):

Hash = sha256(password)

Web link (MD5 with salt): http://asecuritysite.com/encryption/salt

Unfortunately this is weak from both a dictionary attack and brute force, so we add some salt:

Hash = salt + sha256(salt + password)

Now this becomes more difficult as the same password is highly likely to produce a different output. With simple hashing, if one hashed password was cracked, all the other passwords with the same value will also be cracked.

The weakness is that the salt requires to be stored with the hashed password, so the intruder just uses a fast computer – such as using NVIDIA graphics cards on the Amazon Cloud – and tries the most common passwords, and is often able to determine the original password. The reason this happens is become SHA256 has been designed to be fast, so the intruder uses this for their advantage, and can quickly try lots of passwords. If the password is weak and in a dictionary it is relatively easy for the intruder.

MD5 and SHA-1 produce a hash signature, but this can be attacked by rainbow tables. Bcrypt is a more powerful hash generator for passwords and uses salt to create a non-recurrent hash. It was designed by Niels Provos and David Mazières, and is based on the Blowfish cipher. With this link, if you keep pressing the "Generate Hash" button you should get a unique value each time:

📖 **Web link (Bcrypt):** https://asecuritysite.com/encryption/bcrypt

3.5 Common Hashing Methods

There are a wide variety of hashing methods, and can be classified as:

- **General hashes**. This includes the main standardised hashing techniques, such as MD5, SHA1, SHA256 and SHA512.
- **UNIX hashes (with salt)**. This includes ARP1, PBKDF2, PHPASS, DES, MD5, Bcrypt, Sun MD5, SHA1, SHA256 and SHA512.
- **Microsoft Windows hashes**. This includes LM, NTLM, DCC and DCC2.
- **LDAP hashes**. This includes MD5, MD5 (Salted), SHA, SHA (Salted), MD5 (Crypt).
- **Database hashes**. This includes MS SQL 2000, MS SQL 2005, My SQL 323, My SQL 41, Postgres, Oracle 10, and Oracle 11.
- **Others**. This includes Cisco PIX and Cisco Type 7.

3.5.1 LM Hashing

LM Hash is used in many versions of Microsoft Windows operating systems to store user passwords that are fewer than 15 characters long. It is a fairly weak security implementation and can be easily broken using standard dictionary lookups. More modern versions of Windows use SYSKEY to encrypt passwords. The LM hash uses the DES encryption method, and creates an encryption key from the user's password, and encrypting a string of "KGS!+#$%". Its operation is:

- Converting the user's passwords into uppercase, and then NULL-pad to up to 14 bytes. For example "napier" becomes "NAPIER0000 0000" where 0 represents a NULL character (zero value in ASCII).
- The 14-byte password is then split into two 7-byte halves.
- The 7-byte values are used to create two 64-bit DES keys (with the addition of a parity bit for every seven bits.
- Each key uses DES (with ECB) to encrypt the string "KGS!+#$%", which gives two 8-byte cipher values.
- The resulting two values are then concatenated to give a 16-byte value, and thus gives the LM hash.

With NTLM, each of the characters in the input password are converted into Unicode (16-bits character representation – for example an 'a' is 0x61 in ASCII, so its representation in Unicode is 0x0061). An MD4 signature is then taken of this string, and which results in a 128-bit code. While a vast improvement on LM hash, there was no place for a salt value, so once an intruder knows the mapping between the hashed value and the original password, they would easily map them.

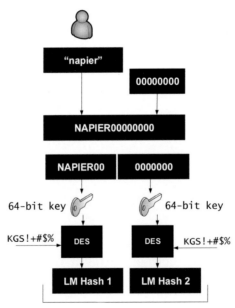

Figure 3.7   LM Hashing.

NT LAN Manager (NTLM) is used in more modern Microsoft Windows systems (Windows XP, Windows Visa, Windows 7 and Windows NT). Rather than using DES it relies on the MD4 hashing algorithm with a series of mathematical calculations. MD4 supports both upper and lower case letters, and does not split the passwords in chunks. Unfortunately, as with LM Hash, it does not use salt.

With NTLM, each of the characters in the input password are converted into Unicode. It uses a Little Endian format for the data, so the "hello" is stored as (Figure 3.8):

```
h\0 e\0 lo \l\0 e0
```

where \0 is 0x00. In C# this is how we convert the password into a Unicode Little Endian format:

```
string GetUnicodeString(string s)
{
    StringBuilder sb = new StringBuilder();
    foreach (char c in s)
    {
        sb.Append((char)c);
        sb.Append((char)0);
    }
    return sb.ToString();
}
```

An **MD4** signature is then taken of this string, and which results in 128-bit code. While a vast improvement on the LM hash, there was no place for a salt value, so once an intruder knew the mapping between the hashed value and the original password, they would easily map them. If you are interested, here is MD2 and MD4:

📖 **Web link (MD2 and MD4):** http://asecuritysite.com/encryption/md2

When you go to this link, select "Unicode Little Endian format" and, for "**hello**", you should get:

```
066DDFD4EF0E9CD7C256FE77191EF43C
```

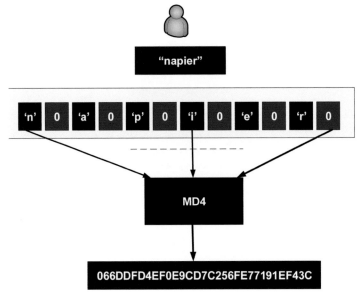

Figure 3.8 NTLM Hashing.

For example for LM Hash:

hashme gives: FA-91-C4-FD-28-A2-D2-57-**AA-D3-B4-35-B5-14-04-EE**
network gives: D7-5A-34-5D-5D-20-7A-00-**AA-D3-B4-35-B5-14-04-EE**
napier gives: 12-B9-C5-4F-6F-E0-EC-80-**AA-D3-B4-35-B5-14-04-EE**

Notice that the right-most element of the hash is always the same, if the password is less than eight characters. With more than eight characters we get:

networksims gives: D7-5A-34-5D-5D-20-7A-00-38-32-A0-DB-BA-51-68-07
napier123 gives: 67-82-2A-34-ED-C7-48-92-B7-5E-0C-8D-76-95-4A-50

For "hello" we get:

LM: FD-A9-5F-BE-CA-28-8D-44-AA-D3-B4-35-B5-14-04-EE
NTLM: 06-6D-DF-D4-EF-0E-9C-D7-C2-56-FE-77-19-1E-F4-3C

We can check these with a Python script:

```
import passlib.hash;
string="hello"
print "LM Hash:"+passlib.hash.lmhash.encrypt(string)

print "NT Hash:"+passlib.hash.nthash.encrypt(string)
```

which gives:

```
LM Hash: fda95fbeca288d44aad3b435b51404ee
NT Hash: 066ddfd4ef0e9cd7c256fe77191ef43c
```

 📖 **Web link (LM/NTLM):** http://asecuritysite.com/encryption/lmhash

Overall a major problem with NTLM, is that the LM Hash is stored along side the NTLM hash. An example of an export of the hashed passwords in Microsoft Windows is (as a pwdump):

```
bill:FDA95FBECA288D44AAD3B435B51404EE:066DDFD4EF0E9CD7C2
56FE77191EF43C:::
```

where both the LM and NTLM hashes are stored for "hello".

3.5.2 APR1 (MD5 with Salt)

The Apache-defined APR1 format addresses the problems of brute forcing an MD5 hash, and basically iterates the hash value 1,000 times. This considerably slows an intruder as they try to crack the hashed value. The resulting hashed string contains "$apr1$" to identify the method and uses a 32-bit salt value. We can use htpassword on Openssl to compute the hashed string (where "bill" is the user and "hello" is the password):

```
# htpasswd -nbm bill hello
bill:$apr1$PkWj6gM4$XGWpADBVPyypjL/cL0XMc1

# openssl passwd -apr1 -salt PkWj6gM4 hello
$apr1$PkWj6gM4$XGWpADBVPyypjL/cL0XMc1
```

We can also create a simple Python program with the passlib library, and add the same salt as the example above:

```
import passlib.hash;

salt="PkWj6gM4"
string="hello"
print "APR1:"+passlib.hash.apr_md5_crypt.encrypt(string, salt=salt)
```

The output is then:

```
APR1:$apr1$PkWj6gM4$XGWpADBVPyypjL/cLOXMc1
```

📖 **Web link (APR1):** http://asecuritysite.com/encryption/apr1

3.5.3 SHA1, SHA256 and SHA512

While APR1 has a salted value, the SHA method has for storing passwords does not have a salted value. SHA produces a 160-bit signature, thus can contain a larger set of hashed value than MD5, but because there is no salt it can be open to rainbow table attacks, and also brute force. The format for the storage of the hashed password on Linux systems is:

```
# htpasswd -nbs bill hello
bill:{SHA}qvTGHdzF6KLavt4PO0gs2a6pQ00=
```

We can also generate salted passwords, and can use the Python script of:

```
import passlib.hash;
salt="8sFt66rZ"
string="hello"
print "SHA1:"+passlib.hash.sha1_crypt.encrypt(string, salt=salt)
print "SHA256:"+passlib.hash.sha256_crypt.encrypt(string, salt=salt)
print "SHA512:"+passlib.hash.sha512_crypt.encrypt(string, salt=salt)

SHA-512 salts start with $6$ and are up to 16 chars long.
SHA-256 salts start with $5$ and are up to 16 chars long.
```

Which produces:

```
SHA1:$sha1$480000$8sFt66rZ$klAZf7IPWRN1ACGNZIMxxuVaIKRj
SHA256:$5$rounds=535000$8sFt66rZ$.YYuHL27JtcOX8WpjwKf2VM876kLTGZHsHwCB
bq9xTD
SHA512:$6$rounds=656000$8sFt66rZ$aMTKQHl60VXFjiDAsyNFxn4gRezZOZarxHaK.
TcpVYLpMw6MnXOlyPQU06SSVmSdmF/VNbvPkkMpOEONvSd5Q1
```

3.5.4 PHPass

Phpass is used as a hashing method by WordPress and Drupal (Figure 3.9). It is public domain software and used within PHP applications. The three main methods used are:

- CRYPT_BLOWFISH. This is the most secure method, and relates to the Bcrypt hashing method.
- CRYPT_EXT_DES. This method uses DES encryption.
- MD5. This is the least preferred method and simply uses an MD5 digest.

The output uses the following to identify the differing types:

- "P". Standard.
- H. Phpbb.
- "1S"; Drupal with SHA-512-like digests.

A sample run with "password" and salt of "ZDzPE45C" for seven rounds gives:

```
$P$5ZDzPE45Ci.QxPaPz.03z6TYbakcSQ0
```

Where it can be seen that the salt value is placed after "P5" ("ZDzPE45C"), and after this there are 22 Base-64 characters (giving a 128-bit hash signature).

We can check the output against the following Python code:

```
import passlib.hash;
string = "password"
salt="ZDzPE45C"
print passlib.hash.phpass.encrypt(string, salt=salt,rounds=7)
```

which should give: P5**ZDzPE45C**i.QxPaPz.03z6TYbakcSQ0.

The code uses seven rounds – to save processor time – but it can vary between 7 and 30. The number of iterations is 2^{rounds}. In this case the hash is "i.QxPaPz.03z6TYbakcSQ0" which is a 128-bit hash signature.

📖 **Web link (phpass):** http://asecuritysite.com/encryption/phpass

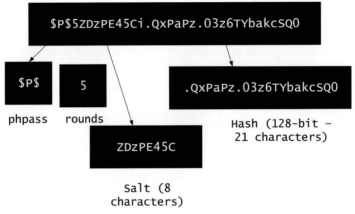

Figure 3.9 Phpass hashing.

Cisco Hashes

Much of the focus of protecting networks is around the server and host infrastructures, but intruders can often use simpler devices to gain a foothold, as these devices will typically have lower security levels and weaker passwords.

This can include a range of Cisco networked devices which store password in a "hashed" format, such as for a Cisco phone, Cisco switch, or Cisco router. An example configuration is:

```
hostname routerA
!
aaa new-model
aaa authentication login default local
aaa authentication ppp default if-needed local
enable secret 5 <hashvalue>
!
username fred password 7 <hashvalue>
username remote password 7 <hashvalue>
username guest password 7 <hashvalue>
```

The problems comes in with the range of passwords that Cisco configuration uses. Two typical password hashing methods are Type 5 and Type 7. For Type 5 we get a simple MD5 hash with a salted value:

```
enable secret 5 $1$iUjJ$rBb3h2PSRAmYieJ.xI12x1
```

where: "$1" identifies that this is a salted MD5 hash; "iUjJ" is the salt used; and where the hash is "rBb3h2PSRAmYieJ.xI12x1".

The privileged password has thus been hashed and salted with MD5, and we can check this with OpenSSL (using a password of "password"):

```
C:\openssl>openssl passwd -1 -salt iUjJ password
$1$iUjJ$rBb3h2PSRAmYieJ.xI12x1
```

and where we can see the hashed passwords match. The salt is small here, and the salt is being stored alongside the hash, so the intruder just pulls off the salt and tries some common passwords, and will easily crack it. For John The Ripper, it takes off the salt, and tries its dictionary of command words, and will have no problems with simple passwords, even which are salted.

With Type 7, there's very little challenge as it's not a hashing method, just an encoded one. If we use:

```
username bill password 7 082949420516
```

It is a fairly easy take to reverse this back with a simple operation.

While new methods use the proper hashing of passwords, with a reasonable challenge, there are often many devices on the network which use weak hashing methods. Type 7 passwords are especially are bad, and the hashed MD5 values are hardly much better.

Type 7 breaks many cryptography rules, including that it gives away the length of the password. For example "hellohowareyou" generates:

```
141F1707000B2224332921303B1C12
```

which gives 30 hex characters ... 14+1 characters, for "hello" we generate "082949420516" which is 6+1 characters.

📖 **Web link (Cisco):** http://asecuritysite.com/encryption/phpass

LDAP Hashes

LDAP (Lightweight Directory Access Protocol) is a protocol where entities can be mapped to a global infrastructure. It is based on X.500 and is included within Active Directory. It uses a number of different hashing methods to hash passwords. The main method uses a number of iterations for:

- {MD5}. MD-5 without salt [RFC1321].
- {SMD5}. Salted MD-5 [RFC1321].
- {SHA}. SHA-1 without salt [FIPS-180-4].
- {SSHA} salted SHA-1 [FIPS-180-4].
- {SHA256}. SHA-256 without salt [FIPS-180-4].
- {SSHA256}. Salted SHA-256 [FIPS-180-4].
- {SHA384}. SHA-384 without salt [FIPS-180-4].
- {SSHA384}. Salted SHA-384 [FIPS-180-4].
- {SHA512}. SHA-512 without salt [FIPS-180-4].
- {SSHA512}. Salted SHA-512 [FIPS-180-4].

Salting the password protects the LDAP hash from a rainbow table attack, but it is still open to brute force attacks. Some of the algorithms hash for a given number of interactions. The greater the number of iterations, the longer the hash will take to crack.

📖 **Web link (phpass):** http://asecuritysite.com/encryption/ldap

PBKDF2

PBKDF2 (Password-Based Key Derivation Function 2) is defined in RFC 2898 and generates a salted hash. Often this is used to create an encryption key from a defined password, and where it is not possible to reverse the password from the hashed value. It is used in TrueCrypt to generate the key required to read the header information of an encrypted drive, and which stores the encryption keys.

PBKDF2 is also used in WPA-2 (Figure 3.10). Its main focus is to produced a hashed version of a password, and includes a salt value to reduce the opportunity for a rainbow table attack. It generally uses over 1,000 iterations in order to slow down the creation of the hash, so that it can overcome brute force attacks. The generalised format for PBKDF2 is:

$$DK = PBKDF2(Password, Salt, MInterations, dkLen)$$

where Password is the pass phrase, Salt is the salt, MInterations is the number of iterations, and dklen is the length of the derived hash.

In WPA-2, the IEEE 802.11i standard defines that the pre-shared key is defined by:

PSK = PBKDF2(PassPhrase, ssid, ssidLength, 4096, 256)

In TrueCrypt, which encrypts file systems, we use PBKDF2 to generate the key (with salt) and which will decrypt the header, and reveal the keys which have been used to encrypt the disk (using AES, 3DES or Twofish). We use:

byte[] result = passwordDerive.GenerateDerivedKey(16,

ASCIIEncoding.UTF8.GetBytes(message), salt, 1000);

which has a key length of 16 bytes (128 bits - dklen), uses a salt byte array, and 1000 iterations for the hash (Minterations). The resulting hash value will have 32 hexadecimal characters (16 bytes).

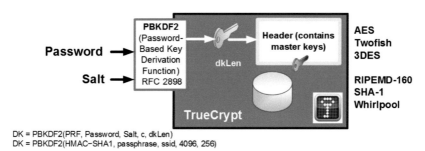

Figure 3.10 PBKDF2.

  ☐ **Web link (PBKDF2):** http://www.asecuritysite.com/encryption/PBKDF2_2

Bcrypt

MD5 and SHA-1 produce a hash signature, but this can be attacked by rainbow tables. Bcrypt (Blowfish Crypt) is a more powerful hash generator for passwords and uses salt to create a non-recurrent hash. It was designed by Niels Provos and David Mazières, and is based on the Blowfish cipher. Bcrypt is used as the default password hashing method for BSD and other systems.

Overall it uses a 128-bit salt value, which gives 22 Base-64 characters (32 hex values). It can use a number of iterations, and which slows down the brute-force cracking of the hashed value. For example, "Hello" with a salt value of "$2a$06$NkYh0RCM8pNWPaYvRLgN9." gives:

$2a$06$NkYh0RCM8pNWPaYvRLgN9. LbJw4gcnWCOQYIom0P08UE ZRQQjbfpy

As illustrated in Figure 3.11, the first part is "$2a$" (or "$2b$"), and then followed by the number of rounds used. In this case is it **6 rounds** which is 2^6 iterations (where each additional round doubles the hash time). The 128-bit (22 character) salt values comes after this, and then finally there is a 184-bit hash code (which is 31 characters).

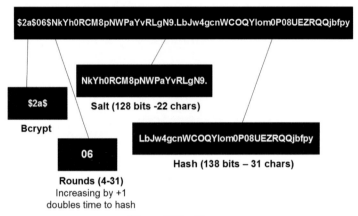

Figure 3.11 Bcrypt.

The slowness of Bcrypt is highlighted with an AWS EC2 server benchmark using hashcat [1]:

- Hash type: MD5 Speed/sec: 380.02 Mwords.
- Hash type: SHA1 Speed/sec: 218.86 Mwords.
- Hash type: SHA256 Speed/sec: 110.37 Mwords.
- Hash type: Bcrypt, Blowfish(OpenBSD) Speed/sec: 25.86 kwords.
- Hash type: NTLM. Speed/sec: 370.22 Mwords.

You can see that Bcrypt is almost 15,000 times slower than MD5 (380,000,000 words/sec down to only 25,860 words/sec). With John The Ripper the benchmarks are:

- md5crypt [MD5 32/64 X2]. 318,237 c/s real, 8881 c/s virtual.
- bcrypt ("$2a$05", 32 iterations). 25,488 c/s real, 708 c/s virtual.
- LM [DES 128/128 SSE2-16]. 88,090 Kc/s real, 2462 Kc/s virtual.

where we can see that Bcrypt is over 3,000 times slower than LM hashes. So, although the main hashing methods are fast and efficient, this speed has a down side, in that they can be cracked easier. With Bcrypt the speed of cracking is considerably slowed down, where each iteration doubles the amount of time it takes to crack the hash with brute force. So to go from 6 to 16 it increases the time by over 1,000 (2^{10}) and from 6 to 26 it increases by over 1 million (2^{20}).

Here is the Python implementation to print hashed values:

```python
import hashlib;
import passlib.hash;

salt="ZDzPE45C"
string="password"
salt2="1111111111111111111111"

print "General Hashes"
print "MD5:"+hashlib.md5(string).hexdigest()
print "SHA1:"+hashlib.sha1(string).hexdigest()
print "SHA256:"+hashlib.sha256(string).hexdigest()
print "SHA512:"+hashlib.sha512(string).hexdigest()

print "UNIX hashes (with salt)"
print "DES:"+passlib.hash.des_crypt.encrypt(string, salt=salt[:2])
print "MD5:"+passlib.hash.md5_crypt.encrypt(string, salt=salt)
print "Bcrypt:"+passlib.hash.bcrypt.encrypt(string, salt=salt2[:22])
print "Sun MD5:"+passlib.hash.sun_md5_crypt.encrypt(string, salt=salt)
print "SHA1:"+passlib.hash.sha1_crypt.encrypt(string, salt=salt)
print "SHA256:"+passlib.hash.sha256_crypt.encrypt(string, salt=salt)
print "SHA512:"+passlib.hash.sha512_crypt.encrypt(string, salt=salt)
```

and which gives:

```
MD5:5f4dcc3b5aa765d61d8327deb882cf99
SHA1:5baa61e4c9b93f3f0682250b6cf8331b7ee68fd8
SHA256:5e884898da28047151d0e56f8dc6292773603d0d6aabbdd62a11ef721d1542d8
SHA512:b109f3bbbc244eb82441917ed06d618b9008dd09b3befd1b5e07394c706a8bb980
b1d7785e5976ec049b46df5f1326af5a2ea6d103fd07c95385ffab0cacbc86

UNIX hashes (with salt)
DES:ZD3yxA4N/XZVg
MD5:$1$ZDzPE45C$EEQHJaCXI6yInV3FnskmF1

Bcrypt:$2a$12$1111111111111111111111uAQxS9vJNRtBb6zeFDV6k7tyB0DZJF0a

Sun MD5:$md5,rounds=34000$ZDzPE45C$$RGKsbBUBhidHsaNDUMEEX0
SHA1:$sha1$480000$ZDzPE45C$gfgoLWRrJHj/ZiXsV101NCX1GfUH
```

In this case we see we are using 12 iterations and a pre-prepared salt of "1111111111111111111111" (22 characters to give a 128-bit salt value):

```
Bcrypt:$2a$12$1111111111111111111111uAQxS9vJNRtBb6zeFDV6k7tyB0DZJF0a
```

We can increase the rounds to 20 with:

```
print ``Bcrypt:"+passlib.hash.bcrypt.encrypt(string, salt=salt2[:22],
rounds=14)
```

to give:

```
Bcrypt:$2a$14$NkYh0RCM8pNWPaYvRLgN9.OcinBT2h.8NWt/KfmHQ5eIr/50zCt8q
```

and which considerably slows down the hashing.

📖 **Web link (Bcrypt):** https://asecuritysite.com/encryption/bcrypt

3.5.5 Non-Cryptographic Hashes

Most of the hashing methods use complex cryptography methods, and which can be time-consuming, and especially focused at microprocessors which have good computing resources. Sometimes we just need a simple checker which does not consume much processing power. Examples of these are:

- Bernstein hash djb2. This is 32 bits long.
- Buzhash. Uses uses a vvariable number of bits.
- CityHash. This is 64, 128, or 256 bits long.
- Fowler–Noll–Vo hash function (FNV Hash). This uses 32-, 64-, 128-, 256-bit, 512-, or 1024-bits signatures.
- Java hashCode() This is 32 bits long.
- Jenkins hash function This is 32 or 64 bits long
- MurmurHash. This is 32, 64, or 128 bits long.
- Numeric hash (nhash). This has a variable length
- Paul Hsieh's SuperFastHash. This is 32 bits long.
- Pearson hashing. This is 8 bits long.
- PJW hash/Elf Hash. This is 32- or 64 bits long.
- SpookyHash. This is 32, 64 or 128 bits long.
- xxHash. This is 32 or 64 bits long.

The Pearson hashing method is designed for 8-bit processes. It produces an 8-bit value which is dependent on the data. With this we use a 256-byte lookup table, and then permulate around the bytes in the data. For example:

```
static const unsigned char T[256] = { 98,  6, 85,150, 36,
23,112,164,135,207,169,  5, 26, 64,165,219, . . .

h := 0
for each c in M loop
index := h xor c
h := T[index]
end loop
return h
```

we can see that we X-OR each byte with a given value (the first one by 98, and the next one by the result of the hash lookup).

 Web link (Pearson): http://asecuritysite.com/encryption/pearson

A message digest provides a fingerprint for data, and is used to prove identity and integrity of messages and entities. The most common ones used are MD5 (128-bit message hash), SHA1 (164-bit message hash), and SHA-256 (256-bit message hash). So for "hello" we get an MD5 signature of "5D41402ABC4B2A76B9719D911017C592" and for SHA-1 it is "AAF4C61DDCC5E8A2DABEDE0F3B482CD9AEA9434D".

Overall the hash signature should be fairly unique, so if we change any part of the data, it will produce a completely different hash signature. The longer the hash signature the less chance we will have of two pieces of data having the same hash signature.

These methods are fairly fast and can run on most computers, but they use cryptography methods for their processing. This makes it difficult to run the hashing method in parallel, so if we had 16 cores on our computer, with these methods we would only be using one of the cores. Along with this the methods such as MD5 and SHA-1 use numbers which are fairly difficult to process on a 32-bit or 64-bit processor.

One of the best ways to improve the throughput is non-crypto hashing. The main methods are typically written in C++, in order that they are fast, and are well mapped to the processor. These include xxHash, Mumur, Spooky, City Hash and FNV. xxHash was created by Yann Collet and is one of the fastest hashing methods, and uses non-cryptographic technique. It works at close RAM limits. The string of "Nobody inspects the spammish repetition" should give the hex value of 0xe2293b2f. With a seed of 1234, we should get a salted value of 0x298ce4e5. The following is a sample run for the speed of the hashes, where it can be seen that xxHash is the fastest of all the hash methods:

Name	Speed	Q.Score	Author
xxHash	5.4 GB/s	10	
MumurHash 3a	2.7 GB/s	10	Austin Appleby
SpookyHash	2.0 GB/s	10	Bob Jenkins
SBox	1.4 GB/s	9	Bret Mulvey
Lookup3	1.2 GB/s	9	Bob Jenkins
CityHash64	1.05 GB/s	10	Pike & Alakuijala
FNV	0.55 GB/s	5	Fowler, Noll, Vo
CRC32	0.43 GB/s	9	
MD5-32	0.33 GB/s	10	Ronald L. Rivest
SHA1-32	0.28 GB/s	10	

Using a Linux Mint 64 bit operating system on a Core i5 3340M at 2.7GHz, we get:

Name	Speed on 64 bits	Speed on 32 bits
XXH64	13.8 GB/s	1.9 GB/s
XXH32	6.8 GB/s	6.0 GB/s

📖 **Web link (Murmur)** http://asecuritysite.com/encryption/murmur
   Web link (xxHash) http://asecuritysite.com/encryption/xxhash
   Web link (Spooky) http://asecuritysite.com/encryption/Spooky
   Web link (CRC) http://asecuritysite.com/encryption/crc32

If a hash method is fast it can be broken easier than a slower one, but one that is fast can be used to quickly hash values. For a sample run of 40 hashes, we can now rank in classifications:

Ulta fast:

Murmur:	545,716 hashes per second

Fast:

SHA-1:	134,412
SHA-256:	126,323
MD5:	125,741
SHA-512:	76,005
SHA-3 (224-bit):	72,089

Medium speed:

LDAP (SHA1):   13,718
MS DCC: 9,582
NT Hash: 7,782
MySQL: 7,724
Postgres (MD5):   7,284

Slow:

PBKDF2 (SHA-256):   5,026
Cisco PIX: 4,402
MS SQL 2000:   4,225
LDAP (MD5):   4,180
Cisco Type 7:   3,775
PBKDF2 (SHA1):   2,348 (five rounds)

Ultra-slow:

LM Hash: 733
APR1:   234
Bcrypt: 103 (five rounds)
DES: 88
Oracle 10: 48

📖 **Web link (Hash test)** https://asecuritysite.com/encryption/htest

We can see, for speed, that Murmur wipes the floor with the rest, with MD5, SHA-1 and SHA-256 all coming in at around the same speed. For the slowcoaches we include Bcrypt, Oracle 10 and ARP1. With Bcrypt and PBKDF2 we have only done five rounds, so in real-life, where we often use more rounds than five, and these methods would be even slower.

3.6 Authenticating the Sender

The next two problems that we have is how to authenticate the sender, and also, how to prove that the message has not been tampered with in any way even by the sender of the message. The main difference between the authentication and verification process from the encryption one, is that when Bob is sending a secret and authenticated email to Alice, Bob uses his **private-key** to encrypt an authentication message (which has been hashed), as illustrated in Figure 3.12. It can be seen that an MD5 hash is taken of the original message, and that this added to the message, and these are then encrypted with Alice's public key (Figure 3.13). This hash signature provides the authentication of Bob, and also that no-one has tampered with the encrypted message (as not even Bob can now decrypt the encrypted message). When received, the encrypted message is then decrypted (Figure 3.14) with the Alice's private-key. This gives the original message, and the encrypted hash signature. The only key which will decrypt this is **Bob's public key**, which will thus authenticate him as the sender, as only he will have the correct private key to initially encrypt the authentication message (Figure 3.15). Alice then computes the MD5 signature for the received message, and check it against the decrypted hash signature that Bob computed. If they are the same, the message has not been tampered with, and that it was really Bob that sent the email.

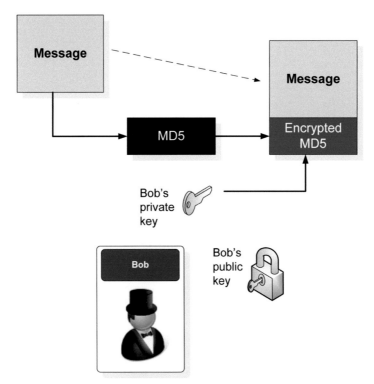

Figure 3.12   Initial part of authentication.

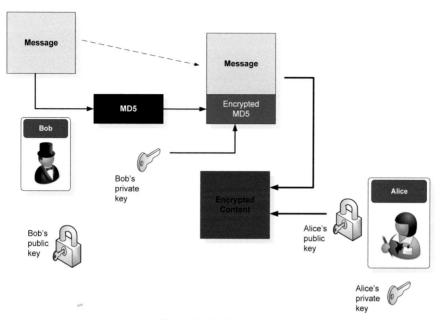

Figure 3.13 Encrypting.

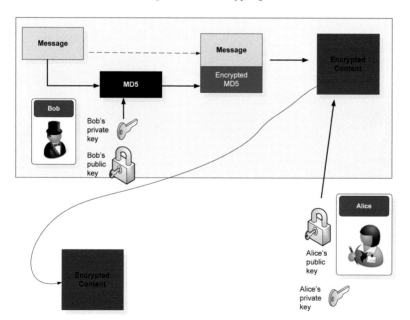

Figure 3.14 Decrypting.

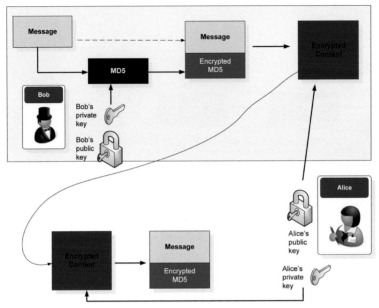

Figure 3.15   Verifying the sender.

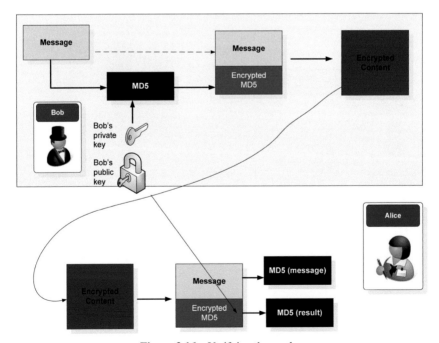

Figure 3.16   Verifying the sender.

3.7 HMAC (Hash Message Authentication Code)

HMAC is a message authentication code (MAC) that can be used to verify the integrity and authentication of the message. It involves hashing the message with a **secret key**, and thus differs from standard hashing, which is purely a one-way function. As with any MAC, it can be used with a standard hash function, such as MD5 or SHA-1, and which results in methods such as HMAC-MD5 or HMAC-SHA-1. Also, as with any hashing function, the strength depends on the quality of the hashing function, and the resulting number of hash code bits. Along with this the number of bits in the secret key is a factor on the strength of the hash. Figure 3.17 outlines the operation, where the message to be sent is converted with a secret key, and the hashing function, to produce an HMAC code. This is then sent with the message. On receipt, the receiver recalculates the HMAC code from the same secret key[1], and the message, and checks it against the received version. If they match, it validates both the sender, and the message (Figure 3.17).

Let's say that the two routers in Figure 3.18 continually challenge each other to answer certain questions. Initially they negotiate a secret shared

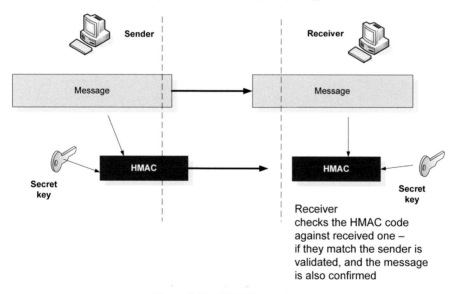

Figure 3.17 HMAC operation.

---

[1]Typically the secret key would either be generated by converting a pass phase into the secret key (such as in some wireless systems) or is passed through the key exchange phase at the start of the connection (such as with Diffie-Hellman).

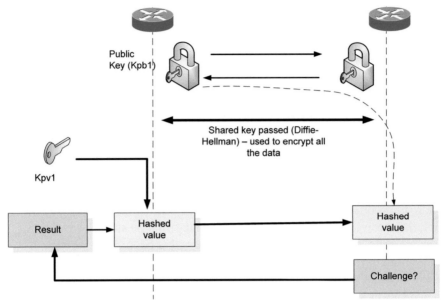

Figure 3.18   Using symmetric encryption and asymmetric authentication.

key, such as "mykey" (or it could be set manually – but this will not be as secure) and negotiate the HMAC type. So a challenge might be to "Multiply 5 and 4?". The answer would be 20, thus using HMAC-MD5, the quizzed device will return back E298452E0 CD44830FEE1DA1C765EB486 (*ref* http://asecuritysite.com/encryption/hmac). The challenger will then do the same conversion, and if it gets the same HMAC code, it will know that the device on the other end is still the same one that it started the connection with. If not, it will disconnect, as it looks as if the original device has been replaced with a spoofed one.

   The following is a simple Python script to cacluate the HMAC hash for a message of "testing123" and a key of "hello"

```
import hashlib
import hmac
def hmacsum(data,key):
        hmachash = hmac.new(key,'',hashlib.sha1)
        hmachash.update(data)
        print(hmachash.hexdigest())
        return
hmacsum("testing123","hello")
```

The following gives some simple .NET code for HMAC conversion:

```
using System;
using System.IO;
using System.Text;
using System.Security.Cryptography;
// Verify with http://hashcalc.slavasoft-inc.qarchive.org/
// Verify: Message="testing123", key="hello" gives
ac2c2e614882ce7158f69b7e3b12114465945d01

namespace hmac
{
   class Class1
   {
   static void Main(string[] args)
   {
       string message = "testing123";
       string key = "hello";

       System.Text.ASCIIEncoding  encoding=new System.Text.ASCIIEncoding();

       byte [] keyByte = encoding.GetBytes(key);

       HMACSHA1 hmac = new HMACSHA1(keyByte);
       byte [] messageBytes = encoding.GetBytes(message);
       byte [] hashmessage = hmac.ComputeHash(messageBytes);
       Console.WriteLine("Hash code is "+ByteToString(hashmessage));
       Console.ReadLine();
   }
   public static string ByteToString(byte [] buff)
   {
      string sbinary="";
      for (int i=0;i<buff.Length;i++)
      {
         sbinary+=buff[i].ToString("X2"); // hex format
      }
      return(sbinary);
   }
  }
 }
}
```

For a key of "hello", and a message of "testing123" gives:

```
The HMAC-SHA-1 hash code is:
AC2C2E614882CE7158F69B7E3B12114465945D01
```

The source code is at:
Source Code link: http://asecuritysite.com/hmac2.zip

In this case we use Python code to generate the HMAC hashes with various hash types:

```
import hashlib;
import hmac;

string = "password"
key="bill"

print "HMAC (MD5): "+hmac.new(key, string,hashlib.md5).hexdigest()
print "HMAC (SHA1): "+hmac.new(key, string, hashlib.sha1).hexdigest()
print "HMAC (SHA224): "+hmac.new(key, string, hashlib.sha224).hexdigest()
print "HMAC (SHA256): "+hmac.new(key, string, hashlib.sha256).hexdigest()
print "HMAC (SHA384): "+hmac.new(key, string, hashlib.sha384).hexdigest()
print "HMAC (SHA512): "+hmac.new(key, string, hashlib.sha512).hexdigest()
```

A sample run is for "password" and a key of "bill" gives:

```
HMAC (MD5): d05e4826b68ade6705b09f50abf8456d
HMAC (SHA1): 6ec0041c6ef1d79dfa8aa6daa79b48d690736a1e
HMAC (SHA224): b9d380c92043a853494643784d454bd516c2446a72c9d37c5bb74934
HMAC (SHA256): 031f225bc62d93beb0543f6d8836e94c98a7c462eb5ba9d9a2606c03e2
452cf1
HMAC (SHA384):4b68107c6358ed3ca06fc924d03ac3480f8fb694ae257c9df09bb4b0d4
c02e2e66f76a0bea9777698592897516c4fa68
HMAC (SHA512): f79e85281fc2bc512f63567fed479ffe19c11c279a124161bd3ab41786
a37752f809cacc8d09f39bf0e401bfe7d7ff46e28c2260e5f7b7e654656a30ad297835
```

With HMAC, the text string is broken-up into blocks of a fixed size, and then are iterated over with a compression function. Typically, such as for MD5 and SHA-1, these blocks are 512 bytes each. With MD5 the output is 128 bits[2] and for SHA-1 it is 160 bits, which is the same as the standard hash functions. HMAC is used in many applications, such as in IPSec and in tunnelling sockets (TLS).

CBC-MAC and CCM

AES is a secret key encryption method, and does not provide authentication of the message. CCM (CCM – Counter with CBC-MAC) can add to AES

---

[2] 128 bits equates to 32 hexadecimal characters (as 4-bits are used for each hex value). For SHA-1, there are 160 bits which gives 40 hexadecimal characters.

by providing an authentication and also encrypt the block cipher mode. It has two parameters: M which indicates the indicates the size of the integrity check value (ICV) and L which defines the size of the length field in octets. To just use CBC-HMAC, the IV is set to zero.

With CBC (Cipher Block Chaining)-MAC (Message Authentication Code) we authenticate messages with a secret shared key. If Bob wants to send some text to Alice, he encrypts it with a shared key and then sends Alice the message digest (or hash of the message) of this. Alice does the same with the secret key and compares the hash that she gets. If they are the same, then Bob has continued to prove his identity (as only he can have the secret key that Bob and Alice share), and that the message has not been changed.

Overall we encrypted the message with the standard form of AES and then throw away everything apart from the last block, and use this as a fixed-length MAC. If the key is not secret the method provides little in the way of security.

For example, it we have a shared key of "test123" and a message of "hello", the CBC-MAC is 9F63F3A838D17066 (16×4 hex character is 64-bits), but with "hellp" is it A2CD2D8CBD8E6DAD. Only by knowing the secret key can we determine the correct hash. In AES Cipher Block Chaining (CBC) encryption we use an IV to make sure that the bits differ for the same message. In CBC-MAC the IV is set to zero, whereas with AES-CCM (Counter with CBC-MAC) the IV value is used to change the message digest.

📖 **Web link (CRC):** https://asecuritysite.com/encryption/ccmaes

3.8 Password Hashing

In order to reduce the opportunity for an intruder determining the password for a given user name, a password is typically hashed using a one-way function. The two main formats involved are with Microsoft Windows and Linux.

Microsoft Windows Hashing

With Microsoft Windows XP, 7 and Visa, the hashed passwords are stored in the Security Account Manager (SAM) database file, and which is stored in a hashed format in a registry hive using an LM or a NTLM hash (Figure 3.19):

```
C:\Windows\System32\config>dir
 Volume in drive C has no label.
 Volume Serial Number is A2B3-7C7A

 Directory of C:\Windows\System32\config
05-Oct-14  05:52 PM          262,144 SAM
05-Oct-14  05:56 PM          262,144 SECURITY
05-Oct-14  08:39 PM      149,946,368 SOFTWARE
05-Oct-14  08:40 PM       15,728,640 SYSTEM
```

LM Hash is used in many version of Windows to store user passwords that are fewer than 15 characters long. It is a fairly weak security implementation and can be easily broken using standard dictionary lookups. More modern versions of Windows use SYSKEY to encrypt passwords.

SAM   Registry:
       HKEY_LOCAL_MACHINE\SAM

Windows 7

- LM Hash (Windows XP, 2003)
- NTLMv2 (Windows 7, 8, etc) – connect to Active Directory
- NTLM (Windows 7, 8, etc) – No salt

```
C:\Windows\System32\config>dir
Volume in drive C has no label.
Volume Serial Number is A2B3-7C7A

Directory of C:\Windows\System32\config
05-Oct-14 05:52 PM          262,144 SAM
05-Oct-14 05:56 PM          262,144 SECURITY
05-Oct-14 08:39 PM      149,946,368 SOFTWARE
05-Oct-14 08:40 PM       15,728,640 SYSTEM
```

- bkhive - dumps the syskey bootkey from a Windows system hive.
- samdump2 - dumps Windows 2k/NT/XP/Vista password hashes.

hashme gives: FA-91-C4-FD-28-A2-D2-57-AA-D3-B4-35-B5-14-04-EE
                 FF2A43841C84518A18795AB6E3C8A62E (NTLM)
napier gives: 12-B9-C5-4F-6F-E0-EC-80-AA-D3-B4-35-B5-14-04-EE
                 307E40814E7D4E103F6A69B04EA78F3D (NTLM)

<user>:<id>:<LM hash>:<NTLM hash>:<comment>:<homedir>:

```
Root@kali:~# cat pw
myuser:500:12B9C54F6FE0EC80AAD3B435B51404EE:307E40814E7D4E103F6A69B04EA78F3D:::
Root@kali:~# john  pw
Loaded 1 password hash (LM DES [128/128 BS SSE2])
NAPIER           (napier)
guesses: 1  time: 0:00:00:00 100% (1)  c/s: 4850  trying: NAPIER - N4PI3R
Use the "--show" option to display all of the cracked passwords reliably
```

Figure 3.19   Windows hashing.

For computers which connect to an Active Directory domain, NTLMv2 is used. John the Ripper and Ophcrack are two of the most widely used tools to crack hashed versions of LM or NTLM passwords (Figure 3.20). Tools such as bkhive and samdump2 can be used to export a password hive into a pwdump format. For example, for "hello" the LM hash is "FD-A9-5F-BE-CA-28-8D-44-AA-D3-B4-35-B5-14-04-EE" and the NTLM hash is "06-6D-DF-D4-EF-0E-9C-D7-C2-56-FE-77-19-1E-F4-3C", and so the pwdump format for a user "bill" is:

```
bill:FDA95FBECA288D44AAD3B435B51404EE:066DDFD4EF0E9CD7C256FE77191EF43C:::
```

where the two formats are defined in their hexadecimal format.

📖 **Web link (LM/NTLM):** http://asecuritysite.com/encryption/lmhash

Figure 3.20 Ophcrack and John The Ripper.

Linux Hashing

Linux can use salted passwords where the username, hashing method and hashed password are stored in a password file. The main methods are:

- **Bcrypt**. This is a secure method but can be slow. The default prefix is $2y$ or $2a$.
- **md5** (APR1). MD5 with salt. The default prefix is $apr1$.
- **crypt()**. At one time this was the default method, but is now seen to be insecure (and limited to eight characters in password). There is no default prefix.
- **salted sha-1**. This is a salted password using SHA-1. The default prefix is {SSHA}.

If we have two users of bill and root, and with passwords of "password" and "redhat", respectively. For APR1 we get:

```
$ cat /etc/shadow
bill:$apr1$oZk9LVLi$mepNMoQbGeN0qp2XIecuj/:15651:0:99999:7:::
root:$apr1$eOzoIRJj$HEwFhY65w0riwDaC5V3G21:15652:0:99999:7:::
```

This contains the user (root), method (APR1), the salt (eOzoIRJj) and the resultant hash (HEwFhY65w0riwDaC5V3G21). When the user logs in, they will enter their password (such as "redhat"). The system then takes the salt value and does a hash of the password with the salt, and checks that it is the same as the stored salted password:

```
C:\openssl>openssl passwd -apr1 -salt eOzoIRJj redhat
$apr1$eOzoIRJj$HEwFhY65w0riwDaC5V3G21
```

We can also generate other formats for the hashed password, such as for SHA-1:

```
bill:{SHA}W6ph5Mm5Pz8GgiULbPgzG37mj9g=
Mike:{SHA}y/2sYAj5yrQIN4TL0YdPdmGNKpc=
```

Figure 3.21 shows an example of the password of "password" and a salt value of "fred", and using MD5 (which is a type of 1). When the user logs in, the system access the password hash file and gains the salt, and uses this with the password to generate the hashed value.

The usage of a salt value considerably improves the strength of the hashed value, but if John the Ripper gets access to the salt value it will try common words with the recoved salt value. In the following case the user (bill) has a password of "password". This is then salted with "ZDzPE45C" to give a hash of "y372GZYCbB1WYtOkbm4/u.". John the Ripper, though, is able to crack the hashed password, as it tries the salt with common words (of which "password" is one of the most common):

```
rootkali:# cat 1.txt
bill:$apr1$ZDzPE45C$y372GZYCbB1WYtOkbm4/u.
rootkali:# john 1.txt
Loaded 1 password hash (FreeBSD MD5 [128/128 SSE2 intrinsics 12x])
password         (bill)
guesses: 1  time: 0:00:00:00 DONE (Mon Jul 27 20:15:28 2015)  c/s: 4866
trying: 123456 - diamond
Use the "--show" option to display all of the cracked passwords reliably
rootkali: # john 1.txt --show
bill:password
```

This takes less than one second to run in Kali. Now, let's check that the salt and the password works using OpenSSL:

```
rootkali:# cat 1.txt
bill:$apr1$ZDzPE45C$y372GZYCbB1WYtOkbm4/u.
rootkali:# openssl passwd -apr1 -salt ZDzPE45C password
$apr1$ZDzPE45C$y372GZYCbB1WYtOkbm4/u.
```

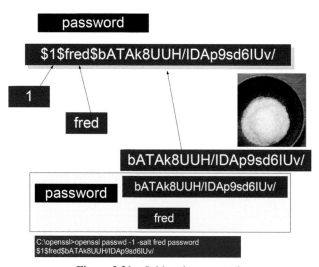

Figure 3.21 Salting the password.

We can also use the crypt library in Python to generate the hashed values. It supports:

1	MD5
2a	Blowfish
5	SHA-256
6	SHA-512

So if we want to compute the MD5 hash for a password of "password" and a salt value of "ZDzPE45C", we use:

```
import crypt;
print crypt.crypt("password","$1$ZDzPE45C$")
```

and the result is:

```
# python 11.py
$1$ZDzPE45C$EEQHJaCXI6yInV3FnskmF1
```

Again if we check with OpenSSL we get:

```
# openssl passwd -1 -salt ZDzPE45C password
$1$ZDzPE45C$EEQHJaCXI6yInV3FnskmF1
```

Unfortunately the crypt() library does not support APR1, so we can use the apr_md5_crypt method:

```
from passlib.hash import apr_md5_crypt;
print apr_md5_crypt.encrypt("password",salt="ZDzPE45C")
```

which gives:

```
$apr1$ZDzPE45C$y372GZYCbB1WYtOkbm4/u.
```

Now we can use this library to create a wide range of hash values using hashlib and passlib:

```
import hashlib;
import passlib.hash;

salt="ZDzPE45C"
string="password"
salt2="1111111111111111111111111"

print "General Hashes"
print "MD5:"+hashlib.md5(string).hexdigest()
print "SHA1:"+hashlib.sha1(string).hexdigest()
```

```
print "SHA256:"+hashlib.sha256(string).hexdigest()
print "SHA512:"+hashlib.sha512(string).hexdigest()

print "UNIX hashes (with salt)"
print "DES:"+passlib.hash.des_crypt.encrypt(string, salt=salt[:2])
print "MD5:"+passlib.hash.md5_crypt.encrypt(string, salt=salt)
print "Bcrypt:"+passlib.hash.bcrypt.encrypt(string, salt=salt2[:22])
print "Sun MD5:"+passlib.hash.sun_md5_crypt.encrypt(string, salt=salt)
print "SHA1:"+passlib.hash.sha1_crypt.encrypt(string, salt=salt)
print "SHA256:"+passlib.hash.sha256_crypt.encrypt(string, salt=salt)
print "SHA512:"+passlib.hash.sha512_crypt.encrypt(string, salt=salt)

print "APR1:"+passlib.hash.apr_md5_crypt.encrypt(string, salt=salt)
print "PHPASS:"+passlib.hash.phpass.encrypt(string, salt=salt)
print "PBKDF2 (SHA1):"+passlib.hash.pbkdf2_sha1.encrypt(string, salt=salt)
print "PBKDF2 (SHA256):"+passlib.hash.pbkdf2_sha256.encrypt(string, salt=salt)
print "PBKDF2 (SHA512):"+passlib.hash.pbkdf2_sha512.encrypt(string, salt=salt)
print "CTA PBKDF2:"+passlib.hash.cta_pbkdf2_sha1.encrypt(string, salt=salt)
print "DLITZ PBKDF2:"+passlib.hash.dlitz_pbkdf2_sha1.encrypt(string, salt=salt)

print "MS Windows Hashes"
print "LM Hash:"+passlib.hash.lmhash.encrypt(string)
print "NT Hash:"+passlib.hash.nthash.encrypt(string)
print "MS DCC:"+passlib.hash.msdcc.encrypt(string, salt)
print "MS DCC2:"+passlib.hash.msdcc2.encrypt(string, salt)

print "LDAP Hashes"
print "LDAP (MD5):"+passlib.hash.ldap_md5.encrypt(string)
print "LDAP (MD5 Salted):"+passlib.hash.ldap_salted_md5.encrypt(string, salt=salt)
print "LDAP (SHA):"+passlib.hash.ldap_sha1.encrypt(string)
print "LDAP (SHA1 Salted):"+passlib.hash.ldap_salted_sha1.encrypt(string, salt=salt)
print "LDAP (DES Crypt):"+passlib.hash.ldap_des_crypt.encrypt(string)
print "LDAP (BSDI Crypt):"+passlib.hash.ldap_bsdi_crypt.encrypt(string)
print "LDAP (MD5 Crypt):"+passlib.hash.ldap_md5_crypt.encrypt(string)
print "LDAP (Bcrypt):"+passlib.hash.ldap_bcrypt.encrypt(string)
print "LDAP (SHA1):"+passlib.hash.ldap_sha1_crypt.encrypt(string)
print "LDAP (SHA256):"+passlib.hash.ldap_sha256_crypt.encrypt(string)
print "LDAP (SHA512):"+passlib.hash.ldap_sha512_crypt.encrypt(string)

print "LDAP (Hex MD5):"+passlib.hash.ldap_hex_md5.encrypt(string)
print "LDAP (Hex SHA1):"+passlib.hash.ldap_hex_sha1.encrypt(string)
print "LDAP (At Lass):"+passlib.hash.atlassian_pbkdf2_sha1.encrypt(string)
print "LDAP (FSHP):"+passlib.hash.fshp.encrypt(string)

print "Database Hashes"
print "MS SQL 2000:"+passlib.hash.mssql2000.encrypt(string)
print "MS SQL 2005:"+passlib.hash.mssql2005.encrypt(string)
print "MySQL:"+passlib.hash.mysql323.encrypt(string)
print "MySQL:"+passlib.hash.mysql41.encrypt(string)
print "Postgres (MD5):"+passlib.hash.postgres_md5.encrypt(string, user=salt)
```

```
print "Oracle 10:"+passlib.hash.oracle10.encrypt(string, user=salt)
print "Oracle 11:"+passlib.hash.oracle11.encrypt(string)


print "Other Known Hashes"
print "Cisco PIX:"+passlib.hash.cisco_pix.encrypt(string, user=salt)
print "Cisco Type 7:"+passlib.hash.cisco_type7.encrypt(string)
print "Dyango DES:"+passlib.hash.django_des_crypt.encrypt(string, salt=salt)
print "Dyango MD5:"+passlib.hash.django_salted_md5.encrypt(string, salt=salt[:2])
print "Dyango SHA1:"+passlib.hash.django_salted_sha1.encrypt(string, salt=salt)
print "Dyango Bcrypt:"+passlib.hash.django_bcrypt.encrypt(string, salt=salt2[:22])
print "Dyango PBKDF2 SHA1:"+passlib.hash.django_pbkdf2_sha1.encrypt(string, salt=salt)
print "Dyango PBKDF2 SHA1:"+passlib.hash.django_pbkdf2_sha256.encrypt(string, salt=salt)
```

It can be seen that in some cases the salt value needs to be at least 22 characters. Also the [:2] modifier will generate a string with 2 characters. When we run it we get:

```
General Hashes
MD5:5f4dcc3b5aa765d61d8327deb882cf99
SHA1:5baa61e4c9b93f3f0682250b6cf8331b7ee68fd8
SHA256:5e884898da28047151d0e56f8dc6292773603d0d6aabbdd62a11ef721d1542d8
. . .
SHA1:pbkdf2_sha256$29000$ZDzPE45C$pd1VbFkOA/VwbhJZhJ+25kHPsKVXika2XsuK
Youdcug=
```

The full output is given in Table 3.1 We can see that our salted APR1 password of "password" and with a salt of "ZDzPE45C" comes out as we expected as

"$apr1$ZDzPE45C$y372GZYCbB1WYtOkbm4/u."

Table 3.1   Sample hash formats

General Hashes	
MD5	5f4dcc3b5aa765d61d8327deb882cf99
SHA1	5baa61e4c9b93f3f0682250b6cf8331b7ee68fd8
SHA256	5e884898da28047151d0e56f8dc6292773603d0d6aabbdd62a11ef72 1d1542d8
SHA512	b109f3bbbc244eb82441917ed06d618b9008dd09b3befd1b5e07394 c706a8bb980b1d7785e5976ec049b46df5f1326af5a2ea6d103fd07 c95385ffab0cacbc86
UNIX Hashes (with salt)	
DES	ZD3yxA4N/XZVg
MD5	1ZDzPE45C$EEQHJaCXI6yInV3FnskmF1
Bcrypt	$2a$12$111111111111111111111uAQxS9vJNRtBb6zeFDV6k7ty B0DZJF0a

Sun MD5	$md5,rounds=34000$ZDzPE45C$$RGKsbBUBhidHsaNDU MEEX0
SHA1	$sha1$480000$ZDzPE45C$gfgoLWRrJHj/ZiXsV101NCX1GfUH
SHA256	5rounds=535000$ZDzPE45C$OuICueKPJYEtr8.A1iZMpZ11 v07uuX/2cXfRrKmF1i6
SHA512	6rounds=656000$ZDzPE45C$uCmjusfwHL378JNZeUuFbTqoe BentVZzoRAVHzke6/mcqJpOppAkVyqn8A41sKXMad3DG 7O2QL/Zn YABfK3j1/
APR1	$apr1$ZDzPE45C$y372GZYCbB1WYtOkbm4/u.
PHPASS	PHZDzPE45Ch4tvOeT9mhtu3i2G/JybR1
PBKDF2 (SHA1)	$pbkdf2$131000$WkR6UEU0NUM$.L1L.AVXTBSsc0FuHR Qz4PNMVXc
PBKDF2 (SHA256)	$pbkdf2-sha256$29000$WkR6UEU0NUM$pd1VbFkOA/VwbhJZhJ.25kHP sKVXika2XsuKYoudcug
PBKDF2 (SHA512)	$pbkdf2-sha512$25000$WkR6UEU0NUM$S.ymDjKjwM9XaQsofRC6 KX1s.pQvZvVmMxdrrLi16pCazREoyJGxe8.Tn6Zhi3S0B6H6r crxITllAEo3rDwBng
CTA PBKDF2	$p5k2$1ffb8$WkR6UEU0NUM=$-L1L-AVXTBSsc0FuHRQz4PNMVXc=
DLITZ PBKDF2	$p5k2$1ffb8$ZDzPE45C$2Cye7ESZt2eO2ouLHuL7h4bJmD 13yGsq
MS Windows Hashes	
LM Hash	e52cac67419a9a224a3b108f3fa6cb6d
NT Hash	8846f7eaee8fb117ad06bdd830b7586c
MS DCC	c531cc9702cbbe9053dfa32d8940c2ca
MS DCC2	920873ab14cffe2420ebf69c6d5f8ee7
LDAP Hashes	
LDAP (MD5)	{MD5}X03MO1qnZdYdgyfeuILPmQ==
LDAP (MD5 Salted)	{SMD5}ZYxs6V7nZOz+ALwZu8nWglpEelBFNDVD
LDAP (SHA)	{SHA}W6ph5Mm5Pz8GgiULbPgzG37mj9g=
LDAP (SHA1 Salted)	{SSHA}Rr2ARpei2FyhmO51IpsE0S1np2BaRHpQRTQ1Qw==
LDAP (DES Crypt)	{CRYPT}fKVEK3hRz/fcU
LDAP (BSDI Crypt)	{CRYPT}_7C/.CcBEsmIgMi.Rbrc
LDAP (MD5 Crypt)	{CRYPT}1HWg9ay6K$oywqsN1iM0M9gG7YeV4C91
LDAP (Bcrypt)	{CRYPT}$2a$12$KzXuLIeRNQJbp38kFQliSOYBA544Tmkj Z1hhQvzrqavmSGMka/1gK
LDAP (SHA1)	{CRYPT}$sha1$480000$Yhd5bf4P$KDtw4NG7r2cnFB4ZNd bugP9Knj6B
LDAP (SHA256)	{CRYPT}5rounds=535000$Uuchrd2YR4h19DA0$cRq2taFl/s RhjB5JQlIVO0IEsn4YULXjI3Os6dOJnjB
LDAP (SHA512)	{CRYPT}6rounds=656000$UbSq4w1iUbiRg9y3$Hkojs.zgu Y4Lg.ZDtiduqmTHSz7nwqKHEPr4fdJhvUIHvPKw/dABSxg BXkcrwU4nRzrS LVjsvetGMKuxO0gv90

(*Continued*)

Table 3.1 Continued

LDAP (Hex MD5)	{MD5}5f4dcc3b5aa765d61d8327deb882cf99
LDAP (Hex SHA1)	{SHA}5baa61e4c9b93f3f0682250b6cf8331b7ee68fd8
LDAP (At Lass)	{PKCS5S2}p5TSmlMqBSCkNCbkPAfgnD+L9jw0OA4XK/B2Uh DwFp8do3TRJjFZgWzQ3FdYs2mM
LDAP (FSHP)	{FSHP1|16|480000}lrI2JoRQqhWCEKI0BuDcGwkbtIDgEOm/ rJ8D9z9NhEXYAW1w5xCf5ePc0ljGGPBI

Database Hashes

MS SQL 2000	0x0100FD7F2FA53B3C77F6247411D6B1178F41ABACCCC 3AC109D46F67A57CD4CC406F1E69F9488992D1C9FB 66DED24
MS SQL 2000	0x01002AE57C4F5923608F88CD76DC41B3114ED77C73C5 52CFBCAB
MS SQL 2000	5d2e19393cc5ef67
MySQL	*2470C0C06DEE42FD1618BB99005ADCA2EC9D1E19
Postgres (MD5)	md5658c6ce95ee764ecfe00bc19bbc9d682
Oracle 10	A8F6239BAE6A967A
Oracle 11	S:43D471EB6253E88ED67C56EFB9BCB0813579C98EE73CA 41474BE963BFCBB

Other Known Hashes

Cisco PIX	NLETddx4AEoSe48z
Cisco Type 7	13151601181B0B382F
Dyango DES	crypt$ZDzPE45C$ZD3yxA4N/XZVg
Dyango MD5	md5ZD32797d3a40d12ed6dc6fa57d0f745ca5
Dyango SHA1	sha1$ZDzPE45C$525954ca97fad2fdb772ebc621bd1d4f846be2d4
Dyango Bcrypt	bcrypt$$2a$12$111111111111111111111uAQxS9vJNRtBb6ze FDV6k7tyB0DZJF0a
Dyango PBKDF2 SHA1	pbkdf2_sha1$131000$ZDzPE45C$+L1L+AVXTBSsc0FuHRQz 4PNMVXc=
Dyango PBKDF2 SHA1	pbkdf2_sha256$29000$ZDzPE45C$pd1VbFkOA/VwbhJZhJ +25kHPsKVXika2XsuKYoudcug=

As with PBKDF2 and Bcrypt, Scrypt is a password-based key derivation function (password-based KDF) which produces a hash with a salt and iterations. The iteration count slows down the cracking and the salt makes pre-computation difficult. The main parameters are: passphrase (P); salt (S); Blocksize (r) and CPU/Memory cost parameter (N – a power of 2). If we use an N value of 16, and r as 1, we get:

```
Phase:  hello
Salt:  test
N:  16
685be7d8bad20c58afbcc7d60fecf9ea4153da4a330af89d01d482cb10ef4f495b
dd004a70d46b7b75bcd24b9e347ebc90681d7c0b06249eca3234a32d70f744
```

We can see that we have 128 hex characters, thus the hash signature is 512 bits long. It intentionally uses a great deal of memory, in order to reduce the risk of GPU-based cracking. The basic method is:

```
Function scrypt(Passphrase,Salt,N,p,dkLen):
    (B0 ... Bp-1) ← PBKDF2(HMAC_SHA256, Passphrase, Salt, 1, p * MFLen)
    for i = 0 to p-1 do
        Bi ← SMix(Bi,N)
    end for
    Output ← PBKDF2(HMAC_SHA256, Passphrase, B0 || B1 ... Bp-1, 1, dkLen)

Function SMix(B,N):
    X ← B
    for i = 0 to N - 1 do
        Vi ← X
        X ← BlockMix(X)
    end for
    for i = 0 to N - 1 do
        j ← Integerify(X) mod N
        X ← BlockMix(X ⊕ Vj)
    end for
    Output ← X

Function BlockMix(B):
    (B0, ... , B2r-1) ← B
    X ← B2r-1
    for i = 0 to 2r - 1 do
        X ← H(X ⊕ Bi)
        Yi ← X
    end for
    Output ← (Y0, Y2, ... , Y2r-2, Y1, Y3, ... , Y2r-1)
```

Integerify() defines a bijective function and maps every entry in a set direct to one in another set.

📖 **Source Code link:** http://asecuritysite.com/Scrypt

3.9 Password Cracking

The strength of a password relates to three major elements:

- **The number of characters in the password**. The more characters that are in the password the stronger the password is likely to be.
- **The range of characters in the password**. The wider the range of characters in a password is likely to increase its strength, especially in using non-alphabet ones (such as "!", "@", and so on).

- **The cracking speed of a brute force generator**. This relates to the speed of the cracker, such as 1,000,000 tries per second.

For example if we have lowercase letters [a–z] we have 26 characters, and add uppercase letters [A–Z], we get 52 characters. If we then have five characters in the password, the range of password combinations will be:

aaaaa to ZZZZZ

which will be 52 to the power of 5 = 380,204,032. If we crack these passwords at a rate of one million per second then it will take around 380 seconds to try all of them (6.23 mins).

Figure 3.22 shows a calculation for [a-zA-Z] with one million password attempts per second. We can see that for a seven character password it takes 11.9 days, and for a 10-character one it takes over 4,000 years.

📖 **Web link (Passwords):** http://asecuritysite.com/encryption/passes

In general terms, as shown in Figure 3.23, we can calculate the number of passwords for a 5-character password with a range of character sets.

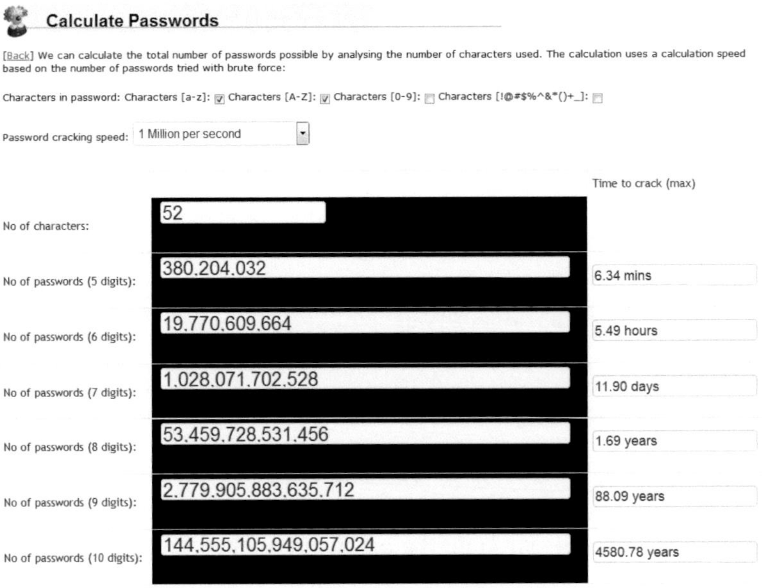

Calculate Passwords

[Back] We can calculate the total number of passwords possible by analysing the number of characters used. The calculation uses a calculation speed based on the number of passwords tried with brute force:

Characters in password: Characters [a-z]: ☑ Characters [A-Z]: ☑ Characters [0-9]: ☐ Characters [!@#$%^&*()+_]: ☐

Password cracking speed: 1 Million per second ▾

		Time to crack (max)
No of characters:	52	
No of passwords (5 digits):	380,204,032	6.34 mins
No of passwords (6 digits):	19,770,609,664	5.49 hours
No of passwords (7 digits):	1,028,071,702,528	11.90 days
No of passwords (8 digits):	53,459,728,531,456	1.69 years
No of passwords (9 digits):	2,779,905,883,635,712	88.09 years
No of passwords (10 digits):	144,555,105,949,057,024	4580.78 years

Figure 3.22   Password strength.

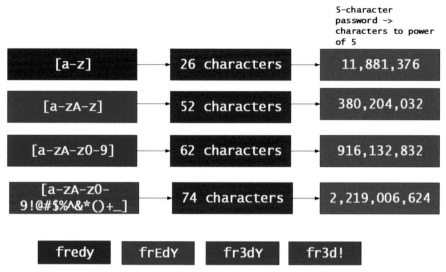

Figure 3.23 Password strength.

Cloud Cracking

Today computer systems are so much more powerful than in the time of Turing, and their speed to try lots of different permutations is the main method that the code crackers use to break encryption. One of the most popular methods for cracking ciphers is to use the NVIDIA CUDA architecture, which runs the cracking algorithms as multiple threads on multiple processor cores, each of which are able to do complex cryptography operations in a minimum number of clock cycles.

Most passwords are stored in a hashed form, which is a one-way function, so that it is not mathematically possible to go backwards from the hash value to the original password. Unfortunately intruders can build massive tables of hashed values, and basically look-up a hashed value, and determine the password which made this. This is known as a rainbow table cracking.

Within the Cloud, rainbow tables are being built for as many common passwords as can be gathered. To illustrate the current state-of-the-art in performance of password cracking methods, if we take a range of characters from a–z, for seven characters in the password we get a password range at "aaaaaaa" to "zzzzzzz". As each character can exist in each position we get over eight billion possible character permutations:

$$26^7 = 8,031,810,176$$

On a typical GPU (Graphical Processing Unit) card, which is multi-threaded and highly parallelized, we could process around 150 million word checked per second. The time to crack any password with this example is 53 seconds, and an average of around 26 seconds.

Now, if we add upper and lower case characters, we go from "aaaa aaa" to "ZZZ ZZZZ", which now gives us around 1 billion billion possible passwords:

$$52^7 = 1,028,071,702,528 \text{ (7 character password [a-zA-Z])}$$

to gives a maximum cracking time of nearly 2 hours (114 minutes), with an average of around one hour. Now we can add the characters of [0-9],[{}/\'".!@#&], and which gives an addition of 20 characters, so that our calculation is:

$$52^7 = 10,030,613,004,288 \text{ (7 character password [a-zA-Z0-9{}/\'".!@#&])}$$

where we get a cracking maximum time of around 18 hours (and an average of around 9 hours). This means it will takes less than a day to crack every one of these passwords:

```
a.{Zi19&
oO!.5pK
LlL1L1L
```

With the addition of one more character, it all gets a little more difficult, as the number of permutations becomes:

$$72^8 = 722,204,136,308,736 \text{ (8 character complex password)}$$

where, with a GPU, and using our examples, it will crack every eight characters password in 1,337 hours or 55 days, which is still acceptable in certain applications. The worrying thing here is that the Cloud is capable of now generating every single MD5 value possible for a 7-digital password. As we require 128-bits to store each one, the total storage will unfortunately be 11,555,266,180,939,776 bytes which is around 11,555 TB.

If we now take over 1,000 of these GPUs, such as by purchasing time on the Amazon GPU Cloud, and allocate each one 1/1000 of the hashes to try, we now only require about one and a half hours to crack an eight character password, with a wide range of characters.

On must remember that for a 128-bit MD5 hash, we can generate 3.4×10^{38} codes, and we are thus using only using 1/471,171,999,457,326, 155,549,623th of the total space, and for SHA-1 (160-bit) we add another 10 zeros to the end.

Salting is the true way to increase the strength of the hashed password and to use more of the MD5 or SHA-1 space, as MD5 can actually deliver 3.4×10^{38} different values. With salt we change the hashing method so that:

$$\text{Hash} = \text{md5 (salt + md5(word))}$$

For example, a common hashing method is APR1, and store the username, salt and hashed password as:

bill:$apr1$UtOa0hnT$17QMSjBPj3urRkG.352kR0

where:UtOa0hn is the salt, and UtOa0hnT$17QMSjBPj3urRkG.352kR0 is the salted password. When a user logs in, the system takes their password (in this case the password is "password"), and takes the method and the salt and recalculates the hash. If they are the same, the password matches. For example:

```
C:\openssl>openssl passwd -apr1 -salt UtOa0hn password
$apr1$UtOa0hn$RZ9RtDAL6mExOKgfbM9sK.
```

The addition of the extra seven upper/lower case characters increases the hashed values of a eight digit password from 722,204,136,308,736 (72^8) to 742,477,635,987,686,000,959,684,608 ($72^8 \times 52^7$), which considerably increases the complexity of cracking process, giving it an equivalent of nearly 15 characters.

3.10 One Time Passwords

Passwords which use a hashed value can be cracked as either with rainbow tables or brute force. An improved method of generating passcode is to generate a different one each time based on an initial seed value, or based on time. The main methods involved are (Figure 3.24):

- **One Time Passwords (OTP)**. This allows a new unique password to be created for each instance, based on an initial seed.
- **Timed One Time Password (TOTP)**. This allows for a new unique passcode to be created for each instance, based on an initial seed and for a given time period.
- **Hashed One Time Password (HOTP)**. This allows a new unique passcode to be created each instance, based on a counter value and an initial seed.

The **One Time Password** protocol has an initial seed, and then a function is applied for each iteration. For example the first time we have f(M), the next time it is f(f(M)), and then f(f(f(M))), and so on. The only way that it is possible to determine the next password is to know the first password. Increasingly OTP is used on systems, where a user is registered and the first password generated. Only by knowing the initial seed of the password generation is it possible to generate each of the following ones, as they cannot be guessed without this seed.

📖 **Web link (OTP):** http://asecuritysite.com/encryption/onetime

TOTP (Timed One Time Password) is a method used to generate single use passwords which are only valid for a certain time period. For example we could have a system which allowed to creation a new account for your mobile phone, but where the password was only valid for a short time. It is also used extensively in two-factor authentication (such as registering with a username/password, and entering the timed password). The time window could be set at one hour, where the user has to register within one hour, or their password would have to be re-generated. The method is defined in RFC6238 and is used in the Google Authenticator service. The following code uses a five second time window to generate a new code, so you can press Generate OTP every five seconds and it should give you a new passcode:

Web link (TOTP): http://asecuritysite.com/encryption/totp

HOTP (Hashed One Time Password) allows a new unique passcode to be created each instance, based on a counter value and an initial seed. For example **"test"** as the passphase will give: **00542354 (0), 00917969 (1), 00493162 (2), 00347259 (3) ...**, and "bill" gives 00578423, 00842117, 00359325, ...:

Web link (HOTP): http://asecuritysite.com/encryption/hotp

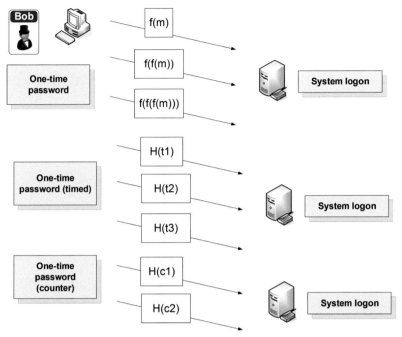

Figure 3.24 Password strength.

3.11 Time Stamp Protocol

TSP (Time-Stamp Protocol) provides a cryptography method to give a verifiable method that a data entity was created at a defined time, and is defined in RFC 3161. It uses a Time Stamping Authority (TSA) that must be trusted for a source of time, and produce a unique time-stamp token, serial number and thumbprint for the data entity.

For a data value of "hello" defined on 23 July 2015 at 8:46pm, we get:

```
Serial Number: 15821993121701984733321759775999
Gen Time: 7/23/2015 8:46:51 PM
Policy 1.2.643.2.2.38.4
Encoded timestamp: 305702010106072A8503020226043021
300906052B0E03021A05000414AAF4C6 1DDCC5E8A2DABEDE0F3B482CD9AEA943
4D020D13F85C98A2000000000024BC7F 180F323031353037323333323034363531
5A3003020164020164
```

This method can be used to verify that an entity was produced at a given time. Any changes in the data will not be verified by the time stamp.

3.12 Winnowing and Chaffing

Ron Rivest, in 1998, proposed a method called Chaffing and Winnowing, and which uses keyless encryption. It involves separating out the useful parts of the transmitted message, and signing these parts with a valid **MAC** (Message Authentication Code), so that they parts can be authenticated. The sender then adds chaff and adds an incorrect MAC for these. At the other end, the receiver will know how to separate them, as the chaff will generate errors. If the codes are signed in some way, Eve cannot tell which is valid from the invalid (Figure 3.25).

With this method, Bob will authenticate each packet with a MAC, of which Alice has the secret key. He can also send bad packets with incorrect MACs, of which Alice will discard. For example, Bob sends:

(1. Hello [Good]) (1.Ankle [Bad]) (2.How [Good]) (2.Ill [Bad]) (3.Are [Good])(3.Also [Bod]) (4.You [Good])(4.Failure [Bad]])

so the message is actually "Hello How Are You"

For example we create a tuple for the data (serial number, data, and MAC). The message that Bob wants to send to Alice is:

{1, Please send me, 12345}
{2, Your details, 43546}
{3, As I am worried about you, 54354}

each of these messages have the correct MAC (12345, 43546 and 54354). Bob can then add some invalid ones:

{1, Please send me, 12345}
{1, Please go away, 8453}
{2, Your details, 43546}
{2, I don't want to see you, 53646}
{3, As I am worried about you, 54354}
{3, And that is the end of it, 44546}

In this case the new MAC codes are invalid (8453, 53646 and 44546), and thus Alice rejects them to give:

Please send me your details as I am worried about you

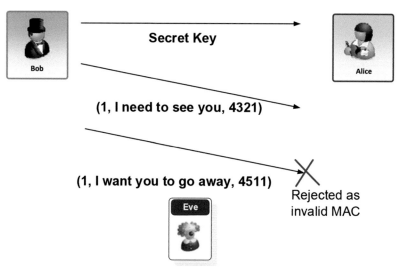

Figure 3.25   Adding chaff.

So, to demonstrate it, you can use:

📖 **Web link (Chaff):** http://asecuritysite.com/encryption/chaff

With this we have a run of:

```
Original data
[(1, 'bill', 'fred123'), (2, 'mike', 'password')]
With chaff
[(1, '\xa1\x8e?0', '0\xcb\x89~6\xdf\x9f'), (1, 'bill',
'fred123'), (2, 'mike', 'password'), (2, '\x06\x05\xefT',
'\x81I\t\xdba\x82\x89L')]
After chaff processed
[(1, 'bill', 'fred123'), (2, 'mike', 'password')]
```

3.13 SHA-3

Keccak won the NIST hash function competition, and is proposed as the SHA-3 standard. It should be noted that it is not replacement SHA-2, which is currently a secure method. Overall Keccak uses the sponge construction where the message blocks are XORed into the initial bits of the state, and then invertibly permuted.

SHA-3 was known as Keccak and is a hash function designed by Guido Bertoni, Joan Daemen, Michaël Peeters, and Gilles Van Assche. MD5 has been shown to be susceptible to attacks, along with theoretical attacks on SHA-1. NIST thus defined that there was a need for a new hashing method which did not use the existing methods for hashing, and setup a competition for competing algorithms.

NIST published the new standard, based on Keccak, on 5 August 2015, and which beat off competition from BLAKE (Aumasson et al.), Grøstl (Knudsen et al.), JH (Hongjun Wu), and Skein (Schneier et al.). After two rounds the final round saw an evaluation of security, performance and hardware space. Generally Blake and Keccah did well in terms of the number of gates which implement the methods. But it was in throughput that Keccak really shone, and beat the others by at least a factor of between three and four. With energy consumption becoming a major factor within mobile devices and in IoT, the energy consumption for Keccak trumped the other finalists. In this case Keccak consumed less than half of the power per bit than Blake.

The sponge function takes a simple function f and involves a number of stages, and where we create a fixed output (dependent on the bit length of the hash function). Simple operations of XOR, AND, and bit shifts are then used, and which leads to a fast generation of the hash function (Figure 3.26). The f permutation function takes a variable-length input and produces an arbitrary output length. A value r is the bit rate, and each f function operates on b bits, and where a capacity is defined as $c = b - r$.

The SHAKE method is useful as it can be used to create a hash method of a variable length. For example the 128-bit version will produce a hash value of 32 hex characters.

NIST has now released the final version of the method as a new standard: Federal Information Processing Standard (FIPS) 202, SHA-3 Standard: Permutation-Based Hash and Extendable-Output Functions. A key factor in the definition of the new standard was that each of the methods submitted required signed statements that the method would be available on a royalty-free basis.

📖 **Web link (SHA-3):** http://asecuritysite.com/Encryption/s3

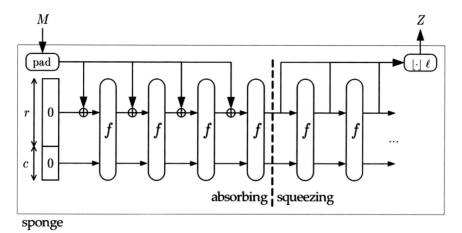

Figure 3.26 SHA-3 [2].

Skein

Skein was a contender for SHA-3 and was created by Bruce Schneier, Niels Ferguson, Stefan Lucks, Doug Whiting, Mihir Bellare, Tadayoshi Kohno, Jon Callas and Jesse Walker. It is based on Bruce's Threefish block cipher and is

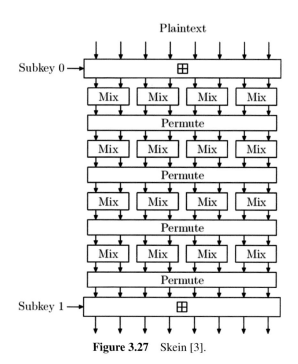

Figure 3.27 Skein [3].

compressed using Unique Block Iteration (UBI). This supports a chaining mode which allows for variable sizes of hashes. It gets its name from the intertwining of the input, which looks like the twining in a skein of yarn.

📖 **Web link (Skein):** http://asecuritysite.com/encryption/sk

Grøstl

Grøstl was designed cryptographers at the Technical University of Denmark (DTU) and TU Graz, and is defined as a new hashing method. Overall it is an iterated hash function, using two fixed and different permutations, along with a compression function. The name Grøstl comes from an Austrian dish of hash.

📖 **Web link (Grøstl):** http://asecuritysite.com/encryption/gro

Blake

BLAKE and BLAKE2 are hash functions based on the ChaCha stream cipher.

📖 **Web link (Grøstl):** http://asecuritysite.com/encryption/blake

3.14 Lab/Tutorial

The lab and tutorial related to this chapter is available on-line at:

http://asecuritysite.com/crypto03

References

[1] "Password Cracking with Amazon Web Services – 36 Cores – Things all the hacking." [Online]. Available: http://blog.nullmode.com/blog/2015/03/22/36-core-aws-john/. [Accessed: 04-Jun-2017].
[2] The Keccak sponge function family: Specifications summary.
[3] Now From Bruce Schneier, the Skein Hash Function. Slashdot, 2008.

4

Public Key

4.1 Introduction

Public key encryption is an asymmetric key method, and uses a public key (which can be distributed) and a private key (which should be kept private). Within the mathematics it should be extremely difficult to determine the private key given the public key, such as the difficulty in factoring a value for its prime number factors. The three main methods that we use for this includes **integer factorization** (such as RSA), **discrete logarithms** (such as ElGamal), and **elliptic curve relationships** (such as the Elliptic Curve).

For integer factorization, if we use a value of 33, then we can see that the prime number factors are 3 and 11, but if we have a 512-bit integer value we generate a value with 148 digits, such as:

13,407,807,929,942,597,099,574,024,998,205,846,127,479,365,820,592,
393,377,723,561,443,721,764,030,073,546,976,801,874,298,166,903,
427,690,031,858,186,486,050,853,753,882,811,946,569,946,433,649,
006,084,096

At the present time 512-bit prime numbers be cracked with high-powered computers, so we often use integers which are 1,024 bits and more.

Public-key encryption is an excellent method of keeping data secure, but it is often too slow for real-time communications. Normally, thus, we use symmetric key encryption to perform the actual encryption of the data or in the encrypting of data within a tunnel, and where public key encryption is used to prove the identity of an entity, or used to protect the passing or agreement of the symmetric key. The two main applications are:

- **Identity checking.** Public key is often used for identity checking, where the entity proving its identity will encrypt a value with the its private key and then other entities can provide the entity by decrypting it with the related public key, as no other entity will have the same key pair. If the private key is thus stolen, other entities can pretend to be the entity which had its private key stolen.

- **Key protection.** The other useful application of public key encryption is in the protection of the symmetric key. This often happens with disk encryption, where the symmetric key which is used to encrypt the files is protected by the public key of an entity. The only key which can then decrypt the symmetric key is the private key.

The key pair and the public key are either stored in an XML format or on a digital certificate. These allow the keys to be processed, stored or transmitted. The usage of digital certificates to identity entities is known as the public key infrastructure (PKI), and where key pairs are generated by trusted entities, such as Verisign. The private key is then kept secret, in order to prove identities, and where the public key is distributed through the PKI infrastructure. The most important element of PKI, is that the private key will always be kept private, and a loss of this would mean that the identity of the entity could be breached, along with any encryption keys that are protected by the key pair.

4.2 RSA

Public-key encryption uses two keys: a public one; and a private one (Figure 4.1). These are generated from extremely large prime numbers, as a value which is the product of two large prime numbers and is extremely difficult to factorize. The two keys are generated, and the public key is passed to the other side, who will then encrypt data destined for this entity using this public key. The only key which can decrypt it is the secret, private key. A well-known algorithm is RSA, and which can be used to create robust keys. Its stages are:

1. Select two large prime numbers, p and q (each will be at least 256 bits long). The factors p and q remain secret and N is the result of multiplying them together. Each of the prime numbers is of the order of 10^{100}. The value at N is known as the modulus.
2. Next, the public-key is chosen. To do this a value e is chosen so that e and $(p-1) \times (q-1)$ are relatively prime. Two numbers are relatively prime if they have no common factor greater than 1 (GCD(a,b)=1). The public-key is then $<e,N>$ and this results in a key which is at least 512 bits long.

3. Next the private key for decryption, d, is computed so that:

$$d = e^{-1} \quad \mod \ [(p-1) \times (q-1)]$$

or:

$$(d \times e) \quad \mod \ [(p-1)(q-1)] = 1$$

where we can define PHI as $(p-1) \times (q-1)$

4. The encryption process to ciphertext, c, is then defined by:

$$c = m^e \quad \mod \ N$$

5. The message, m, is then decrypted with:

$$m = c^d \quad \mod \ N$$

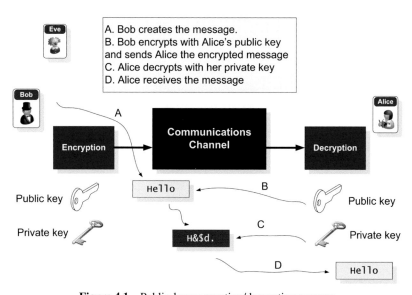

Figure 4.1   Public-key encryption/decryption process.

Sample values

We need to watch the values of e (the exponent) is not too small, especially if m^e is less then N (as we can easily discover the message), so a typical value that we use for e is 010001 (in hex), which is 65,635 (as an integer). In this case we have an N value (the modulus) which has 256 hex characters (1,024 bits), and which is created from two 512-bit prime numbers, such as:

```
e=010001,
n=9C7A2D4655B25862026DEB341403B5EB081C67DB343F18E430C2A975AB97578D
   B9DAEDC9B589CBDB7B53521380A98307106348E84684BE04E4B66661B60B3B55
   163DD067F31792A9390D57FFF12F3A67ACCD8DAD22E945AA2AAB98BAC53EF9AD
   45C8DADA107601FEE3C12F965EF012494292E77621DC6CB50CDCD402AED903C7
```

If we look at the decryption key, we see that the *d* value is much larger than the *e* value

```
d=30AEFA83158A856108EC75EF2002FF887E5F8818752AE46DAA9605EF2C51CBD
   5B66B5CEC12E52A5DC102ED5850016D58B74A8C9667CA48EC70D4270A637F1F
   181EF2000CDD5DA1E6C714CEF1F4B9E853B98D8301FB5E28FF6FB536F85B7BC
   3BB1BD7EBF43C8363A9BE00CFB9FECC27396D55D42EDD9FE995FA022B142990
   41A1,
n=9C7A2D4655B25862026DEB341403B5EB081C67DB343F18E430C2A975AB97578
   DB9DAEDC9B589CBDB7B53521380A98307106348E84684BE04E4B66661B60B3B
   55163DD067F31792A9390D57FFF12F3A67ACCD8DAD22E945AA2AAB98BAC53EF
   9AD45C8DADA107601FEE3C12F965EF012494292E77621DC6CB50CDCD402AED
   903C7
```

📖 **Web link (RSA keys):** http //asecuritysite.com/encryption/rsa2
   Web link (RSA): http //asecuritysite.com/encryption/rsa

Examples

Example 1
Let's select P = 11 Q = 3

The calculation of *n* and PHI is:

$n = P \times Q = 11 \times 3 = 33$
$PHI = (P - 1)(Q - 1) = 20$

The factors of PHI are 1, 2, 4, 5, 10 and 20. Next the public exponent *e* is generated so that the greatest common divisor of *e* and PHI is 1 (*e* is relatively prime with PHI). Thus, the smallest value for *e* is:

$e = 3$

Next we can calculate *d* from:

$(3 \times d) \bmod (20) = 1$

Thus the smallest value of *d* will be:

$d = 7$

And the keys will be:
Encryption key [33,3]
Decryption key [33,7]

Then, with a message of 4, we get:

Cipher = $(m)^e$ mod *N*
Cipher = $(4)^3$ mod 33 = 31

Decoded = $(cipher)^d$ mod *N*
Decoded = 31^7 mod 33 = 4

Example 2
Let's select the same P and Q, but we'll pick a different *e* value:

P = 11 Q = 3

The calculation of *N* and PHI is:

$N = P \times Q = 11 \times 3 = 33$
PHI = $(P - 1)(Q - 1) = 20$

We can select *e* as:

$e = 7$

Next we can calculate *d* from

$7 \times d$ mod (20) = 1
$d = 3$

And the keys will be:
Encryption key [33,7]
Decryption key [33,3]

Then, with a message of 2, we get:

Cipher = $(2)^7$ mod 33 = 31

Decoded = 313 mod 33 = 2

Example 3
Let's select P = 13 Q = 11

The calculation of N and PHI is

N = P × Q = 13 × 11 = 143
PHI = (P − 1)(Q − 1) = 120

We can select *e* as:

e = 7

Next we can calculate *d* from:

(7 × *d*) mod (120) = 1

d = 103

And the results keys will be:
Encryption key [143,7]
Decryption key [143,103]

Then, with a message of 7, we get:

Cipher = $(7)^7$ mod 143 = 6

Decoded = $(6)^{103}$ mod 143 = 7

📖 **Web link (P and Q values):** http //asecuritysite.com/encryption/rsa_2

XML keys

Within public key, an XML format can be used to define the Modulus (N) and the Exponent (e):

```
<RSAKeyValue>
<Modulus>
mtNFzSrQKBXi3NJs118He2Eir8pIFuTXnsQS0U7BWxkRGoGF/qK0FD
CPx7VbrJMZb7gttXInANnpj/SNKIxxsQ==</Modulus>
<Exponent>AQAB</Exponent>
</RSAKeyValue>
```

The following is example of the key pair which has a Modulus (*N*), an Exponent (*e*), prime numbers (P and Q), DP (*d* mod P-1), DQ (*d* mod Q-1), and InverseQ (INV(Q) mod P):

```
<RSAKeyValue>
<Modulus>
mtNFzSrQKBXi3NJs118He2Eir8pIFuTXnsQS0U7BWxkRGoGF/qK0FD
CPx7VbrJMZb7gttXInANnpj/SNKIxxsQ==</Modulus>
<Exponent>AQAB</Exponent>
<P>yIelVYRqHEHy+lJdAeb6baCAduADPj1ya1k4mB3Xr+0=</P>
<Q>xacYXOwF7A4cuq1QrTbPPO+aqATqFsHvJAqKQNv6KFU=</Q>
<DP>bgoZjRrzi3wZFIo75X5Vb/ECbbkxrmbTsdqs9rRxlmU=</DP>
<DQ>I807lYFPJU39GDdSmL2H1lLUYcDaIhso1Q9vsYXnDy0=</DQ>
<InverseQ>eMmd366oBE4kguzx4cUH+4Ei69+7GRVSifAMU5FxgvQ=
</InverseQ>
<D>hiRzH9XuUCzWSFkQ8HFnfCCm+wQZ/av8nZRocWz43kG6rycWDug
cJmwI4rKzcWtZYukjQssxYRCzoALYiHwoIQ==</D>
</RSAKeyValue>
```

 📖 **Web link (XML keys):** http //asecuritysite.com/encryption/rsa3

The following is some .NET code to generate 1024-bit public and private keys in an XML format:

```
System.Security.Cryptography.RSACryptoServiceProvider
RSAProvider;
RSAProvider = new System.Security.Cryptography.RSACryp
toServiceProvider(1024);
publicAndPrivateKeys = RSAProvider.ToXmlString(true);
justPublicKey = RSAProvider.ToXmlString(false);
StreamWriter fs = new StreamWriter("c\\public.xml");
fs.Write(justPublicKey);
fs.Close();
fs = new StreamWriter("c\\private.xml");
fs.Write(publicAndPrivateKeys);
fs.Close();
```

It converts the keys into an XML format, such as given in Figure 4.2 (which contains both the private and public key). A sample output for the public key is:

```
<RSAKeyValue>
<Modulus>
1NtbP2f+I/3AiwKd+QeHhhsnlTkfufLKS4muFruJ8CwIRFhsyo9yoC
IVydb6v0VdDtfg3F10iTGQw6waXy4QQ2LB4utIqASRumqU2cVNBLYk
B/p7eHByTm3GAhxvyTOGWPidcbVCrIrYor9ck9M79syetG7ZEpHd8h
y4Qm6BuP8=
</Modulus>
<Exponent>AQAB</Exponent>
</RSAKeyValue>
```

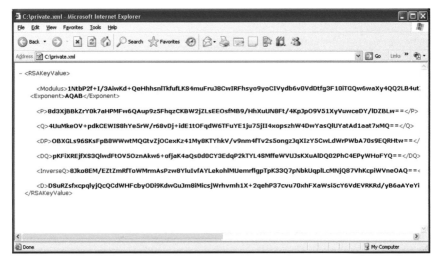

Figure 4.2   XML-based private key.

The code to then read the keys is:

```
XmlTextReader xtr = new XmlTextReader("c\\private.xml");
publicAndPrivateKeys=""; // reset keys
justPublicKey="";
while (xtr.Read())
{
    publicAndPrivateKeys += xtr.ReadOuterXml();
}
xtr.Close();
```

```
xtr = new XmlTextReader("c\\public.xml");
while (xtr.Read())
{
    justPublicKey += xtr.ReadOuterXml();
}
xtr.Close();
```

and then to encrypt a message (txt) with the public key:

```
RSACryptoServiceProvider rsa = new RSACryptoServiceProvider();
string txt= tbTxtEncrypt.Text;
rsa.FromXmlString(justPublicKey);
byte[] plainbytes = System.Text.Encoding.UTF8.GetBytes(txt);
byte[] cipherbytes = rsa.Encrypt(plainbytes,false);
this.tbTxtEncrypted.Text=Convert.ToBase64String(cipherbytes);
```

and then to decrypt with the private key:

```
RSACryptoServiceProvider rsa = new RSACryptoServiceProvider();
string txt=tbTxtEncrypted.Text;
rsa.FromXmlString(publicAndPrivateKeys);
byte[] cipherbytes = Convert.FromBase64String(txt);
byte[] plainbytes = rsa.Decrypt(cipherbytes,false);
System.Text.ASCIIEncoding enc = new System.Text.ASCIIEncoding();
this.tbTxtDecrypt.Text = enc.GetString(plainbytes);
```

OpenSSL

The OpenSSL library is often used to perform and check cryptography operations. In the following we generate a 1,204-bit key pair, and then export the public key to **mykey.pub:**

```
$ openssl version
OpenSSL 1.0.1f 6 Jan 2014

$ openssl genrsa -out mykey.pem 1024
Generating RSA private key, 1024 bit long modulus
.............................................................
.........++++++...++++++
e is 65537 0x10001

$ openssl rsa -in mykey.pem -pubout > mykey.pub
writing RSA key
```

```
$ cat mykey.pub
-----BEGIN PUBLIC KEY-----
MIGfMA0GCSqGSIb3DQEBAQUAA4GNADCBiQKBgQDXv9HSFkpM+ZoOQcpdHBZiUwX8EzIK
m0nsgjc5ZTYVaF9CMLtmKoTzep7aQX9o9nKepFt1kQ73Ta9vOPd6CX61/cgYXy2tShw0
imrtFaVDFjX+7kLmcOuWbFFCoZMtJxIaXaa9SV2kARxOCTJ2uOjRTCCeXU09IJGHnIhS
NJeIJQIDAQAB
  -----END PUBLIC KEY-----
```

Fermat's Little Theorem

Fermat's little theorem is used to justify RSA. It states that:

$$a^{p-1} \bmod p = 1$$

where p is a prime number and a has no common factors in p. Let's take an easy one, at a = 4, p = 5

a^{p-1} gives $4^3 = 256$
$a^{p-1} \bmod 1$ gives 256 mod 5 = 1

📖 **Web link (Fermat's Little Theorem):** http//asecuritysite.com/encryption/ fermat

Commutative encryption (SRA)

Commutative encryption allows us to decrypt in any order. For this we can use SRA (Shamir, Rivest and Aldeman) and generate encryption keys which share P, Q and N. With maths, operators such as multiplication are commutative, such as:

$$3 \times 5 \times 4 = 4 \times 5 \times 3$$

In encryption, most operations are non-commutative, so we need to modify the methods. One way is to use RSA, but generate two keys which have shared P, Q and N values. So we generate Bob and Alice's keys using the same two prime numbers (P and Q), so that they share the same N value (modulus).
    So let's start with Bob:

Let's select P = 7, Q = 13

The calculation of N and PHI is:

$N = 7 \times 13 = 91$

$PHI = (P - 1)(Q - 1) = 72$

We need to make sure that our encryption key (e) does not share any factors with PHI (gcd(PHI,e)=1). We can select e as:

$e = 5$

Next we can calculate d from

$(d \times 5) \bmod (72) = 1$

One answer for this is 29. Thus:

$d = 29, e = 5, N = 91$

Encryption key [91,5]

Decryption key [91,29]

Now for Alice. We have:

$N = 7 \times 13 = 91$
$PHI = (P - 1)(Q - 1) = 72$

We can select e as (and should not share any factors with PHI):

$e = 7$

Now we must solve:

$(7 \times d) \bmod (72) = 1$

For this we get 31. Alice's keys are then:

$d = 31, e = 7, N = 91$

Encryption key [91,7]
Decryption key [91,31]

A demo at the process is given here:

📖**Web link (SRA):** http //asecuritysite.com/encryption/comm2

RSA – partially homomorphic cryptosystem

With homomorphic encryption, we can perform mathematical operations with ciphered values. For RSA, we have a partially homomorphic cryptosystem, where we can take two values and the cipher them. Next we multiply the results together, and the deciphered result will be the multiplication of the two values. RSA is a partially homomorphic crypto system. If we have two values (V_1 and V_2) and an exponent of e and modulus of N:

$\text{Cipher}_1 = V_1{}^e \ (\text{mod } N)$
$\text{Cipher}_2 = V_2{}^e \ (\text{mod } N)$

Then:

$\text{Cipher}_1 \times \text{Cipher}_2 = V_1{}^e \ V_2{}^e \ (\text{mod } N) = (V_1 V_2)^e \ (\text{mod } N)$

If we decrypt this value we will get the result of the multiplication of V_1 and V_2. If we take an example, we have:

$e = 79, d = 1019, N = 3337, V_1 = 5, V_2 = 6$

Then we can calculate the ciphers:

$\text{Cipher}_1 \text{ is} = 5^{79} \ (\text{mod } 3337) = 270$
$\text{Cipher}_2 \text{ is} = 6^{79} \ (\text{mod } 3337) = 2086$
$\text{Cipher}_1 \times \text{Cipher}_2 = 270 \times 2086 \ (\text{mod } 3337) = 2604$

We now decrypt:

$\text{Decrypt} = 2604^{1019} \ (\text{mod } 3337) = 30$

Web link (RSA with homomorphic): http//asecuritysite.com/encryption/h_rsa

4.3 Elliptic Curve Ciphers (ECC)

RSA has a heavy overhead on processor loading, and is not well suited to embedded systems (as the power drain can be high, along with heavy requirements for processing and memory). An improved solution over RSA is Elliptic Curve which is often used in key exchange methods (such as with Elliptic Curve Diffie Hellman – ECDH) and for the creation of digital signatures (Elliptic Curve Digital Signature Algorithm – ECDSA). Within key exchange we use Elliptic Curve methods to generate a shared key, where as within digital signatures we use a private key to encrypt an specific object, and then the public key is used to decrypt it, and thus proving that the object was signed by the private key. In 2000, the patents related to RSA timed-out, but some patents still exist around Elliptic Curve methods.

The main advantages of Elliptic Curve methods are:

- Much smaller keys. The prime number P is normally only 160 bits, and much smaller than in RSA. This considerably speeds up the encryption process.
- Creation of the curves are more difficult than generating prime numbers, which makes it more difficult to crack than RSA.
- They can be used to factorise values, such as finding the prime number factors within RSA.

An elliptic curve takes the form of:

$$y^2 = x^3 + ax + b$$

A plot of $y^2 = x^3 - 3x + 5$ is shown in Figure 4.3.

📖 **Web link (Elliptic curve):** http //asecuritysite.com/comms/plot05

Overall elliptic curve is seen as a replacement for RSA, especially for embedded systems which would struggle to cope with the processing requirements of RSA. The Elliptic Curve equation is in the form of:

$$y^2 = x^3 + ax + b \ (\mathrm{mod} \ p)$$

where y, x, a and b are all within Fp, (and are integers modulo p). The values of *a* and *b* are coefficients of the curve. The curve must fulfill one condition:

$$4a^3 + 27b^2 \neq 0$$

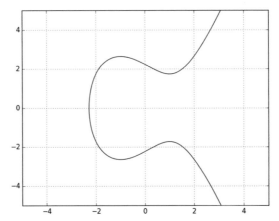

Figure 4.3   Public-key encryption/decryption process.

which guarantees that the curve will not contain any singularities. There are two interesting properties of elliptic curves. The first is horizontal symmetry, where a point on the curve is reflected over the x-axis and remains on the same curve. Another interesting property is that a non-vertical line intersects the curve in three places.

If we analyse the graph in Figure 4.4 if we select any two points on the curve and then draw a straight line between them. For P and Q on the graph, we get $nP = Q$, and where n is a scalar. So it we have P and Q, it is not computationally feasible to determine n, if n is large enough. The trapdoor problem involved with Elliptic Curve Ciphers (ECC) involves using the following curve and a prime number p:

$$y^2 = x^3 + \mathrm{a}x + \mathrm{b} \pmod{p}$$

We also have points on the curve (P and nP). We then need to find the value of n. If P is extremely large and n is also an extremely large number, it is then easy to determine nP. But if we only know P and nP, it is extremely difficult to determine n. While we could easily find out the value of P if we have 2P (as n will be two), it becomes extremely difficult to find n when n is extremely large. n is the discrete logarithm between P and nP, and that the main operation is point multiplication, which differs from prime number factorization used in RSA.

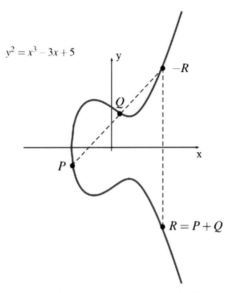

Figure 4.4 Elliptic curve property [1].

The following provides examples of the keys produced by Elliptic curve:

```
++++Keys++++
Bob's private key   02da0024026c6aeccf91e869dac11b5dc6b
5fd9d2b532d12ba63926cf59a8d0c2b5d9100b69be9e7
Bob's public key    02da00240727903797335fd45d88ddb16253
b8e17448f6fbf138c7dc37fa2869145e4eaaf02aae08002349ee16
4660dc9d55eb90e4d2d7e77b4176b16ff0e879a38520ec83b71c78
c13456c0cf
Alice's private key   02da002329e3ca0ca45072e54c287a79
a11fbe0e9571fd88c251ccbb8c5b474d6c016356886ba0
Alice's public key   02da002406f215ffee3b640145792655e
53d00ca24cd013c5e135814c1c71859fdb687b5f605f985002403e
555632044d13d57565bed9e91924cb2f711c4fdeef6ad8116abfb2
3722d71c6ef537e
```

Bitcoins use Elliptic Curve cryptography with 32 byte private keys (which is a random number) and 64 byte public keys, on a secp256k1 curve. A private key is a 32-byte number chosen at random, which selects an extremely large value at P. In OpenSSL, we can create a random number with:

```
C \ > openssl ecparam –name secp256k1 –genkey –out priv.pem

C \ > type ec–priv.pem
-----BEGIN EC PARAMETERS-----
```

```
BgUrgQQACg==
-----END EC PARAMETERS-----
-----BEGIN EC PRIVATE KEY-----
MHQCAQEEIEa56GG2PTUJyIt4FydaMNItYsjNj6ZIbd7jXvDY4ElfoAcGBSuBBAAK
oUQDQgAEJQDn8/vd8oQpA/VE3ch0lM6VAprOTiV9VLp38rwfOog3qUYcTxxX/sxJ
11M4HncqEopYIKkkovoFFi62Yph6nw==
-----END EC PRIVATE KEY-----
```

Next we can generate the public key based on the private key:

```
C \ > openssl ec -in priv.pem -text -noout
read EC key
Private-Key  (256 bit)
priv
       46 b9 e8 61 b6 3d 35 09 c8 8b 78 17 27 5a 30
       d2 2d 62 c8 cd 8f a6 48 6d de e3 5e f0 d8 e0
       49 5f
pub
       04 25 00 e7 f3 fb dd f2 84 29 03 f5 44 dd c8
       74 94 ce 95 02 9a ce 4e 25 7d 54 ba 77 f2 bc
       1f 3a 88 37 a9 46 1c 4f 1c 57 fe cc 49 97 53
       38 1e 77 2a 12 8a 58 20 a9 24 a2 fa 05 16 2e
       b6 62 98 7a 9f
ASN1 OID   secp256k1
```

The public key has 64 bytes (512 bits), and is made up of two 32 byte values
(x,y) and is a point on the secp256k1 elliptic curve function of:

$$y^2 = x^3 + 7 \ (\text{mod p})$$

and relates to an (x,y) point in relation to the private key (n) and a generator
(G). With the private key (32 bytes – 256 bits), we have a random number. In
this case it is in the form of:

```
       46 b9 e8 61 b6 3d 35 09 c8 8b 78 17 27 5a 30
       d2 2d 62 c8 cd 8f a6 48 6d de e3 5e f0 d8 e0
       49 5f
```

With Bitcoins, the private key defines our identity and we use it to sign for
transactions, and prove our identity to others with the public key. For the
public key we have an (x,y) point and is defined in a raw form starting with a
0x04 value and then followed by the x co-ordinate and then the y-co-ordinate:

```
04
25 00 e7 f3
fb dd f2 84
29 03 f5 44
dd c8 74 94
ce 95 02 9a
```

```
ce 4e 25 7d
54 ba 77 f2
bc 1f 3a 88

37 a9 46 1c
4f 1c 57 fe
cc 49 97 53
38 1e 77 2a
12 8a 58 20
a9 24 a2 fa
05 16 2e b6
62 98 7a 9f
```

📖 **Web link (ECC key generation):** http //asecuritysite.com/encryption/ecc

We can also use OpenSSL to view the details of the curve:

```
C:> openssl ecparam -in priv.pem -text -param_enc explicit
-noout
Field Type: prime-field
Prime:
    00:ff:ff:ff:ff:ff:ff:ff:ff:ff:ff:ff:ff:ff:ff:
    ff:ff:ff:ff:ff:ff:ff:ff:ff:ff:ff:ff:fe:ff:
    ff:fc:2f
A:   0
B:   7 (0x7)
Generator (uncompressed):
    04:79:be:66:7e:f9:dc:bb:ac:55:a0:62:95:ce:87:
    0b:07:02:9b:fc:db:2d:ce:28:d9:59:f2:81:5b:16:
    f8:17:98:48:3a:da:77:26:a3:c4:65:5d:a4:fb:fc:
    0e:11:08:a8:fd:17:b4:48:a6:85:54:19:9c:47:d0:
    8f:fb:10:d4:b8
Order:
    00:ff:ff:ff:ff:ff:ff:ff:ff:ff:ff:ff:ff:ff:ff:
    ff:fe:ba:ae:dc:e6:af:48:a0:3b:bf:d2:5e:8c:d0:
    36:41:41
Cofactor:  1 (0x1)
```

Overall we have a prime number (p), and fixed point G (the generator), which on the curve. We then multiply the generator (G) by the scalar private key n. This operation is extremely difficult to reverse in modular arithmetic. The result is the public key P which is:

$$P = n \times G$$

It should not be computationally possible, within a reasonable time period, to determine the scalar (the private key value) between the generator and the public key value. Within Bitcoins, we use the private key to sign a transaction, and then which is proven by the public key (Elliptic Curve Digital Signature Algorithm). More details on elliptic curve ciphers here:

📖 **Web link (Elliptic Curve):** http //asecuritysite.com/encryption/elc

The strength of the keys depends on the encryption key. NIST defines that an 80-bit symmetric key is equivalent to a 1,024 bit RSA key, and a 160 bit Elliptic Curve key. The other examples of recommended key sizes (in bit length) are

Symmetric Keys	RSA/Diffie-Hellman Key	Elliptic Curve Key
80	1,024	160
112	2,048	224
128	3,072	256
192	7,680	384
256	15,360	521

4.4 ElGamal

ElGamal is a public key method that is used in both encryption and digital signing. It is used in many applications and uses discrete logarithms. At the root is the generation of p which is a prime number and G (which is a value between 1 and $p - 1$, and must be a safe value – as defined in the link given next). At the core of discrete logarithms we have:

$$Y = G^x \mod p$$

where p is a prime number and G is a generator, with x being a random number. What we want is the each value of x that we choose should give us a unique value of Y (obviously between 0 and $p - 1$), so:

3^1 mod 5 gives 3, while 3^2 mod 5 gives 4, and so on.

This process allows us to pick a G value from a cyclic group:

Web link (Picking G): http //asecuritysite.com/encryption/pickg

Let's illustrate ElGamal with an example. First Bob generates a prime number (P) and a number (G), which is in the cyclic group for values between 1 and ($P - 1$):

P = 3191
G = 1118

Bob select a random number (*x*) which will be his private key:

x = 101

He then calculates:

Y = 1983 (Y = G^x mod *P*)

Bob's public key is now [P, G, Y] and he sends these values to Alice. The private key is *x*. Alice then creates a message.

Message (to send) = 43

and then she selects a random value (*k*), and calculates two new values (*a* and *b*)

k (random value) = 191
a value = 2,890 (a = G^k mod *P*)
b value = 1,549 (b = y^k M mod *P*)

These values are then sent to Alice, and she decrypts them with:

Message (decrypted)   43 $\left(\frac{b}{a^x} \bmod P\right)$

Note that the calculation for $\left(\frac{b}{a^x} \bmod P\right)$ is implemented as:

$$a^{(P-1-x)}b \bmod P$$

📖 **Web link (El Gamal):** https //asecuritysite.com/encryption/elgamal

4.5 Cramer-Shoup

Cramer-Shoup is a public key encryption method that is an extension of ElGamal but adds a one-way hashing method which protects against an adaptive chosen ciphertext attack.

📕 **Web link (Cramer-Shoup):** https //asecuritysite.com/encryption/cramer

Key generation:

- Alice generates two random generators in the range 1 to $p-1$ (g_1, g_2).
- Alice select five random values (x_1, x_2, y_1, y_2, z).
- Alice computes $c = g_1{}^{x1} g_2{}^{x2}$, $d = g_1{}^{y1} g_2{}^{y2}$, $h = g_1{}^{z}$
- Alice publishes (c, d, h), and shares g_1, g_2, p. She keeps (x_1, x_2, y_1, y_2, z) secret.

Encryption:

- Bob creates a message (m) and uses Alice's public key.
- Bob creates a random number (k).
- Bob calculates $u_1 = g_1{}^{k}$, $u_2 = g_2{}^{k}$.
- Bob calculates $e = h^k\, m$

Decryption

- Alice decrypts with $m = e/(u_1{}^{z})$.

Web link (Cramer-Shoup): https://asecuritysite.com/encryption/cramer

4.6 Paillier Cryptosystem

With homomorphic encryption, defined by Craig Gentry in 2010, we can operate on data without even decrypting it. Craig defined a scenario where Alice had a jewelry box, which she locked with her key, and where her workers could not gain access to the gems contained within it. Then when they wanted to work on the gems, they could do so with special gloves, but couldn't remove them from the box.

Homomorphic encryption allows ciphered values to be moved to wherever they are required, and then processed, without giving away the original data. Data could thus traverse across the Internet and move to places that it is required, and then used to calculate results. For your tax return we might see:

Sales (Web)   &*X43=%
Sales (Print)   *65tfd1=
_____

Total Sales   64,532 (=B1+B2)

In this case the sales values are ciphered, but we can still process the addition of the two values. We could also apply subtraction, multiplication and division.

📖 **Web link (Paillier):** https //asecuritysite.com/encryption/pal

4.7 Knapsack Encryption

RSA is just one way of doing public key encryption. Knapsack is an alternative where we can create a public key and a private one. The knapsack problem defines a problem where we have a number of weights and then must pack our knapsack with the minimum number of weights that will make it a given weight. In general the problem is:

- Given a set of numbers A and a number b.
- Find a subset of A which sums to b (or gets nearest to it).

So imagine you have a set of weights of 1, 4, 6, 8 and 15, and we want to get a weight of 28, we could thus use 1, 4, 8 and 15 (1+4+8+15=28).

So our code would become 11011 (represented by '1', '4', no '6', '8' and '15').

Then if our plain text is 10011, with a knapsack of 1, 4, 6, 8, 15, we have a cipher text of 1+4+8+15 which gives us 28.

A plain text of 00001 will give us a cipher text of 15.

With public key cryptography we have two knapsack problems. One of which is easy to solve (private key), and the other difficult (public key).

Creating a public and a private key

We can now create a super-increasing sequence with our weights where the current value is greater than the sum of the preceding ones, such as {1, 2, 4, 9, 20, 38}. Super-increasing sequences make it easy to solve the knapsack problem, where we take the total weight, and compare it with the largest weight, if it is greater than the weight, it is in it, otherwise it is not.

For example with weights of {1, 2, 4, 9, 20, 38} with a value of 54, we get

Check 54 for 38? Yes (smaller than 54). [1] We now have a balance of 16.
Check 16 for 20? No. [0].
Check 16 for 9? Yes. [1]. We now have a balance of 5.
Check 5 for 4? Yes. [1]. We now have a balance of 1.
Check 1 for 2? No. [0].
Check 1 for 1? Yes [1].

Our result is 101101.

If we have a non-super-increasing knapsack such as {1, 3, 4, 6, 10, 12, 41},
and have to make 54, it is much more difficult. So a non-super-increasing
knapsack can be the public key, and the super-increasing one is the
private key.

Making the Public Key

We first start with our super-increasing sequence, such as {1, 2, 4, 10, 20, 40}
and take the values and multiply by a number n, and take a modulus (m) of
a value which is greater than the total, for example we could select 120. For
n we make sure that there are no common factors with any of the numbers.
Let's select an n value of 53, so we get:

$1 \times 53 \bmod(120) = 53$
$2 \times 53 \bmod(120) = 106$
$4 \times 53 \bmod(120) = 92$
$10 \times 53 \bmod(120) = 50$
$20 \times 53 \bmod(120) = 100$
$40 \times 53 \bmod(120) = 80$

So the public key is {53, 106, 92, 50, 100, 80} and the private key is {1, 2,
4, 10, 20, 40}. The public key will be difficult to factor while the private key
will be easy. Let's try to send a message that is in binary code:

111010 101101 111001
We have six weights so we split into three groups of six weights:

$111010 = 53 + 106 + 92 + 100 = 351$
$101101 = 53 + 92 + 50 + 80 = 275$
$111001 = 53 + 106 + 92 + 80 = 331$

Our cipher text is thus 351 275 331.

The two numbers known by the receiver is thus 120 (m modulus) and 53 (n multiplier).

We need n^{-1}, which is a multiplicative inverse of n mod m, i.e. $n(n^{-1}) = 1$ mod m. For this we find the inverse of n:

$n^{-1} = 53^{-1}$ mod 120
$(53 \times n)$ mod $120 = 1$

So we try values of n^{-1} in ($53 \times n^{-1}$ mod 120) in order to get a result of 1:

n^{-1}	Result
1	53
2	106
3	39
...	
75	15
76	68
77	1

So the inverse is 77.

The coded message is 351 275 331 and is now easy to calculate the plain text:

351×77 mod$(120) = 27 = 111010$ (1+2+4+20)
275×77 mod$(120) = 55 = 101101$
331×77 mod$(120) = 47 = 111001$

The decoded message is thus

111010 101101 110001
which is the same as our original message.

The decrypting is easy and the only thing that was difficult to find the inverse value at n, which is not too difficult.

 Web link (Knapsack): http //asecuritysite.com/encryption/knap

4.8 Identity-Based Encryption

Identity-based Encryption (IBE) is an alternative to PKI (Public Key Infrastructure), and involves generating the encryption key from a piece of the identity of the recipient. For example we could use the email address of the recipient to generate the public key of the receiver.

For this we have some shared parameters with a trust center that both Bob and Alice trust. If Alice wants to send Bob an email, she takes the parameters from the trust center, and then uses Bob's email address to generate his public key (Figure 4.5). When Bob receives the encrypted email, he contacts the trust center and the center generates the private key required to decrypt the email.

📖 **Web link (IBE):** http //asecuritysite.com/encryption/ibe

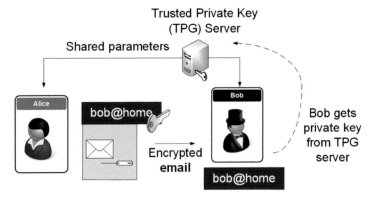

Figure 4.5   Public-key encryption/decryption process.

4.9 Lab/Tutorial

The lab and tutorial related to this chapter is available on-line at:

http //asecuritysite.com/crypto04

Reference

[1] "An introduction to elliptic curve cryptography I Embedded." [Online]. Available http //www.embedded.com/design/safety-and-security/439 6040/An-Introduction-to-Elliptic-Curve-Cryptography. [Accessed 05-Jun-2017].

5

Key Exchange

5.1 Introduction

The major problem of secret-key encryption is how to pass the key between Bob and Alice, without Eve listening (Figure 5.1). The two main methods for this is to either use a key exchange protocol (such as Diffie-Hellman) or to encrypt the key with a public key and pass it to the other side, and use the private key to decrypt it. With Diffie-Hellman we use the difficulty of solving discrete logarithms, and where we have to solve for x (which is the discrete logarithm of y with respect to a base g modulo p):

$$y = g^{x} \mod p$$

With public key methods, we use the difficulty of the factorising a value into its prime number factors, or use elliptic curve methods. Overall for Bob and Alice to generate a symmetric key (a secret key), they assume that Eve is listening to their communications, and talk openly, and where at the end of the key negotiation they will have the same secret key, but Eve, even though she has been listening, will not.

Bob could thus communicate openly Alice, and end up with an agreed secret key, but how does Bob actually know that he is communicating with Alice, and that the key they negotiate is the same? As we will see in a later chapter, we often have to use key pairs as part of the identification process, as we need to check the identity of one or more of the entities involved in the key exchange process. These key pairs (a public and a private key) can either static, and which are created from a trusted source, such as from a trusted digital certificate, or which can be generated for each connection. From a trust point-of-view the ones created from the trusted source are more likely to be trusted, but as they are static, a leakage of the private key could compromise any of the key exchanges created with the key pair. Another method is to create the key pairs when a secret key is required, and where a new key is created for each key exchange. This is typically defined as a session key.

In some situations we need to make sure that we are connecting to a trusted end source, as Eve could break the communication process, and become a man-in-the-middle. We thus often validate at least one of the entities involved in the negotiation. This is normally achieved through one side passing its public key, and to sign some data with their private key, and where the other side verifies that it has signed the data with the required private key. Normally it is the server which proves its identity to the client.

An important concept within key exchange is the usage of **forward secrecy** (FS), which means that a comprise of the long-term keys will not compromise any previous session keys. For example if we send the public key of the server to the client, and then the client sends back a session key for the connection which is encrypted with the public key of the server, then the server will then decrypt this and determine the session key. A leakage of the public key of the server would then cause all the sessions which used this specific public key to be compromised. FS thus aims to overcome this by making sure that all the sessions keys could not be compromised, even though the long-term key was compromised.

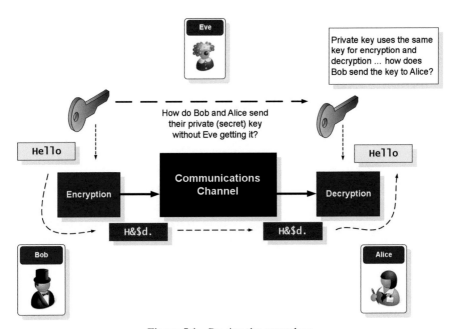

Figure 5.1 Passing the secret key.

Another major concept is where the key is **ephemeral**. With some key exchange methods the same key will be generated if the same parameters are used on either side. This can cause problems as an intruder could guess the key, or even where the key was static and never changed. With ephemeral methods, a different key is used for each connection, and, again, the leakage of any long-term key would not cause all the associated session keys to be breached.

5.2 Diffie-Hellman Key Exchange

The problem of creating a shared symmetric key over a public network was solved by Whitfield Diffie in 1975, who created the Diffie-Hellman method. With this method, Bob and Alice generate two random values, and perform some calculations (Figure 5.2 and Figure 5.3), and pass the result of the calculations to each other. We first use two shared values (g – a generator, and p – a prime number). Bob then generates a random number (x) and Alice generates a random number (y). Next:

$$\text{Bob computes A} = g^x \bmod p$$
$$\text{Alice computes B} = g^y \bmod p$$

Bob sends A to Alice, and Alice sends B to Bob. The result becomes:

$$\text{Bob computes Key} = B^x \bmod p$$
$$\text{Alice computes Key} = A^y \bmod p$$

This will give be the same shared key (which is $g^{xy} \bmod p$).

The basics of the operation is that we agree on the generator (g) and a prime number (p), which are agreed by both Bob and Alice (Figure 5.4). Alice and Bob generate their values (a and b), and where Alice passes g^a (mod p) and Bob passes g^b (mod p). Once they raise the received values to their random value that they have created, they will end up with the same shared key (g^{ab} (mod p)).

Once these values have been received at either end, Bob and Alice will have the same secret key, which Eve cannot compute (without extensive computation). Diffie-Hellman is used in many applications, such as in VPNs (Virtual Private Networks), SSH, and secure FTP. The following shows a trace of a connection to a secure FTP site:

```
STATUS:> Initializing SFTP21 module...
STATUS:> Resolving host name mysite.com...
STATUS:> Host name mysite.com resolved: ip = 1.2.3.4.
STATUS:> Connecting to SFTP server ftp1.napier.ac.uk:22 (ip = 1.2.3.4)
                          Key Method: Diffie-Hellman-group1-SHA1
                          Host Key Algorithm: SSH-RSA
                          Session Cipher: 192 bit TripleDES-cbc
                          Session MAC: HMAC-MD5
                          Session Compressor/Decompressor: ZLIB
STATUS:> Getting working directory...
STATUS:> Home directory: /home/test
```

Where it can be seen that this is a secure FTP transaction, the **encryption** being used is **3DES** (TripleDES), the message **authentication** method is **HMAC-MD5** and the **key exchange** is Diffie-Hellman. Overall Diffie-Hellman has three groups: Group 1, Group 3 or Group 5, which vary in the size of the prime number used.

  📖 **Web link (Diffie-Hellman):** http://asecuritysite.com/encryption/diffie
       Web link (Diffie-Hellman real): http://asecuritysite.com/encryption/diffie2

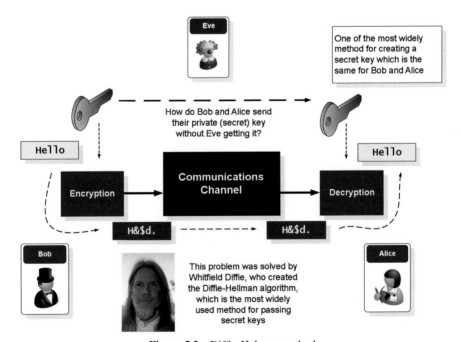

Figure 5.2   Diffie-Helman method.

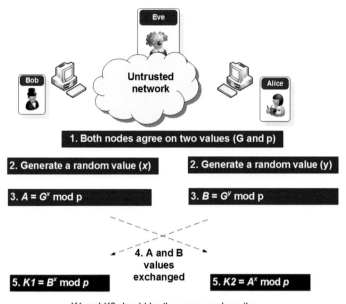

Figure 5.3 Diffie-Hellman process.

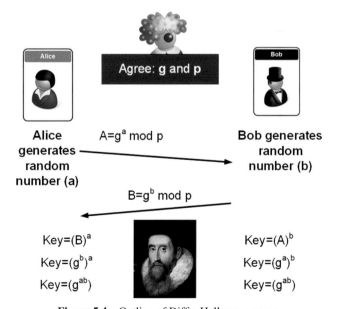

Figure 5.4 Outline of Diffie-Hellman process.

A simple Python program to calculate small values of G and n is:

```
import random
import base64
import hashlib

g=11
p=1001

x=random.randint(5, 10)
y=random.randint(10,20)

A=(g**x) % p

B=(g**y) % p

print 'g: ',g,' (a shared value), n: ',p, ' (a prime number)'

print '\nAlice calculates:'
print 'a (Alice random): ',x
print 'Alice value (A): ',A,' (g^a) mod p'

print '\nBob calculates:'
print 'b (Bob random): ',y
print 'Bob value (B): ',B,' (g^b) mod p'

print '\nAlice calculates:'
keyA=(B**x) % p
print 'Key: ',keyA,' (B^a) mod p'
print 'Key: ',hashlib.sha256(str(keyA)).hexdigest()

print '\nBob calculates:'
keyB=(A**y) % p
print 'Key: ',keyB,' (A^b) mod p'
print 'Key: ',hashlib.sha256(str(keyB)).hexdigest()
```

which gives a sample run of:

```
g:  11  (a shared value), n:  1001  (a prime number)

Alice calculates:
a (Alice random):  7
Alice value (A):  704  (g^a) mod p

Bob calculates:
b (Bob random):  16
Bob value (B):  627  (g^b) mod p

Alice calculates:
Key:  627  (B^a) mod p
```

```
Key:   9a35532c7499c19daeacafc961657409c7280ce59d7ae1a3606dd638
       ac3d99ec

Bob calculates:
Key:   627   (A^b) mod p
Key:   9a35532c7499c19daeacafc961657409c7280ce59d7ae1a3606dd638
       ac3d99ec
```

5.3 Creating the Generator

The value of g must be selected so that every value of x gives a unique value (Y) for a given prime number (p):

$$Y = g^x \mod p$$

What we want is that each value of x that we choose should give us a unique value of Y (obviously between 0 and p–1), so if we have g = 3 and p = 5, then we can compute the Y values of:

$$3^1 \mod 5 -> 3$$
$$3^2 \mod 5 -> 4$$
$$3^3 \mod 5 -> 2$$
$$3^4 \mod 5 -> 1$$

and which is known as a cyclic group (Zp) where each value is unique as an output (up to a value at p–1). We can create these values with the following Python code (for possible g values up to p–1):

```
import sys
import random
p=11
for x in range (1,p):
        rand = x
        exp=1
        next = rand % p

        while (next <> 1 ):
                next = (next*rand) % p
                exp = exp+1

        if (exp==p-1):
                print rand
```

For g values with $p = 11$, we get possible generators of 2, 6, 7 and 8. We can get this by also calculating the values for g^x mod p for different generator values and x values, but using $p = 11$. In this case we see that $g = 3$, $g = 4$, $g = 5$ and $g = 9$ have repeated values as an output, so they cannot be used:

p	11							
Generator	2	3	4	5	6	7	8	9
x	g^x mod p	g^x mod p	g^x mod p	g^x mod p	g^x mod p	g^x mod p	g^x mod p	g^x mod p
2	4	9	5	3	3	5	9	4
3	8	5	9	4	7	2	6	3
4	5	4	3	9	9	3	4	5
5	10	1	1	1	10	10	10	1
6	9	3	4	5	5	4	3	9
7	7	9	5	3	8	6	2	4
8	3	5	9	4	4	9	5	3
9	6	4	3	9	2	8	7	5
10	1	1	1	1	1	1	1	1

The strength of the Diffie-Hellman method normally relates to the size of the prime number bases which are used in the key exchange, where Group 5 uses a 1,536-bit prime number, Group 2 uses a 1,024-bit prime, and Group 1 uses a 768-bit prime number. In the following we use Openssl to create a 768-bit key:

```
C:\> openssl dhparam -out dhparams.pem 768 -text

C:\> type dhparams.pem
Diffie-Hellman-Parameters: (768 bit)
    prime:
        00:d0:37:c2:95:64:02:ea:12:2b:51:50:a2:84:6c:
        71:6a:3e:2c:a9:80:e2:65:b2:a5:ee:77:26:22:31:
        66:9e:fc:c8:09:94:e8:9d:f4:cd:bf:d2:37:b2:fb:
        b8:38:2c:87:28:38:dc:95:24:73:06:d3:d9:1f:af:
        78:01:10:6a:7e:56:4e:7b:ee:b4:8d:6b:4d:b5:9b:
        93:c6:f1:74:60:01:0d:96:7e:85:ca:b8:1f:f7:bc:
        43:b7:40:4d:4e:87:e3
    generator: 2 (0x2)
-----BEGIN DH PARAMETERS-----
MGYCYQDQN8KVZALqEitRUKKEbHFqPiypgOJlsqXudyYiMWae/MgJlOid9M2/0jey
+7g4LIcoONyVJHMG09kfr3gBEGp+Vk577rSNa021m5PG8XRgAQ2WfoXKuB/3vEO3
QE1Oh+MCAQI=
-----END DH PARAMETERS-----
```

In this case we use the value of 2 for the generator (which is often the default value for the generator), but can also use a value of 5 (using the "-5" option). For a g value of 2, and a prime number (p) of 11, we get (safe g values are 2, 6, 7 and 8):

$2^1 \bmod 11 = 2$ $2^2 \bmod 11 = 4$
$2^3 \bmod 11 = 8$ $2^4 \bmod 11 = 5$
$2^5 \bmod 11 = 10$ $2^6 \bmod 11 = 9$
$2^7 \bmod 11 = 7$ $2^8 \bmod 11 = 3$
$2^9 \bmod 11 = 6$ $2^{10} \bmod 11 = 1$

If we use a generator (g) of 5, and generate a 768-bit prime number, we get:

```
Diffie-Hellman-Parameters: (768 bit)
    prime:
        00:8b:48:7b:80:7c:fe:69:6a:a6:30:29:08:3b:e7:
        2b:c6:90:8b:68:63:6b:ff:ba:29:5a:52:9e:98:a7:
        d8:4a:1b:2f:fe:e6:35:e8:af:de:51:6b:5f:e8:2f:
        79:aa:6a:65:ed:85:64:99:ce:84:e3:b3:0c:37:77:
        47:78:d3:33:45:da:4e:0b:49:82:83:c1:7b:2a:c7:
        8d:11:8e:e2:7b:93:2c:85:46:62:6c:93:a5:25:88:
        3a:83:fd:fd:10:e5:f7
    generator: 5 (0x5)
-----BEGIN DH PARAMETERS-----
MGYCYQCLSHuAfP5paqYwKQg75yvGkItoY2v/uilaUp6Yp9hKGy/+5jXor95Ra1/o
L3mqamXthWSZzoTjsww3d0d40zNF2k4LSYKDwXsqx40RjuJ7kyyFRmJsk6UliDqD
/f0Q5fcCAQU=
-----END DH PARAMETERS-----
```

If we take a simple example of a generator of 5 we cannot use a prime number of 11, or 13, so let's use 17 (where safe generate values are 3, 5, 6, 7, 10, 11, 12 and 14):

$5^1 \bmod 17 = 5$ $5^2 \bmod 17 = 8$
$5^3 \bmod 17 = 6$ $5^4 \bmod 17 = 13$
$5^5 \bmod 17 = 14$ $5^6 \bmod 17 = 2$
$5^7 \bmod 17 = 10$ $5^8 \bmod 17 = 16$
$5^9 \bmod 17 = 12$ $5^{10} \bmod 17 = 9$
$5^{11} \bmod 17 = 11$ $5^{12} \bmod 17 = 4$
$5^{13} \bmod 17 = 3$ $5^{14} \bmod 17 = 15$
$5^{15} \bmod 17 = 7$ $5^{16} \bmod 17 = 1$

5.4 Diffie-Hellman Examples

Let's select a prime number of:

$p = 11$

Safe values for g are 2, 6, 7 and 8, so let's select $g = 7$. Bob and Alice generate random numbers (x and y):

$$x = 3 \qquad y = 4$$

Bob calculates A:

$$A = g^x \ (\text{mod } p) = 7^3 \ \text{mod } 11 = 343 \ (\text{mod } 11) = 2$$

Alice calculates B:

$$B = g^y \ (\text{mod } p) = 7^4 \ \text{mod } 11 = 2401 \ (\text{mod } 11) = 3$$

They swap values and they generate the key:

Key (Bob) $= B^x \ (\text{mod } p) = 3^3 \ \text{mod } 11 = 27 \ (\text{mod } 11) = 5$
Key (Alice) $= A^y \ (\text{mod } p) = 2^4 \ \text{mod } 11 = 16 \ (\text{mod } 11) = 5$

This is their shared key. As another example, let's select:

$$p = 3049$$

A safe value of p for 3,049 is $g = 282$. Bob and Alice generate random numbers (x and y):

$$x = 21 \qquad y = 6$$

Bob calculates A:

$$A = 282^{21} \ \text{mod } 3,049 = 438$$

Alice calculates B:

$$B = 282^6 \ \text{mod } 3,049 = 1,924$$

They swap values and they generate the shared key:

Key (Bob) $= 1924^{21} \ (\text{mod } 3,049) = 2,736$
Key (Alice) $= 438^6 \ (\text{mod } 3,049) = 2,736$

5.5 Ephemeral Diffie-Hellman with RSA (DHE-RSA)

The problem with DH is that if Bob and Alice generate the same values each time, they will always end up with the same secret key. With Ephemeral Diffie-Hellman (DHE) a different key is used for each connection, and a leakage of the public key would still mean that all of the communications were secure. Within DHE-RSA, the server signs the Diffie-Hellman parameter (using a private key from an RSA key pair) to create a pre-master secret, and where a master key is created which is then used to generate a shared symmetric encryption key.

Normally when we create a shared key we create a tunnelled connected between a client and a server. This is normally defined through SSL (Secure Socket Layer) or TLS (Transport Layer Security), and where a client initially connects to a server. We then define the tunnel type (such as TLS or SSL), the key exchange method (such as DHE-RSA), a symmetric key method to be used for the encryption process (such as 256-bit AES with CBC) and a hashing method (such as SHA). This can be defined as a string as:

TLS_DHE_RSA_WITH_AES_256_CBC_SHA

and is contained in a ClientHello message that goes from the client to the server. A ServerHello is then returned with the digital certificate of the server and which contains the public key of the server. A simplified handshake is defined in Figure 5.5 where the client sends the definition for a TLS cipher suite. In this case we are using a handshaking methods of DHE-RSA, a 256-bit AES-CBC shared key, and with a SHA hash signature. The server will then generate a random value (x) and create a value for g (the generator) and p (the prime number). Along with this the server will then generate g^x and take the g, p and g^x parameters and encrypt them with the private key of the server. This creates a signature for the server.

Next the server will create a message with g, p, g^x, and the signature of nonces (random values) and the Diffie-Hellman parameters (g, p, and g^x). The server then sends this message with a digital certificate containing its public key. When the client receives it, it will check the certificate for its validity, and then extract the public key. The client then checks the signature by decrypting the signed value and checks it against the parameters already contained in the message (g, p, and g^x). If the values are the same, the server has been validated. Now, as with DH, the client will create a random value of y, and sends the value of g^y back to the server. The g^{xy} value will then be the pre-master secret. In this way the client knows that it has received valid values of the DH parameters, and can trust that the connection does not have a man-in-the-middle.

We now have a pre-master secret, as illustrated in Figure 5.6, which is shared by the client and server, and which can then be used to create a master key by using a PRF (Pseudorandom Function). In TLS 1.2 this is created using an HMCA-SHA256 hashed value (and which will generate a 256-bit key). To create the actual key used we feed the master key and the nonce into the PRF and generate the shared key for the session.

📖 **Web link (DHE-RSA):** http://asecuritysite.com/dhe

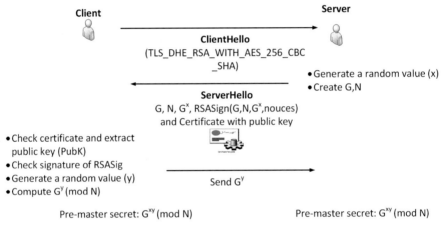

Figure 5.5   Example DHE-RSA process.

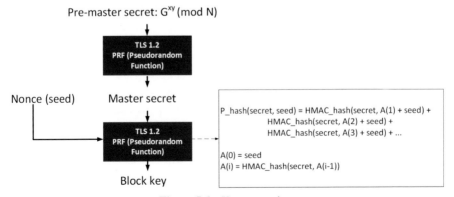

Figure 5.6   Key generation.

5.6 (Ephemeral) Elliptic Curve Diffie-Hellman (ECDHE)

ECDHE can be used to create a shared key between Bob and Alice. Initially, as with the Elliptic Curve method, we define some domain parameters, such as p, a, b, and G (see the previous chapter). From this Bob and Alice each generate an elliptic curve key pair, either with a static version where the keys have been generated by a trusted source, or with a dynamic method. We then have a private key (d) which is a random number and where the public key (P) is equal to dG (where G is a point on a defined elliptic curve). This gives us a public key for Alice (P_A), and one for Bob (P_B), along with a private key for Alice (d_A) and Bob (d_B).

$$\text{Bob computes } (x_k, y_k) = d_B P_A$$
$$\text{Alice computes } (x_k, y_k) = d_A P_B$$

The value of x_k then becomes the shared secret. The reason we can assume this is that:

$$d_A P_B = d_A d_B G = d_B d_A G = d_B P_A$$

The keys can either come from a digital certificate (which uses static keys) or can be ephemeral, and where the public and private keys are generated for each connection. Normally, too, the secret values (x_k) are hashed in order to remove weaknesses within the Diffie-Hellman key exchange process.

5.7 Diffie-Hellman Weaknesses

Netscape first defined SSL (Secure Socket Layer) Version 1.0 in 1993, and eventually, in 1996, released a standard which is still widely used: SSL 3.0. While many in the industry used it, it did not become an RFC standard until 2011 (which was assigned RFC 6101). SSL has now been exposed by many problems including FREAK ("Factoring RSA Export Keys") and was introduced to comply with US Cryptography Export Regulations, where the keys used for exportable software were limited to 512-bits or less (and were defined as RSA EXPORT keys – DHE_EXPORT). The RFC states:

> The server key exchange message is sent by the server if it has no certificate, has a certificate only used for signing (e.g., DSS [DSS] certificates, signing-only RSA [RSA] certificates), or FORTEZZA KEA key exchange is used. This message is not used if the server certificate contains Diffie-Hellman [DH1] parameters.

Note: According to current US export law, RSA moduli larger than 512 bits may not be used for key exchange in software exported from the US. With this message, larger RSA keys may be used as signature-only certificates to sign temporary shorter RSA keys for key exchange.

In 2015, a paper entitled *Imperfect Forward Secrecy: How Diffie-Hellman Fails in Practice* – showed that it was fairly easy to precompute on values for two popular Diffie-Hellman parameters (and which use the DHE_EXPORT cipher set). The research team found that one of them was used as a default in the around 7% of the Top 1 million web sites and was hard coded into the Apache httpd service. Overall, at the time, it was found that over 3% of Web sites were still using the default. The parameters were (where we see a generator value of 2):

```
Diffie-Hellman-Parameters: (512 bit)
prime:
   00:9f:db:8b:8a:00:45:44:f0:04:5f:17:37:d0:ba:
   2e:0b:27:4c:df:1a:9f:58:82:18:fb:43:53:16:a1:
   6e:37:41:71:fd:19:d8:d8:f3:7c:39:bf:86:3f:d6:
   0e:3e:30:06:80:a3:03:0c:6e:4c:37:57:d0:8f:70:
   e6:aa:87:10:33
generator: 2 (0x2)
```

Another group was found within the OpenSSL library (dh512.pem), and defined with:

```
Diffie-Hellman-Parameters: (512 bit)
prime:
   00:da:58:3c:16:d9:85:22:89:d0:e4:af:75:6f:4c:
   ca:92:dd:4b:e5:33:b8:04:fb:0f:ed:94:ef:9c:8a:
   44:03:ed:57:46:50:d3:69:99:db:29:d7:76:27:6b:
   a2:d3:d4:12:e2:18:f4:dd:1e:08:4c:f6:d8:00:3e:
   7c:47:74:e8:33
generator: 2 (0x2)
```

The DHE_EXPORT Downgrade attack then involves forcing the key nego-tiation process to default to 512-bit prime numbers. For this the client only offers DHE_EXPORT for the key negotiation, and the server, if it is setup for this, will accept it. The precomputation of 512-bit keys with *g* values of 2 and 5 (which are common) are within a reasonable time limits. The ways to overcome the problems are to:

- **Disable Export Cipher Suites**. With this we disable any negotiation of the key using export grade ciphers. This will have no effect on connections, as no existing version of Web browsers actually depend on export level ciphers.
- **Use (Ephemeral) Elliptic-Curve Diffie-Hellman (ECDHE)**. This method uses a key exchange method based on an Elliptic-Curve Diffie-Hellman (ECDH) key exchange.
- **Uses a strong group**. With this we make sure that we are using the strong generation of a prime number which cannot be precomputing. Currently 2,048-bit prime numbers are recommended. A strong prime number is generated with "openssl dhparam -out dhparams.pem 2048". Normally this will take a few minutes to compute, as it involves a random process, so can only be used to statically assign the parameters.

5.8 Using the Public Key to Pass a Secret Key

Diffie-Hellman methods have been used extensively to create a shared secret key, but suffers from a man-in-the-middle attack, where Eve sits in-between and passes the values back and forward, and negotiates two keys: one between Bob and Eve, and the other between Alice and Eve. An improved method is to use public key encryption, where Alice passes her public key to Bob, and then Bob creates an encryption key and encrypts this with Alice's public key. Alice then receives this and decrypts the key with her private key, to reveal the shared key. As Alice is the only one to have the private key to match the public key, so the method is secure (Figure 5.7). The major problem with this method is that a breach of Alice's private key would compromise all the previous key exchanges. Eve may also trick Bob with a fake public key for Alice.

With key exchange we typically have a time-out for the key to be used, after which time the key is renegotiated. This allows a smaller time window for Eve to determine the key. The FREAK (Factoring RSA Export Keys) vulnerability caused many problems as the negotiation used a 512-bit public key, where the 512-bit private key can be determined using graphic processors running in the Cloud. If Eve determines the private key associated with the public key, she can read all of the communications sent using the secret key. The key pair can thus be static (such as from a digital certificate which has been created from a trusted provider), or can be generated for each connection.

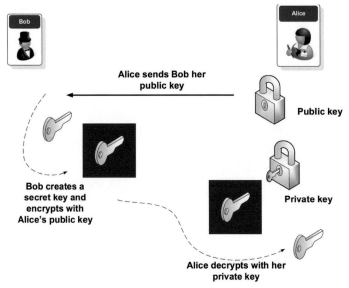

Figure 5.7   Sharing a key using public-key encryption.

5.9 Lab/Tutorial

The lab and tutorial related to this chapter is available on-line at:

http://asecuritysite.com/crypto05

6

Authentication and Digital Certificates

6.1 Introduction

Eve will often try many things to trick Bob into revealing information. One method is for her to steal Bob's identity and then pretend to be him. She may also spoof a device and get Bob to connect to it, and thus get him to reveal some of his data (Figure 6.1). Eve could also use a spoof identity in order to get him to connect to a fake Web site with valid looking content. This is defined as trap-door impersonation. In a modern world, proving identity is just as important as keeping things secret, and where we see identity checking every time we connect to a secure Web site (Figure 6.2). For secure communications using HTTPS, the digital certificate is used to prove the identity of the server for which the Web browser connects to. Normally symmetric encryption is used to secure the communications (such as with 256-bit AES) and a hashing method (such as SHA-1) to prove integrity of the communications.

The previous chapter outlined the way data can be encrypted so that it cannot be viewed by anyone other than those it is intended for. With symmetric encryption, Bob and Alice use the same secret key to encrypt and decrypt the message. This can be generated using a key exchange method (such as using the Diffie-Hellman method). With public-key encryption, though, Bob and Alice do not have this problem, as Alice can advertise her public key so that Bob can use it to encrypt communications to her. The only key that can decrypt the communications is Alice's private key (which, hopefully, Eve cannot get hold off). In most cases for identity we use public key encryption to sign something with a private key, and then other entities can prove it with the associated public key. We now, though, have three further problems:

- How can we tell that the message has not been tampered with?
- How does Bob distribute his public key to Alice, without having to post it onto a Web site or for Bob to be on-line when Alice reads the message?
- Who can we *really* trust to properly authenticate Bob? Obviously we can't trust Bob to authenticate that he really is Bob.

183

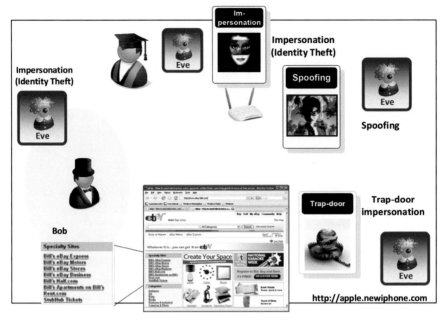

Figure 6.1   Impersonation.

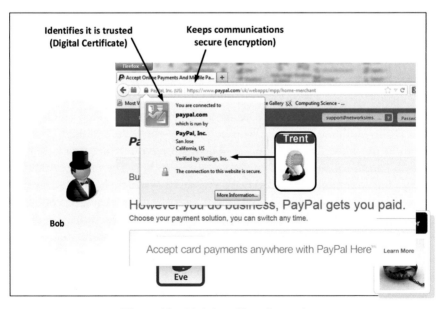

Figure 6.2   Digital certificate integration.

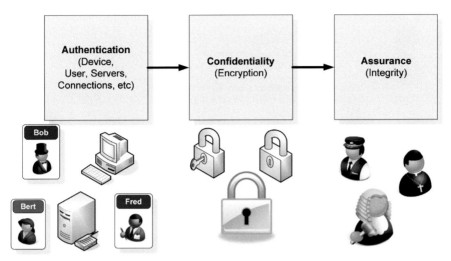

Figure 6.3 Authentication, confidentiality and assurance.

This chapter will show the importance of authentication and assurance, along with confidentiality (Figure 6.3). A key concept in authentication is the way that different entities authenticate themselves. The main methods are: one-way server authentication; one-way client authentication; and mutual authentication (Figure 6.4). With one-way server authentication, the server sends its authentication credentials to the client, such as with a digital certificate. The client then checks this and will verify that it has been created by an entity which it trusts. This is the method used by SSL when a connection is made, and which is used by secure application protocols such as HTTPS, FTPS, SSH, and so on. With one-way client authentication, the client proves its identity to the server. This might be though a hardware address, a nonce, or an IP address. With two-way authentication, both the client and the server identify themselves to each other, and is thus the most secure method, as we reduce the risk of a spoof device on either end.

Another important concept in authentication is that of end-to-end authentication, where the user authenticates themselves to the end service (Figure 6.5) or with intermediate authentication, where only part of the conversation between the entities is authenticated. The major problem with intermediate authentication is that it tends to authenticate only part of the connection, and where the user is not properly authenticated to the end service. It is also possible to have both intermediate and end-to-end authentication,

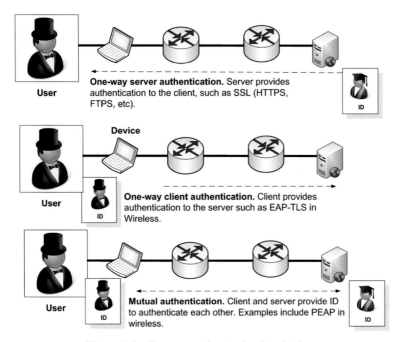

Figure 6.4   One-way and mutual authentication.

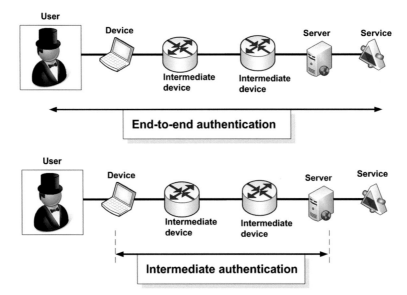

Figure 6.5   End-to-end authentication.

where intermediate devices can authenticate themselves to each other, and where the client might also authenticate themselves to the server/service. This has the advantage of making sure that the route taken for the data packets goes through a valid route, such as for data packets between two organisational sites.

6.2 Methods of Authentication

There are many ways to authenticate devices, applications, and users, each with their strengths and weaknesses. These include:

- **Network/physical addresses**. These are simple methods of verifying a device. The network address, such as an IP address, though, can be easily spoofed, but the physical address is less easy and is a more secure implementation. Unfortunately, the physical address can also be spoofed, either through software modifications of the data frame, or by reprogramming the network interface card. Methods of authentication include DHCP, in which an IP address is granted to a host based on a valid MAC address.
- **Username and password**. The use of usernames and passwords are well known but are often open to security breaches, especially from dictionary attacks on passwords, and from social engineering attacks. In wireless networks, methods such as LEAP include a username and password for authentication, but this also is open to dictionary-type attacks.
- **Authentication certificate**. This verifies a user or a device by providing a digital certificate which can be verified by a reputable source. In wireless networks such methods include EAP-TLS and PEAP. Sometimes it is the user/requester that has to provide a certificate (to validate the user), whereas in other protocols it is the server that is required to present a certificate to the user (to validate the server).
- **Tokens/Smart cards**. With this method a user can only gain access to a service after they have inserted their personal smart card into the computer and, typically, enter some other authentication details, such as their PIN code. In wireless networks, methods include RSA SecurID Token Card and Smartcard EAP.
- **Pre-shared keys**. This uses a pre-defined secret key. In wireless networks, methods include EAP-Archie.

- **Biometrics**. This is an improved method over a physical token where a physical feature of the user is scanned. The scanned parameter requires to be unchanging, such as fingerprints or retina images.
- **OpenID**. This type of authentication uses a URL (or XRI – Extensible Resource Identifier) to authenticate themselves from an trusted identity provider.

Often, there is often a trade-off between the robustness and authenticity of the method versus the ease of use, as illustrated in Figure 6.6. Generally many systems are moving is towards multiple methods of authentication, such as (Figure 6.7):

- Something you know?
- Something you have?
- Something you are?
- Somewhere you are?

Increasingly public key encryption is used to prove identity, as the private key can be used to sign an entity and then be proven with the public key. The main issue is then having a trusted agent who can distribute the public key.

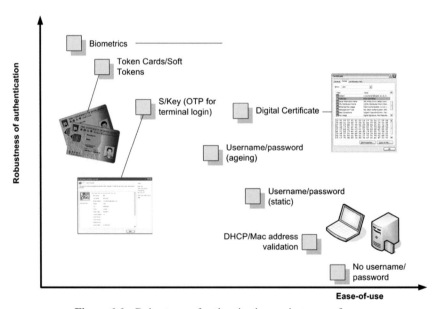

Figure 6.6   Robustness of authentication against ease-of-use.

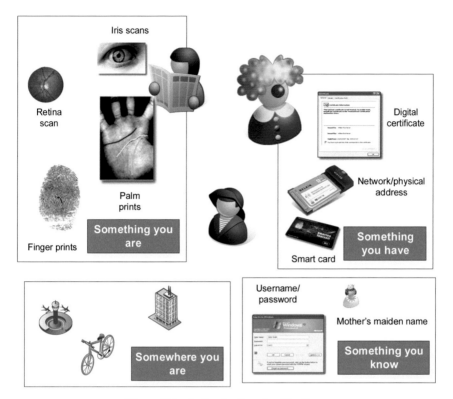

Figure 6.7   Authentication classifications.

6.3 Digital Certificates and PKI

We have seen that it is possible for Bob to sign a message with his private key, and that this is then decrypted by Alice with his public key. There are many ways that Alice could get Bob's public key, but a major worry for her is that who should she trust to receive his public key? One way would be for Bob to post his public key on his web site, but what happens if the web site is down, or if it is a fake web site that Alice uses. Also if Alice asked Bob for his public key by email, how does she really know that Bob is the one who is responding? Thus we need a method to pass public keys in the verifiable way. One of the best ways is to provide a digital certificate which contains, amongst other things, the public key of the entity which is being authenticated. Obviously anyone could generate one of these certificates, so there are two ways we can create trust. One is to setup a server on our own network which provides the digital certificates for the users and

Figure 6.8   Sample certificate.

devices within an organization, or we could generate the digital certificate from a trusted source from well-known Certificate Authorities (CAs), such as Verisign, GlobalSign Root, Entrust and Microsoft. These are generated by trusted parties and which has their own electronic thumbprint to verify the creator, and thus can be trusted by the recipient, or not. Figure 6.8 shows a sample certificate, and Figure 6.9 shows issued details.

6.3.1 PKI and Trust

The major problem that we now have is how to determine if the certificate we get for Bob is credible, and can be trusted. The method used for this is to setup a PKI (Public Key Infrastructure), where digital certificates are generated by a trusted root CA (Certificate Authority), and which is trusted by both parties. As seen in Figure 6.10, Bob asks the root CA for a certificate, for which the CA must check his identity, after which, if validated, it will grant Bob with a certificate. This certificate is digitally signed with the private key of the CA, so that the public key of the CA can be used to check the validity of it.

In most cases, the CA's certificate is installed as a default as a Trusted Root Certificate on the device, and is used to validate all other certificate issued by them. Thus when Bob sends his certificate to Alice, she checks the creditability of it (Figure 6.11), and if she trusts the CA, she will

Figure 6.9 Issued certificate.

accept it[1]. Unfortunately, the system is not perfect, and there is sometimes a lack of checking of identities from CA, and where Eve could thus request a certificate, and be granted one (Figure 6.12). The other method is to use a self-signed certificate, which has no creditability at all, as anyone can produce a self-signed certificate and where there is no validation of it. An example of this is shown in Figure 6.9, where a certificate has been issued to Bill Buchan (even though the user is Bill Buchanan).

Thus our trusted root CA, which we will call Trent, is trusted by both Bob and Alice, but at what level of trust? Can we trust the certificate for authenticating emails, or can we trust it for making secure network connections? Also, can we trust it to digital sign software components? It would be too large a job to get every entity signed by Trent (the root authority), so we introduce Faythe, who is trusted by Trent to sign on his behalf for certain things, such as that Faythe issues the certificates for email signing and nothing else. Thus we get the concept of an intermediate authority, which is trusted to sign certain applications (Figure 6.13), such as for document authentication, code signing, client authentication, user authentication, and so on.

---

[1]Unfortunately many people when faced with a certificate will not actually know if the CA is a credible one, or not, and this is the main weakness of the PKI/digital certificate system. There are many cases of self-signed certificate, and of certificates which are not valid, faking the user.

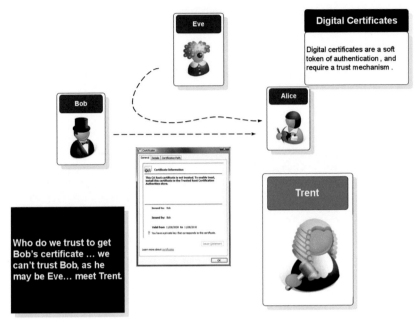

Figure 6.10   Getting a certificate.

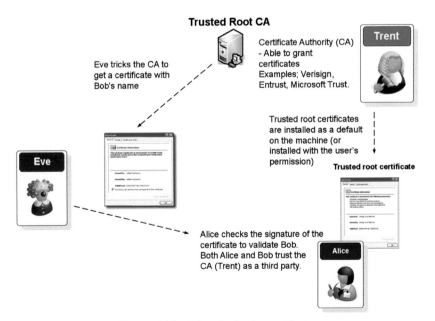

Figure 6.11   Alice checks the certificate.

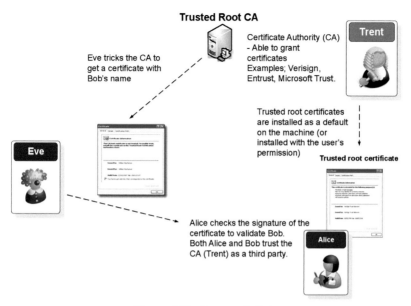

Figure 6.12 Eve spoofs Bob.

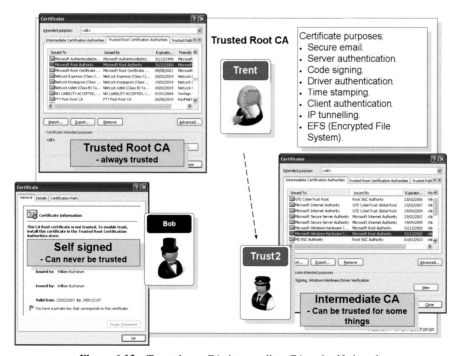

Figure 6.13 Trusted root CA, intermediate CA and self-signed.

We can then have two different types of certificates. The first contains the key pair (the public and the private key) and which is created by the signing CA. This certificate must be kept securely as it contains the private key. In order to distribute the public key, we then export the public key onto a distributable certificate. The owner of the certificate with the key pair can thus export a signed certificate which contains only the public key.

6.3.2 Digital Certificate Types

Digital certificates are used by many devices, such as for servers to prove their identity, and for smart cards to provide their public keys. Typical digital certificate types are:

- IKE.
- PKCS #7.
- PKCS #10.
- RSA signatures.
- X.509v3 certificates. These are exchanged at the start of a conversion to authenticate each device.

A key factor in integrated security is the usage of digital certificates, and are a way of distributing the **public key** of the entity. The file used is typically in the form of X.509 certificate files. Figure 6.14 and Figure 6.15 shows an example export process to a CER file, while Figure 6.16 shows the actual certificate. The standard output is in a binary format, but a Base-64 conversion can be used as an easy way to export/import on a wide range of systems, such as for the following:

```
-----BEGIN CERTIFICATE-----
MIID2zCCA4WgAwIBAgIKWHROcQAAAABEujANBgkqhkiG9w0BAQUFADBgMQswCQYD
VQQGEwJHQjERMA8GA1UEChMIQXNjZXJ0aWExJjAkBgNVBAsTHUNsYXNzIDEgQ2Vy
dGlmaWNhdGUgQXV0aG9yaXR5MRYwFAYDVQQDEw1Bc2NlcnRpYSBDQSAxMB4XDTA2
MTIxNzIxMDQ0OVoXDTA3MTIxNzIxMTQ0OVowgZ8xJjAkBgkqhkiG9w0BCQEWF3cu
YnVjaGFuYW5AbmFwaWVyLmFjLnVrMQswCQYDVQQGEwJVSzEQMA4GA1UECBMHTG90
aGlhbjESMBAGA1UEBxMJRWRpbmJ1cmdoMRowGAYDVQQKExFOYXBpZXIgVW5pdmVy
. . .
H+vXhL9yaOw+Prpzy7ajS4/3xXU8vRANhyU9yU4qDA==
-----END CERTIFICATE-----
```

The CER file format is useful in importing and exporting single certificates, while other formats such as the Cryptographic Message Syntax Standard – PCKS #7 Certificates (.P7B), and Personal Information Exchange – PKCS #12 (.PFX, .P12) can be used to transfer more than one certificate. The main information for a distributable certificate will thus be:

- The entity's public key (Public key).
- The issuer's name (Issuer).
- The serial number (Serial number).
- Start date of certificate (Valid from).
- End date of certificate (Valid to).
- The subject (Subject).
- CRL Distribution Points (CRL Distribution Points).
- Authority Information (Authority Information Access). This will be shown when the recipient is prompted to access the certificate, or not.
- Thumbprint algorithm (Thumbprint algorithm). This might be MD5, SHA1, and so on.
- Thumbprint (Thumbprint).

The certificate, itself, can then be trusted to verify a host of applications (Figure 6.17), such for:

- Server authentication.
- Client authentication.
- Code signing.
- Secure email.
- Time stamping.
- IP security.
- Windows hardware driver verification.
- Windows OEM System component verification.
- Smart card logon.
- Document signing.

Overall an intermediate CA should have some expertise that allows them to validate the functions they have been signed. A particular weakness of PKI is where a user is tricked into installing a root CA certificate or an intermediary certificate authority, and which are then used to validate fake certificates.

📖 **Web link (Digital Certificates):** https://asecuritysite.com/encryption/digitalcert

Figure 6.14   Exporting digital certificates.

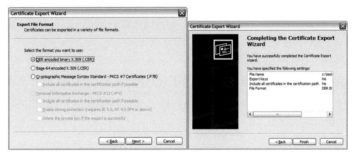

Figure 6.15   Exporting digital certificates.

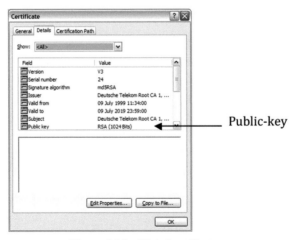

Figure 6.16   Digital certificates.

Figure 6.17 Options for signing.

6.3.3 Digital Certificate Reader

The C# code to read an X509 cer file is:

```
using System;
using System.Security;
using System.Net;
using System.Security.Cryptography.X509Certificates;
namespace ConsoleApplication3
{
  class Class1
  {
    static void Main(string[] args)
    {
      X509Certificate cer = X509Certificate.CreateFromCertFile("c:\\test.cer");

      System.Console.WriteLine("Serial Number: {0}",cer.GetSerialNumberString());
      System.Console.WriteLine("Effective Date: {0}",
            cer.GetEffectiveDateString());
      System.Console.WriteLine("Name: {0}",cer.GetName());
      System.Console.WriteLine("Public key: {0}",cer.GetPublicKeyString());
      System.Console.WriteLine("Public key algorithm: {0}",
            cer.GetKeyAlgorithm());
      System.Console.WriteLine("Issuer: {0}",cer.GetIssuerName());
      System.Console.ReadLine();
    }
  }
}
```

And the output from this is:

```
Serial Number: C0DD5E19983C6F575EFE454E7E66AD02
Effective Date: 08/11/1994 16:00:00
Name: C=US, O="RSA Data Security, Inc.", OU=Secure Server Certification Authority
Public key: 308185027E0092CE7AC1AE833E5AAA898357AC2501760CADAE8E2C37CEEB35786454
03E5844051C9BF8F08E28A8208D216863755E9B12102AD7668819A05A24BC94B256622566C88078FF7
81596D840765701371763E9B774CE35089569848B91DA7291A132E4A11599C1E15D549542C733A6982
B197399C6D706748E5DD2DD6C81E7B0203010001
Public key algorithm: 1.2.840.113549.1.1.1
Issuer: C=US, O="RSA Data Security, Inc.", OU=Secure Server Certification Authority
```

It can be seen that this digital certificate defines the public key for the owner, and is thus a way for a user or organisation to distribute their public key. Thus if a user sends an authenticated message, they sign it with their private key, and the only key which will be able to decrypt it will be the public key contained within the digital certificate. For example, the Microsoft .NET framework includes the digital signing for software components, and which involves the creator signing them with their private key, and only the public key will be able to authenticate them. If software component is changed, it will not be authenticated or authorized.

We can also use OpenSSL to view a certificate (in this case it is a Google certificate):

```
$ openssl x509 -in google.cer -noout -text
 Certificate:
     Data:
         Version: 3 (0x2)
         Serial Number:
             31:c5:76:fa:67:35:ff:66
         Signature Algorithm: sha256WithRSAEncryption
         Issuer: C=US, O=Google Inc, CN=Google Internet Authority G2
         Validity
             Not Before: Jul 15 12:26:30 2015 GMT

             Not After: Oct 13 00:00:00 2015 GMT
         Subject: C=US, ST=California, L=Mountain View, O=Google Inc,
CN=www.google.co.uk
         Subject Public Key Info:
             Public Key Algorithm: rsaEncryption
             RSA Public Key: (2048 bit)
                 Modulus (2048 bit):
                     00:90:1d:d8:b9:e3:46:17:9d:fc:bc:f5:7f:30:86:
                     18:8f:c5:3f:9c:66:ac:49:de:44:c2:c0:fb:7f:8e:
                     2d:b1:6a:eb:2e:2a:1f:ff:c5:da:75:43:ad:1f:a9:
                     96:82:df:02:1e:a3:c8:e7:7e:e1:ec:3e:6c:94:bf:
..
                     c4:fe:47:ae:c4:e0:fa:b2:05:ec:2c:51:97:e2:af:
```

```
              10:89:f9:8c:ab:c7:25:02:71:d1:a1:70:41:68:1f:
              83:e7
          Exponent: 65537 (0x10001)
  X509v3 extensions:
      X509v3 Extended Key Usage:
          TLS Web Server Authentication, TLS Web Client Authentication
      X509v3 Subject Alternative Name:
          DNS:www.google.co.uk
      Authority Information Access:
          CA Issuers - URI:http://pki.google.com/GIAG2.crt
          OCSP - URI:http://clients1.google.com/ocsp

      X509v3 Subject Key Identifier:
          F4:36:65:05:AC:FE:B6:E3:CD:F5:2C:06:90:75:62:34:78:E7:B9:5E
      X509v3 Basic Constraints: critical
          CA:FALSE
      X509v3 Authority Key Identifier:
          keyid:4A:DD:06:16:1B:BC:F6:68:B5:76:F5:81:B6:BB:62:1A:BA:
              5A:81:2F

      X509v3 Certificate Policies:
          Policy: 1.3.6.1.4.1.11129.2.5.1

      X509v3 CRL Distribution Points:
          URI:http://pki.google.com/GIAG2.crl

  Signature Algorithm: sha256WithRSAEncryption
      70:b7:ea:da:21:58:42:b2:c4:6b:ed:b8:22:72:21:4e:f0:43:
      31:65:ff:4f:9d:ef:8c:6e:e6:a2:a9:2e:aa:b7:63:45:87:f9:

      50:c7:86:40:0c:fc:c4:9f:a7:ce:cc:04:4c:33:f8:9e:9e:84:
      df:4a:81:6e
```

📖 **Web link:** https://asecuritysite.com/encryption/certopenssl

6.4 Key and Certificate Management

The main stages of key/certificate management are:

- Initialisation. This includes registration, key pair generation, certificate creation and certificate/key distribution, certificate dissemination, and key backup.
- Issued. This includes certificate retrieval, certificate validation, key recovery and key update.
- Cancellation. This includes certificate expiration, certificate revocation, key history and key archiving.

A digital certificate has a start and end date, and which will define the valid period of the certificate. After the end date, the certificate cannot be trusted. Sometimes, though, the certificate might be breached, or where a fake certificate has been issued. In this case the certificate must be revoked, and where browsers are informed that they cannot trust the certificate. RFC 5280 thus defines "Internet X.509 Public Key Infrastructure Certificate and Certificate Revocation List (CRL) Profile", and includes two main states for revocation:

- **Revoked**. This is where a certificate has been revoked, and cannot be reversed, and often occurs when a certificate is defined as having its private key breached.
- **Hold**. In this case the certificate's trust level is on hold, and can be reversed at some time in the future. It could relate to a private key being thought to be compromised, but where an investigation has show that it has not been breached.

Within the revocation request the reasons given are:

- **Key Compromise**. This defines that the private key has been compromised.
- **CA Compromise**. This defines that the CA has been compromised.
- **Affiliation Changed**. This defines that the certificate affiliation defined within the certificate has changed.
- **Superseded**. This defines that there is an updated certificate, and that this certificate is not valid any more.
- **Cessation Of Operation**. This defines a generic reason of a termination of the certificate, such as where a company has gone in liquidation.
- **Certificate Hold**. This defines where a hold is placed on a certificate.
- **Remove from CRL**. This is where a remove is defined from the list.
- **Privilege Withdrawn**. This defines where a privilege to sign certificates has been removed.
- **AA Compromise**.

In order to keep track of the certificates which have been revoked or on hold, a CRL (Certificate revocation list) is published at defined time or periods, or is generated when a certificate has been revoked. The CRL must be published by the CA who originally generated the targeted certificates, and is only valid for a given amount of time (which is typically less than 24 hours). These CRLs are signed by the CA, in order that they can be validated, and thus signed by the private key of the CA, and then checked against its public key (which is stored in a root certificate folder or preinstalled within a Web browser). Normally a CRL is defined in an X.509 format and encoded in DER (binary) or PEM (text) formats.

Apart from CRLs files becoming large over time, the major problem with CRLs:

- **Lack of checking**. Many systems do not continually check the list whenever a certificate is used. It also requires an online service to be available for validating the certificate, which defeats the main advantage of PKI, in that the certificate is self-authenticating.
- **Revoking error**. This problem occurs when a certificate is revoked through a mistake, and where applications can fail to operate.
- **Denial of service on the CA**. As systems must check the CRL, a denial-of-service against the CRL provider will thus cause problems in accessing information on the certificates which have been revoked. If an application, such as a Web browser, cannot get a validation of the certificate, it may then fail to load a Web site.

An alterative to CRL is to use Online Certificate Status Protocol (OCSP), and which is a light-weight online service to checks the validity of a certificate. Some browsers, such as Firefox, use OCSP to validate certificates.

The core part of PKI is the concept of the root certificate. These are self-signed certificates from a root CA, and where all the certificates signed by a root certificate are trusted. These certificates are normally embedded into the operating system when it is installed on a system. A breach of the root certificate will thus cause major problems, as each of the certificates signed by the root certificate will not be trusted any more.

6.5 Creating a Signed Certificate

The basic process of creating a certificate signed by a CA is defined in Figure 6.18. Initially the organization creates a key-pair and then creates a CSR (Certificate Signing Request) which contains the details of the organization (such as the organization name, the domain-name, the contact email address, the locality, and the unit name). The CA then takes its private key from a key pair and its digital certificate, and signs a new digital certificate for the organization.

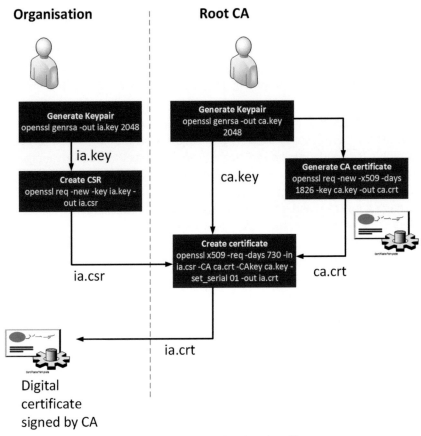

Figure 6.18   Creating a CA signed certificate.

The start of the process is the creation of a key pair. This can be achieved using Openssl:

```
$ openssl genrsa -out ca.key 2048
Loading 'screen' into random state - done
Generating RSA private key, 2048 bit long modulus
.................................................+++
.......................................................+++
e is 65537 (0x10001)
```

In this case we have created a 2,048-bit key pair, and which is contained in the ca.key file:

```
-----BEGIN RSA PRIVATE KEY-----
MIIEogIBAAKCAQEA7g1oS54PiY/6h2wzN0eY8yzCANa26y6BNBfmK1RKJ9Mc/rXt
CofBfunBR5kXhOpQnkmDo9eo4Uu3tPz9qor3zAdaUIOoo8jQw4faQjY35tDTrl9X
dBtzkilR6Qnce6VJrLT/t/uL5fIVvxitnRJSd1QEWPoyT5LOvZsw/39qtXv/6v+W
wVfTgnWGce99Wpt/PvtYD9u9EfbFUZvbrcl4APokMfbQJfBPwhpX/XKfskKZgt0M
3Km2Ik2kmohKJ5M37KDC8pMoyV2vJ0iZsWAaxPvjaaXkX36XUyL+CbmjtyaQMLXr
vNGwxy6OvtYjla/PR1JlPEeKSaCQI/O9/5xnYQIDAQABAoIBAGTsqEAO5hVzRkrt
05TnNPA8FJAYd/qjf8GfNEVAeiQCPDO8259wSNfOsNPzEuaWFNHW5wmqn/3MhTkl
. . .
2GIIwnnAYGlJ2Q/aJAKtR1j58ygW/++n99+l5/2J9Rw3g3Eap5V3QLJ1qchOAvEq
WmrLAoGAEpIYa1atB2Atv0FRravY86HmlWHbfFrfs5ZkBAKzCpNqvo7m/ih69U1v
7DV+b0ejF+lW4Jww5q8htdVln9UgUiLQ8O8HJMwOxa4wGB0KM96nIqRJSmX4DB4r
r7VsAT6lLlald/plFO/D/evZe4lTWz6C3n/RgttHueDGj8YqckI=
-----END RSA PRIVATE KEY-----
```

Next create a self-signed root CA certificate ca.crt for MegaCorp:

```
$ openssl req -new -x509 -days 1826 -key ca.key -out ca.crt
You are about to be asked to enter information that will be
incorporated into your certificate request.
What you are about to enter is what is called a Distinguished Name
or a DN.
There are quite a few fields but you can leave some blank
For some fields there will be a default value,
If you enter '.', the field will be left blank.
-----
Country Name (2 letter code) [AU]:UK
State or Province Name (full name) [Some-State]:None
Locality Name (eg, city) []:Edinburgh
Organization Name (eg, company) [Internet Widgits Pty Ltd]:MegaCorp
Organizational Unit Name (eg, section) []:None
Common Name (e.g. server FQDN or YOUR name) []:None
Email Address []:none
```

Next we will create a subordinate CA (My Little Corp), and which will be used for the signing of the certificate. First, generate the key:

```
$ openssl genrsa -out ia.key 2048
Generating RSA private key, 2048 bit long modulus
.................................................+++
.........................................................+++
e is 65537 (0x10001)
```

Next we will request a certificate for our newly created subordinate CA and create a code signing request (CSR):

```
$ openssl req -new -key ia.key -out ia.csr
You are about to be asked to enter information that will be incorporated
into your certificate request.
What you are about to enter is what is called a Distinguished Name
 or a DN.
There are quite a few fields but you can leave some blank
For some fields there will be a default value,
If you enter '.', the field will be left blank.
-----
Country Name (2 letter code) [AU]:UK
State or Province Name (full name) [Some-State]:None
Locality Name (eg, city) []:Edinburgh
Organization Name (eg, company) [Internet Widgits Pty Ltd]:My Little Corp
Organizational Unit Name (eg, section) []:MLC
Common Name (e.g. server FQDN or YOUR name) []:MLC.none
Email Address []:

Please enter the following 'extra' attributes
to be sent with your certificate request
A challenge password []:Qwerty123
An optional company name []:
```

The code signing request has the form of:

```
-----BEGIN CERTIFICATE REQUEST-----
MIICyTCCAbECAQAwajELMAkGA1UEBhMCVUsxDTALBgNVBAgTBE5vbmUxEjAQBgNV
BAcTCUVkaW5idXJnaDEXMBUGA1UEChMOTXkgTGl0dGxlIENvcnAxDDAKBgNVBAsT
A01MQzERMA8GA1UEAxMITUxDLm5vbmUwggEiMA0GCSqGSIb3DQEBAQUAA4IBDwAw
. . .
k1b4DqOvInWLOs+yuWT7YYtWdr2TNKPpcBqbzCYzrWL6UaUN7LYFpNn4BbqXRgVw
iMAnUh9fvLMe7oreYfTaevXT/506Sj9WvQFXTcLtRhs+M30q22/wUK0ZZ8APjpwf
rQMegvzXXEIO3xEGrBi5/wXJxsawRLcM3ZSGPu/Ws950oM5Ahn8K8HBdKubQ
-----END CERTIFICATE REQUEST-----
```

We can then create a certificate from the subordinate CA certificate and signed by the root CA.

```
$ openssl x509 -req -days 730 -in ia.csr -CA ca.crt -CAkey ca.key -
set_serial 01 -out ia.crt
Signature ok
subject=/C=UK/ST=None/L=Edinburgh/O=My Little Corp/OU=MLC/CN=MLC.none
Getting CA Private Key
```

If we want to use this certificate to digitally sign files and verify the signatures, we need to convert it to a PKCS12 file:

```
$ openssl pkcs12 -export -out ia.p12 -inkey ia.key -in ia.crt -chain
-CAfile ca.crt
Enter Export Password: Qwerty123
Verifying - Enter Export Password: Qwerty123
```

The crt format is encoded in binary. If we want to export to a Base64 format, we can use DER:

```
$ openssl x509 -inform pem -outform pem -in ca.crt -out ca.cer
```

and for My Little Corp:

```
$ openssl x509 -inform pem -outform pem -in ia.crt -out ia.cer
```

A view of the cer file shows that it is in Base64 format:

```
-----BEGIN CERTIFICATE-----
MIIESzCCAzOgAwIBAgIJAJh3rnD4llQKMA0GCSqGSIb3DQEBBQUAMHYxCzAJBgNV
BAYTAlVLMQ0wCwYDVQQIEwROb25lMRIwEAYDVQQHEwlFZGluYnVyZ2gxETAPBgNV
BAoTCE1lZ2FFDb3JwMQ0wCwYDVQQLEwROb25lMQ0wCwYDVQQDEwROb25lMRMwEQYJ
KoZIhvcNAQkBFgRub251MB4XDTE3MDYyNTEyNTUxMloXDTIyMDYyNTEyNTUxMlow
. . .
RT5OLx5sHfl+Dr5CrV0WM5zyt3SrF/vyAMVCBZDzonioPi0mSfCtf0CHPNXEow9v
jAxNExKpicVWW+eiT7ZdMzIT1ulaYtgO7T9OCsmIqym/zxZzadvA+3jYjzugfq1W
iPivmvWHCq5aiAvyzdqlFTt2AE55Ym16T2vVFbr4kDb9j1+Wuo5Gk+B0o/4rq7A=
-----END CERTIFICATE-----
```

The Base64 format is often used to distribute the certificate through an email.

In the above example, we created a self-signed CA, but if we use a root CA, we will receive a CA Certificate Signing Request (CSR) after the key pair has been created, and which is then sent to a CA in order to create a digital identity certificate. This normally involves passing the public key along with identity information (such as for a related domain name) and a digital signature. If the request is successful, the CA sends back a signed certificate and which has been signed by the private key of the CA. In the CSR given in the previous example, we can use Python to view:

```
import OpenSSL.crypto
from OpenSSL.crypto import load_certificate_request, FILETYPE_PEM

csr = '''-----BEGIN NEW CERTIFICATE REQUEST-----
MIICyTCCAbECAQAwajELMAkGA1UEBhMCVUsxDTALBgNVBAgTBE5vbmUxEjAQBgNV
BAcTCUVkaW5idXJnaDEXMBUGA1UEChMOTXkgTG10dGxlIENvcnAxDDAKBgNVBAsT
. . .
```

```
k1b4DqOvInWLOs+yuWT7YYtWdr2TNKPpcBqbzCYzrWL6UaUN7LYFpNn4BbqXRgVw
iMAnUh9fvLMe7oreYfTaevXT/506Sj9WvQFXTcLtRhs+M30q22/wUK0ZZ8APjpwf
rQMegvzXXEIO3xEGrBi5/wXJxsawRLcM3ZSGPu/Ws950oM5Ahn8K8HBdKubQ
-----END NEW CERTIFICATE REQUEST-----'''

req = load_certificate_request(FILETYPE_PEM, csr)
key = req.get_pubkey()
key_type = 'RSA' if key.type() == OpenSSL.crypto.TYPE_RSA else 'DSA'
subject = req.get_subject()
components = dict(subject.get_components())
print "Key algorithm:", key_type
print "Key size:", key.bits()print "Common name:", components['CN']
print "Organisation:", components['O']
print "Orgainistional unit", components['OU']
print "City/locality:", components['L']
print "State/province:", components['ST']
print "Country:", components['C']
```

A sample run gives:

```
Key algorithm: RSA
Key size: 2048
Common name: MLC.none
Organisation: My Little Corp
Orgainistional unit MLC
City/locality: Edinburgh
State/province: None
Country: UK
```

📖 **Web link (Digital Certificates):** https://asecuritysite.com/encryption/csr

6.6 Digital Certificate Passing

Figure 6.19 outlines that we have Bob, Alice, Trent and Eve. Bob and Alice both have public and private keys, and both of them have agreed to trust MegaCorp for proving the identity of both Alice and Bob. The private key on either side must be protected against accesses from malicious parties, as it will be possible to both decrypt the communications, and also pretend to be the entity who owns the private key. If Bob wants to send something to Alice, he must first get Alice's public key. Normally the public key is passed through a distributable digital certificate, and which does not contain the private key (Figure 6.20).

Bob then creates the message and then creates a signature (in the real case this is likely to be the hash of the data in the message (Figure 6.21). Bob then take the signature and encrypts it with his private key (Figure 6.23), and then will take the message and the encrypted signature, and encrypt the whole lot with Alice's public key (Figure 6.24). Alice then receives the encrypted content (Figure 6.25), and takes the private key off her non-distributable digital certificate and uses that to decrypt the encrypted content (at this stage only she can decrypt the message, as Bob will not have the right key to decrypt).

She can now read the message, but wants to know that it was Bob who sent it, as Eve could have used Alice's public key and sent the message, pretending to be Bob. Next she receives Bob's distributable digital certificate, which has his public key, and which has been checked by MegaCorp (Figure 6.25). She then takes Bob's public key and decrypts the signature, and then takes her own signature of the message and compares them. If they are the same, see knows that Bob sent the message, and also that the message is unchanged. So using public key, we have implemented privacy (with Alice's public key), integrity (proven with Bob's public key) and identity (proven with Bob's public key). In most real life cases we would use symmetric key encryption for the secrecy part.

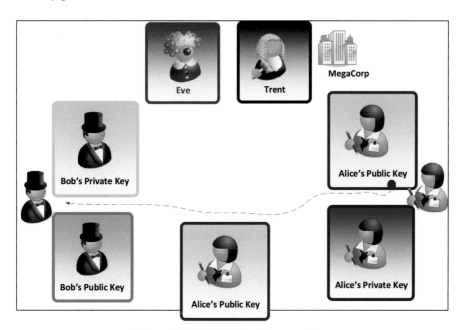

Figure 6.19 Digital Certificate passing.

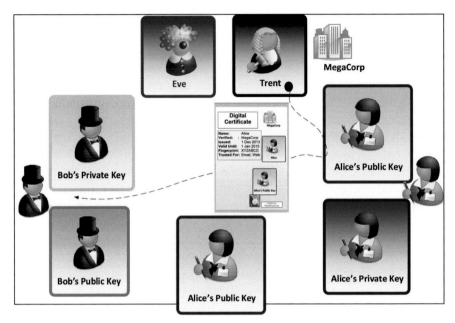

Figure 6.20 Digital Certificate passing.

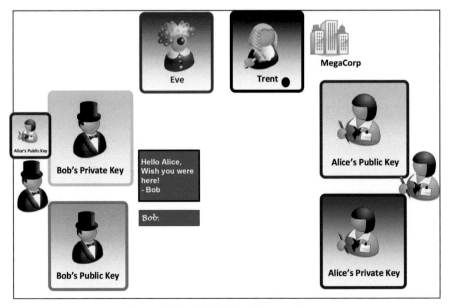

Figure 6.21 Digital Certificate passing.

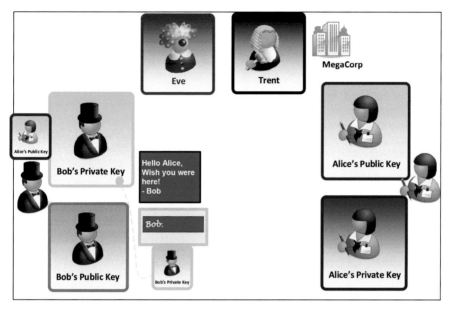

Figure 6.22 Digital Certificate passing.

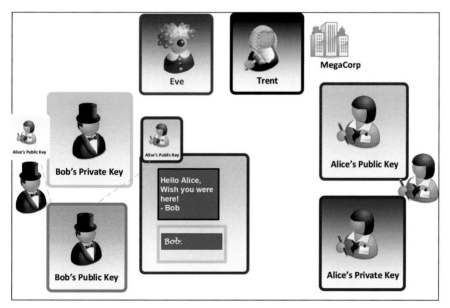

Figure 6.23 Digital Certificate passing.

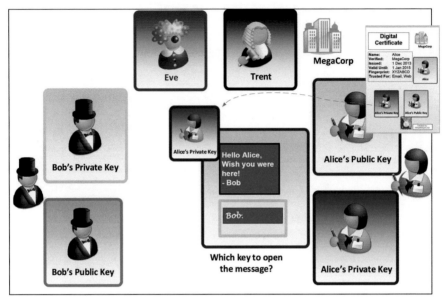

Figure 6.24   Digital Certificate passing.

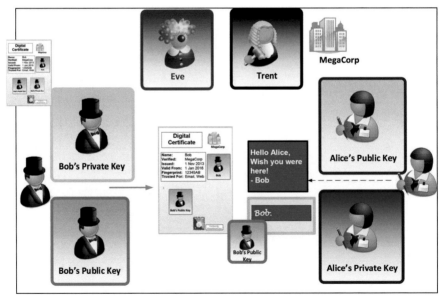

Figure 6.25   Digital Certificate passing.

6.7 Email Encryption

A popular type of email encryption is PGP (Pretty Good Privacy) which uses asymmetric encryption to encrypt the data, and adds the private key of the sender to provide authentication. It can be seen, in Figure 6.26, that the first stage takes the message and produces an MD5 hash, which is encrypted, using RSA, with the sender's private key. As the Alice has the Bob's public key, she should be able to decrypt it, and compare with the hash of the decrypted message. After a ZIP stage, a session key (using the IDEA encryption method, in this example) will be used to encrypt the output from the ZIP process, and Alice's public key will be used to encrypt the session key. Only Alice will then be able to determine the session key. The output is then be converted to ASCII characters using Base-64 (as required in standard email transmission). Alice will then use her private key to decrypt the session key. After which she will determine its contents. She will then use Bob's public key to decrypt the hashed value. This will then be compared with the hashed value from the message. If they are the same, then the message and the sender have been authenticated.

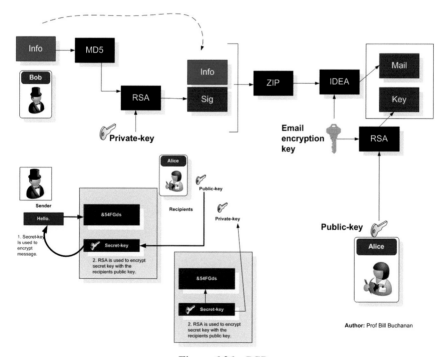

Figure 6.26   PGP.

The true genius of PGP is the usage of unique key to encrypt the email message. The email is thus encrypted using IDEA and with a randomly generated key. Next the encryption key is encrypted with Alice's public key. At the receiver, all Alice has to do is to decrypt the IDEA key, and then decrypt the ciphered message with it. The great advantage of this is that symmetric encryption/decryption is much faster and less process intensive than asymmetric methods. This is similar to someone locking up all the doors in a house, and the placing all the keys in a safe deposit box, that only one person holds the secret code for. Once the person has closed the door on the keys, even they cannot then get access to them, and only the person with the correct combination can get access to them. Each time we might create new keys, but the combination can stay the same.

6.8 Kerberos

The major problem with current authentication systems is that they are not scalable, and they lack any real form of proper authentication. A new authentication architecture is now being proposed, which is likely to be the future of scalable authentication infrastructures – Kerberos. It uses tickets which are gained from an Identity Provider (IP – and also known as an Authentication Server), and which are trusted to provide an identity to a Relying Party (RP). The basic steps are:

Client to IP:
- A user enters a username and password on the client.
- The client performs a one-way function on the entered password, and this becomes the secret key of the client.
- The client sends a cleartext message to the IP requesting services on behalf of the user.
- The IP checks to see if the client is in its database. If it is, the IP sends back a session key encrypted using the secret key of the user (MessageA). It also sends back a ticket which includes the client ID, client network address, ticket validity period, and the client/TGS (Ticket Granting Server) session key encrypted using the secret key of the IP (MessageB).
- Once the client receives messages A and B, it decrypts message A to obtain the client/TGS session key. This session key is used for further communications with IP.

Client-to-RP:

- The client now sends the ticket to the RP, and an authentication message with the client ID and timestamp, encrypted with the client session key (MessageC).
- The RP then decrypts the ticket information from the secret key of the IP, of which it recovers the client session key. It can then decrypt MessageD, and sends back a client-to-server ticket (which includes the client ID, the client network address, validity period, and the client/server session key). It also sends the client/server session key encrypted with the client session key.

The Kerberos principle is well-known in many real-life authentication, such as in an airline application, where the check-in service provides the authentication, and passes a token to the passenger (Figure 6.27). This is then passed to the airline security in order to board the plane. There is thus no need to show the form for the original authentication, as the passenger has a valid ticket. Figures 6.28 and 6.29 show the detail of the Kerberos protocol which involves an Authentication Server, and a Ticket Grant Server.

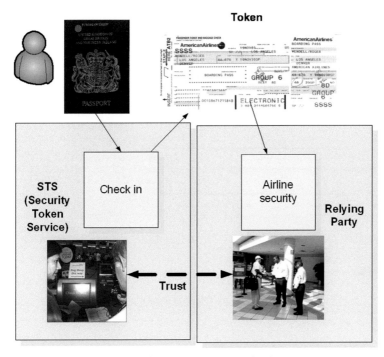

Figure 6.27   Ticketing authentication.

AS_REQ is the initial user authentication request This message is directed to the KDC component known as Authentication Server(AS).

AS_REQ = (
Principal$_{Client}$,Principal$_{Service}$,IP_list,Lifetime)

Eg Principal$_{Client}$ = Principal for user (such as fred@home.com), IP_list = all IP address which will use the ticket(may be null if behind NAT), lifetime = require life of the ticket

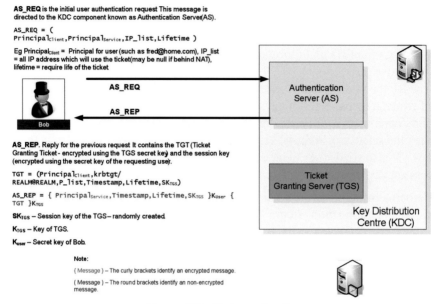

AS_REP. Reply for the previous request It contains the TGT (Ticket Granting Ticket - encrypted using the TGS secret key) and the session key (encrypted using the secret key of the requesting use).

TGT = (Principal$_{Client}$,krbtgt/
REALM@REALM,P_list,Timestamp,Lifetime,SK$_{TGS}$)

AS_REP = { Principal$_{Service}$,Timestamp,Lifetime,SK$_{TGS}$ }K$_{User}$ {
TGT }K$_{TGS}$

SK$_{TGS}$ – Session key of the TGS– randomly created.

K$_{TGS}$ – Key of TGS.

K$_{user}$ – Secret key of Bob.

Note:
{ Message } – The curly brackets identify an encrypted message.
(Message) – The round brackets identify a non-encrypted message.

Figure 6.28   Kerberos (Part I).

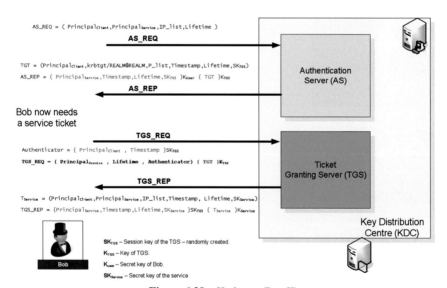

Figure 6.29   Kerberos (Part II).

6.9 Kerberos Key Sharing

In order to understand the Kerberos key sharing method, let's relate this to real life. Bob and Alice trust Trent, but want a way to identify each other and communicate in a secret way. So Alice goes to Trent and says that she has to prove her identity to Bob, and vice-versa. For this Trent will make a special key for a box, and will make a copy for Bob and Alice (he might also keep a copy for himself, just in case they lose them). Trent will then take a photograph of Alice, and write down the date and time on it, and the amount of time he can verify Alice for. He will then put it into the box, and gives the box to Alice, along with the key. Along with this he will give her a sealed letter for the attention of Bob which has his stamp on it. Inside will be a photograph of Alice that he took, and the secret key, along with the date/time that he created the key.

Alice goes home, and then puts her photograph in the box, and locks it with the secret key. She then passes the box, without the key, along with the sealed letter to Bob. Bob opens the sealed letter, which has a key inside to open up the box, and which has the photograph that Trent took of Alice. Bob then opens the box with the secret key provided by Trent, and takes out the photograph that Alice has provided. If it is the same as the one that Trent put in the sealed letter, Bob thus verifies Alice's identity.

Bob and Alice now have the same key to open and close the secret box, and can now use it to send secret messages to each other. No-one else will have that unique key, thus any messages in there must have been provided by Bob and Alice.

First we determine the ID for Alice, Bob, and key to be used by Alice to communicate with Trent, and for Bob to communicate with Trent:

Bob's ID (B)	Bob
Alice's ID (A)	Alice
Alice Shared Key (with Trent) – E_A	49287e4abf2276a94cf66e652d012dad6397 de8c752afa40e5a5edb91c590f25
Bob Shared Key (with Trent) – E_B	5dea5e3be847720df8924a06e933edb1d0eef aca4fd65cc053f4a48eeb2ecae4

Now Trent calculates a timestamp, a Life Time (L), a random session key (K) and Bob's identity (B), and encrypts with Alice's Key:

Timestamp (T)	8/4/2015 4:22:44 PM
Lifetime (L)	100
Random Key (E_K)	91e6ec6c8ca8a54bfc9a0f3a759c7a8f0c4478 7575fbbe6e65101157256129876
Bob's ID	Bob
Trent sends to Alice: $E_A(T,L,K,B),E_B(T,L,K,A)$	2c84240f87d11c5212887c3728de5400,c3 3696b185a7f8cf99ae816e197d048e

Next Alice can now decrypt the first part ($E_A(T,L,K,B)$) as she has the encryption key for this. She can then determine the session key (K). Next she encrypts her identity (A) and the timestamp (T) with the session key (E_K) and sends to Bob:

Timestamp (T)	8/4/2015 4:22:44 PM
Alice sends to Bob: $E_K(A,T),E_B(T,L,K,A)$	de3c27656fad3e9e09aa2a12c0f5a7d7,c336 96b185a7f8cf99ae816e197d048e

Bob can now decrypt $E_B(T,L,K,A)$ with E_B, and will thus determine K (which is the session key). After this he can then decrypt $E_K(A,T)$, to determine Alice's identity (A) and the Timestamp (T). He will then increment the T stamp by one, and encrypt with the session key and send back to Alice:

Timestamp (T+1)	8/4/2015 4:22:45 PM
Bob sends to Alice: $E_K(T+1)$	cf610ccd8eb56dcadd485018d7b137c5

Alice will then receive this, and decrypt with the session key E_K, and determine that it has the correct time stamp, and thus proves that Bob has sent it back. Alice and Bob now have a shared key, and can now use it to send encrypted content.

📖 **Web link (Kerberos):** http://asecuritysite.com/encryption/ker

6.10 Lab/Tutorial

The lab and tutorial related to this chapter is available on-line at:

http://asecuritysite.com/crypto06

7

Tunneling

7.1 Introduction

There are many risks to data packets as they travel over networks, especially over untrusted networks. This involves the risks of someone sniffing the network packets and reading their contents. Along with this we can have problems around data packets being changed as the travel over networks, and where an intruder could setup a spoofed gateway. It is thus important to integrate privacy (typically using encryption), the authentication of devices, and the integrity of the data packets. This is typically performed with a tunnel where an encryption key is used on either end of the tunnel, and where all the data packets are encrypted.

7.2 SSL/TLS Connections

In order to secure higher level protocols, the Secure Socket Layer (SSL) was inserted between the application layer and the transport layer. It has since evolved through SSL v2, SSL v3 (TLS 1.0), and onto TLS 1.1 and TLS (Transport Layer Sockets) 1.2. One of the greatest flaws of SSL v2 was the usage of "export-grade ciphersuites" – which were created to comply with US Export regulations, and which made sure that the keys were crackable. This included a small key size, such as using a 40-bit session key for a connection.

SSL allows for the data above the transport layer to be encrypted, and where the TCP server ports are moved from the defaults for the application protocols to the SSL supported versions. For example HTTP moves from a default TCP Port of 80 to a TCP Port of 443 for HTTPS. For email, the default port for sending email (SMTP) is TCP Port 25, whereas SMTPs typically moves this to TCP Port 465. Figure 7.2 shows the initial connection for HTTPS where the client sends a SYN TCP segment to the server for TCP

217

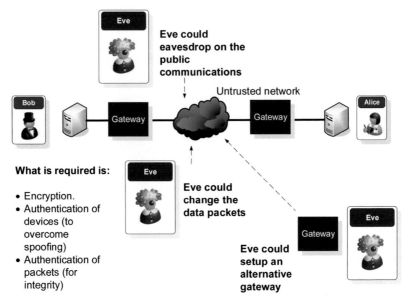

Figure 7.1   Encryption, authentication and integrity.

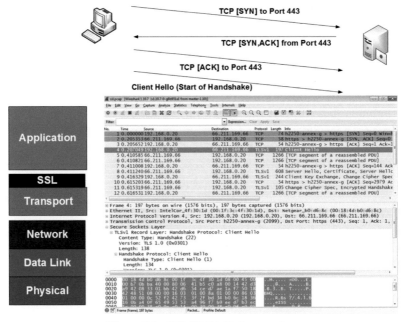

Figure 7.2   SSL/TLS initial connection.

Port 443. The server sends back a SYN/ACK from TCP Port 443, and the client responds with an ACK TCP segment to the server. If successful the connection is then continued into the SSL/TLS handshaking phase, where the client sends a Client Hello.

The first two bytes of the SSL/TLS data message contains the connection type. For 0x03 0x00 it is SSL 3.0, for 0x03 0x01 it is TLS 1.0 and for TLS 1.1 it is 0x03 0x02 (Figure 7.3).

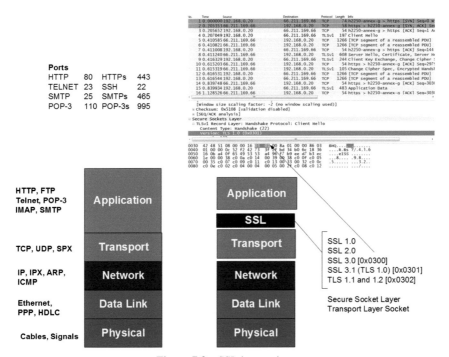

Figure 7.3   SSL integration.

7.3 SSL/TLS Handshaking

With SSL/TLS, as illustrated in Figure 7.4, the tunnel is created with a symmetric key method (such as with RC4 or AES), and then a signature is created with a defined hashing method (such as SHA-1 or MD5). Normally the client creates a session key which will be used by the symmetric key method. In the case of RSA key exchange the client will receive the public key of the server, and then the client encrypts the session key with this public key. Once received, the server will then decrypt the encrypted session key with its private key, and thus both the client and the server will have the same session key for the tunnel. The main stages involved are normally:

- **Client Hello.** This is sent from the client to the server and defines the cipher suites that the client supports.
- **Server Hello.** This sends back the digital certificate from the server and the selected cipher suite from the list that the client sent.
- **Client Key Exchange.** This is sent from the client and contains the information required to generate the session key.

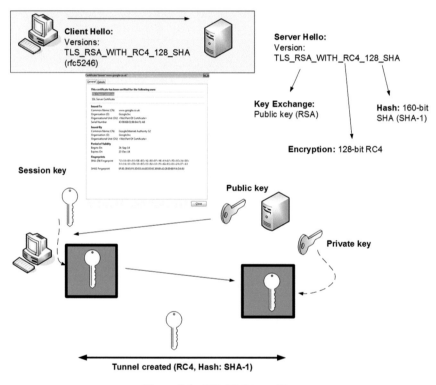

Figure 7.4   SSL/TLS tunnelling.

The initial phase involves the exchange of handshaking messages, and has the format of:

```
enum {
    hello_request(0), client_hello(1) server_hello(2),
    certificate(11), server_key_exchange (12),
    certificate_request(13), server_hello_done(14),
    certificate_verify(15), client_key_exchange(16),
    finished(20)
    (255)
} HandshakeType;

struct {
    HandshakeType msg_type;
    uint24 length;
    select (HandshakeType) {
        case hello_request:        HelloRequest;
        case client_hello:         ClientHello;
        case server_hello:         ServerHello;
        case certificate:          Certificate;
        case server_key_exchange:  ServerKeyExchange;
        case certificate_request:  CertificateRequest;
        case server_hello_done:    ServerHelloDone;
        case certificate_verify:   CertificateVerify;
        case client_key_exchange:  ClientKeyExchange;
        case finished:             Finished;
    } body;
} Handshake;
```

Figure 7.5 outlines the connection involved with SSL and TLS. Initially the client connects to the server on a given port (such as port 443 for HTTPS), and which contains a list of the cipher suits that it can support. The client initially sends a Client Hello Request and which contains a list of the ciphers suites that it supports. These are defined in RFC5246, and define the connection type (SSL or TLS), the symmetric encryption method (such as RC4, 3DES or AES), the key exchange method (such as RSA or Diffie Hellman) and the hashing method (such as MD5 or SHA-1). Examples are:

- TLS_RSA_WITH_RC4_128_MD5 {0x00,0x04}. Uses a TLS connection, with 128-bit RC4 as the symmetric encryption key for the tunnel, RSA for the key exchange method, and MD5 for the hashing method.
- TLS_RSA_WITH_AES_128_CBC_SHA {0x00,0x2F}. Uses a TLS connection, with 128-bit AES and CBC as the symmetric encryption key for the tunnel, RSA for the key exchange method, and SHA-1 for the hashing method.
- TLS_RSA_WITH_AES_256_CBC_SHA256 {0x00,0x3D}. Uses a TLS connection, with 256-bit AES and CBC as the symmetric

encryption key for the tunnel, RSA for the key exchange method, and SHA-256 for the hashing method.

- TLS_DHE_RSA_WITH_AES_128_CBC_SHA {0x00,0x33}. Uses a TLS connection, with 128-bit AES and CBC as the symmetric encryption key for the tunnel, Diffie-Hellman Ephemeral for the key exchange method, and SHA-1 for the hashing method.
- TLS_DH_anon_WITH_AES_256_CBC_SHA256 {0x00,0x6D}. In this we have a Diffie-Hellman key exchange, and where neither side is authenticated (anon), and uses a symmetric encryption key of 256-bit AES and CBC, and using a SHA-256 for hashing. This method is open to a Man-in-the-Middle (MITM) attack.

In some connections, the client may also send its digital certificate, but for HTTPS, only the server sends its digital certificate. In the example in Figure 7.6 we see that the client has sent 56 cipher suites within the Client Hello (1), and where the server selects TLS, with RSA key exchange, 3DES encryption with CBC and SHA-1 for hashing. We can see also that the server's digital certificate is contained within the Server Hello (2).

📖 **Web link (View an SSL connection):** http://asecuritysite.com/log/ssl.zip

Figure 7.7 outlines an example of using an RSA key exchange, where the client sends back an RSA Encrypted PreMaster Secret, and which is the session key (which is a random value of the required key size) which has

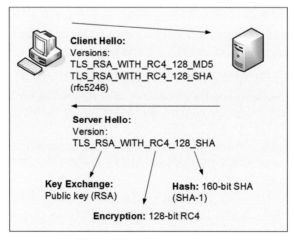

Figure 7.5   SSL connections.

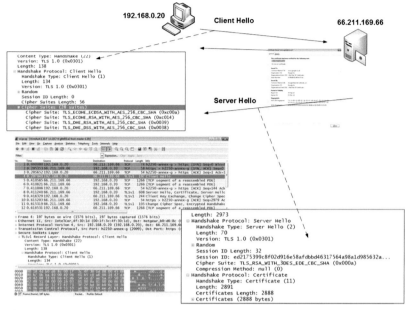

Figure 7.6 SSL connections.

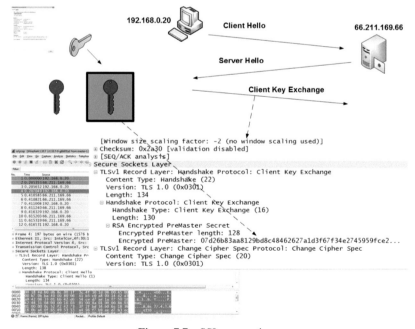

Figure 7.7 SSL connections.

been encrypted with the public key of the server. The server will then decrypt this with its private key. After this, the client and server will have the same shared key.

We can use OpenSSL to view the details of the connection. In the following we see a connection with www.google.com on TCP Port 443: It can be seen that the server has a 2,048-bit public key, and has selected ECDHE-RSA-AES128-GCM-SHA256. In this case Google is using 128-bit AES with GCM (Galois/Counter Mode) and which creates an AES cipher stream using the CTR method, rather than using a block cipher. GCM is thus faster and can recover easier from errors. The master key is also defined in the handshake:

```
$ openssl s_client -connect www.google.com:443
CONNECTED(00000003)
depth=2 C = US, O = GeoTrust Inc., CN = GeoTrust Global CA
verify error:num=20:unable to get local issuer certificate
verify return:0
---
Certificate chain
 0 s:/C=US/ST=California/L=Mountain View/O=Google Inc/CN=www.google.com
   i:/C=US/O=Google Inc/CN=Google Internet Authority G2
 1 s:/C=US/O=Google Inc/CN=Google Internet Authority G2
   i:/C=US/O=GeoTrust Inc./CN=GeoTrust Global CA
 2 s:/C=US/O=GeoTrust Inc./CN=GeoTrust Global CA
   i:/C=US/O=Equifax/OU=Equifax Secure Certificate Authority
---
Server certificate
-----BEGIN CERTIFICATE-----
MIIEdjCCA16gAwIBAgIISVyALWN+akUwDQYJKoZIhvcNAQEFBQAwSTELMAkGA1UE
---
SOx4I5L0D0jZYqKfJuImGcFwdIETq0EpCmkhJfGNHjVdzC/h/T61TmaY
-----END CERTIFICATE-----
subject=/C=US/ST=California/L=Mountain View/O=Google Inc/CN=www.google.com
issuer=/C=US/O=Google Inc/CN=Google Internet Authority G2
---
No client certificate CA names sent
---
SSL handshake has read 3719 bytes and written 446 bytes
---
New, TLSv1/SSLv3, Cipher is ECDHE-RSA-AES128-GCM-SHA256
Server public key is 2048 bit
Secure Renegotiation IS supported
Compression: NONE
Expansion: NONE
SSL-Session:
    Protocol  : TLSv1.2
    Cipher    : ECDHE-RSA-AES128-GCM-SHA256
    Session-ID: 9D92CEC32FA9F86C6D902081EE186C4FC68234FFF7B903D6621A86C980
                92BD51
    Session-ID-ctx:
    Master-Key: B8A14DB1D3021E80B53F30EA94D2EEA155A995B926879B08E3D971EB16
                873D16F62929899E2FA368D374716DB14A412B
```

```
Key-Arg    : None
PSK identity: None
PSK identity hint: None
SRP username: None
TLS session ticket lifetime hint: 100800 (seconds)
TLS session ticket:
0000 - fa 8d cb 50 53 3d 99 c8-b4 11 20 0c ca 53 e9 bd    ...PS=.... ..S..
0010 - f8 8e 15 14 ec 82 c1 56-ab d9 9b 36 c2 56 b0 db    .......V...6.V..
0020 - 2b d4 07 56 a5 02 ac 1f-34 fa 72 21 fd 7c ba 97    +..V....4.r!.|..
0030 - 2a ae e9 20 04 ef 8a e5-a0 57 28 3a c7 67 04 ac    *.. .....W(:.g..
0040 - 7d 14 bf b0 6d 96 9f cb-eb 0c 0a 40 07 5f a6 84    }...m......@._..
0050 - e2 3b 98 0b e7 f4 b1 e1-04 be 15 6b 36 a5 57 b3    .;.........k6.W.
0060 - 11 98 f2 f4 20 fe b5 7f-6b 10 4e 7a f9 b5 6d 02    .... ...k.Nz..m.
0070 - 30 ec 07 e6 f0 c0 49 81-31 6b 30 f9 b0 d3 c4 25    0.....I.1k0....%
0080 - 62 f3 92 33 e8 25 cc 22-32 84 54 e6 0e 76 b1 45    b..3.%."2.T..v.E
0090 - 3a 60 83 cf 1b b0 97 7d-05 03 47 20 29 12 d9 8d    :`.....}..G )...
00a0 - 6f 5a b4 f2                                        oZ..

Start Time: 1413136351
Timeout   : 300 (sec)
Verify return code: 20 (unable to get local issuer certificate)
```

 📖 **Web link (View an SSL with site):** http://asecuritysite.com/encryption/ssl

7.4 SSL Risks

The DROWN (Decrypting RSA using Obsolete and Weakened eNcryption) [1] vulnerability focused on Web servers running SSLv2, and where the attack does not actually involve an SSLv2 connection, but on the legacy "export" ciphersuites. In 2011, Juliano Rizzo and Thai Duong demonstrated BEAST (Browser Exploit Against SSL/TLS) [2] and showed a weakness in CBC based ciphers, for SSL Version 3/TLS 1.0.

In 2014, Bodo Möller (along with Thai Duong and Krzysztof Kotowicz) from Google [3] announced a major vulnerability in SSLv3, and where the plaintext of the encrypted content could be revealed by an intruder. The flaw itself had been speculated on for a while, but the announcement showed that it could actually be used to compromise secure communications. It was named POODLE (Padding Oracle On Downgraded Legacy Encryption) [4], and it highlighted a method where Web servers deal with older versions of the SSL (Secure Socket Layer) protocol.

The FREAK ("Factoring RSA Export Keys") [5] vulnerability suffered from the weakness introduced to comply with US Cryptography Export Regulations, and where the keys used for exportable software was limited to 512-bits or less (and were defined as RSA EXPORT keys). The related RFC on SSL states:

Note: According to current US export law, RSA moduli larger than 512 bits may not be used for key exchange in software exported from the US. With this message, larger RSA keys may be used as signature-only certificates to sign temporary shorter RSA keys for key exchange.

By 2015, it was fairly easy to get a cloud-based instance to factorize 512-bit keys (which involves finding the matching key to a 512-bit public key). The exploit could thus involve a man-in-the-middle, and who could downgrade the keys used to 512 bits. With Logjam [6] the weakness of 512-bit Diffie-Hellman key exchange method was exposed, and where 1,024 bits can be cracked by nation states. The Diffie-Hellman key exchange method has been known to be weak for many years, but in [7] the authors outlined a man-in-the-middle to downgrade connections for "export-grade" Diffie-Hellman. For this they found 82% of vulnerable servers that supported the 512-bit group, and for 7% of Alexa Top Million HTTPS sites.

Within a tunnel, we typically use symmetric encryption (such as AES). Other ciphers include RC4 and DES, and which are seen to be weak in their implement. DES, for example, supports a 56-bit key, and which can easily be cracked by brute force. For RC4, while was popular for many years, researchers found a major statistical weakness [8].

With Perfect Forward Secrecy (PFS) we create assurances that the session keys will not be compromised, even if the private key of the server is compromised. By generating a unique session key for every session a user initiates, even the compromise of a single session key will not affect any data other than that exchanged in the specific session protected by that particular key.

7.5 VPN Tunnels

Within a VPN (Virtual Private Network) tunnel we aim to create a connection from a host machine to a trusted network, and which is tunneled through a public network. This can support privacy (though encryption), authentication, and integrity checking. As illustrated in Figure 7.8, the most common tunnel protocols are PPTP (Point-to-point Tunneling Protocol), L2TP (Layer 2 Tunneling Protocol) and IPSec.

While SSL/TLS suffers from a range of security threats, IPSec is seen to be one of the most robust methods in connecting a computer system to a remote network, and in a secure and robust manner. Within the tunneling mode, the connection is tunneled over a public network, but the network traffic is unprotected on either side of the connection. This mode allows for the inspection of network packets on either side. Normally this is a typical method used within a corporate network, as IDS and virus scanners can be applied onto the network traffic. With transport mode, we have end-to-end tunneling, where the encryption scope spans across of the network, and where no intermediate scanning is possible for the content stored within the packets (Figure 7.9). Figure 7.10 illustrates the scope of the tunnel. Note that if the network connection uses SSL, it is often not possible to fully examine the contents of the network stream, even within tunneling mode.

The IPSec protocol includes two mechanisms which can be used separately or together (Figure 7.11):

- ESP (Encapsulated Security Protocol). ESP takes the original data packet, and breaks off the IP header. The rest of the packet is encrypted, with the original header added at the start, along with a new ESP field at the start, and one at the end. It is important that the IP header is not encrypted as the data packet must still be read by routers as it travels over the Internet. Only the host at the other end of the IPSec tunnel can decrypt the contents of the IPSec data packet.
- AH (Authentication Header). This encrypts the complete contents of the IP data packet, and adds a new packet header. ESP has the weakness that an intruder can replay previously sent data, whereas AH provides a mechanism of sequence numbers to reduce this problem.

IPSec traffic can be identified from information within the network and transport layers. Overall the handshaking process uses UDP Port 500 for key exchange, and if this is blocked, the tunnel will not be created. Along with this the value of 50 is defined for the IP Protocol field in the IP header for ESP and 51 for AH (Figure 7.12).

There are two main phases in setting up an IPSec connection. The first phase defines the IKE (Internet Key Exchange) phase, where the hashing method, the encryption and key exchange methods are defined (Figure 7.13) and the second phase defines the polices to be used for the tunnel. This includes the lifetime of the SA (Security Association) and whether we are using AH and/or ESP. A sample of a connection between two Cisco routers is shown in Figure 7.14.

In an aggressive mode, the phases are merged into one exchange, where the first packet from the client to the server defines IKE SA values: the Diffie-Hellman public key; a nonce for the other party to sign; and an identity packet. The server then sends back the details it has selected.

The following shows an example of the key exchange phases, where a
UDP data packet is sent from UDP Port 500 to a destination UDP Port
500. In this example the host is at 192.168.0.20 and the destination is at
146.176.210.2. The first key exchange packet contains a list of the Transform
Payloads (#1 is shown), and supports AES-CBC and SHA-1, with a lifetime
of 2,147,483 seconds (24.86 days):

```
No.    Time       Source         Destination     Protocol    Length    Info

2      0.631802   192.168.0.20   146.176.210.2   ISAKMP      918       Aggressive

Frame 2: 918 bytes on wire (7344 bits), 918 bytes captured (7344 bits)
Ethernet II, Src: IntelCor_4f:30:1d (00:1f:3c:4f:30:1d), Dst: Netgear_b0:d6:8c
(00:18:4d:b0:d6:8c)
Internet Protocol Version 4, Src: 192.168.0.20 (192.168.0.20), Dst:
146.176.210.2 (146.176.210.2)
User Datagram Protocol, Src Port: 65341 (65341), Dst Port: 500 (500)
Internet Security Association and Key Management Protocol
    Initiator SPI: 0490174339c81264
    Responder SPI: 0000000000000000
    Next payload: Security Association (1)
    Version: 1.0
    Exchange type: Aggressive (4)
    Flags: 0x00
    Message ID: 0x00000000
    Length: 860
    Type Payload: Security Association (1)
        Next payload: Key Exchange (4)
        Payload length: 556
        Domain of interpretation: IPSEC (1)
        Situation: 00000001
        Type Payload: Proposal (2) # 1
            Next payload: NONE / No Next Payload  (0)
            Payload length: 544
            Proposal number: 1
            Protocol ID: ISAKMP (1)
            SPI Size: 0
            Proposal transforms: 14
            Type Payload: Transform (3) # 1
                Next payload: Transform (3)
                Payload length: 40
                Transform number: 1
                Transform ID: KEY_IKE (1)
                Transform IKE Attribute Type (t=1,l=2) Encryption-Algorithm :
                  AES-CBC
                Transform IKE Attribute Type (t=2,l=2) Hash-Algorithm : SHA
                Transform IKE Attribute Type (t=11,l=2) Life-Type : Seconds
                Transform IKE Attribute Type (t=12,l=4) Life-Duration : 2147483
                Transform IKE Attribute Type (t=14,l=2) Key-Length : 256
```

An example of the initial packet sent is shown in Figure 7.13. The reply that
comes back from the server (Figure 7.14) then defines the selected transform.

In this case the encryption method chosen is 3DES-CBC and the hashing method is MD5.

📖 **Web link (View IPSec trace):**

https://asecuritysite.com/forensics/pcap?infile=ipsec.pcap

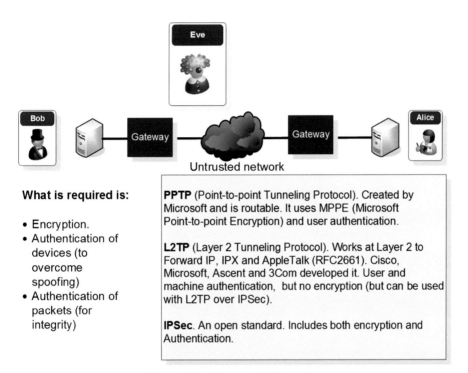

What is required is:

- Encryption.
- Authentication of devices (to overcome spoofing)
- Authentication of packets (for integrity)

PPTP (Point-to-point Tunneling Protocol). Created by Microsoft and is routable. It uses MPPE (Microsoft Point-to-point Encryption) and user authentication.

L2TP (Layer 2 Tunneling Protocol). Works at Layer 2 to Forward IP, IPX and AppleTalk (RFC2661). Cisco, Microsoft, Ascent and 3Com developed it. User and machine authentication, but no encryption (but can be used with L2TP over IPSec).

IPSec. An open standard. Includes both encryption and Authentication.

Figure 7.8 Common VPN tunnels.

The decision for a host to tunnel traffic is normally setup within the routing table. The host in Figure 7.15 does not have a tunnel setup, and will thus send all the traffic to the 192.168.0.1 interface. Once we have created the tunnel (as shown in Figure 7.16), we now see that new destination routes have been setup for 146.176.1.0/24, 146.176.2.0/24, and so on. The traffic for these networks are set for the 146.176.0.1 gateway, and through the 146.176.212.218 network interface (which has been setup by the VPN client as the Cisco Systems VPN Adapter).

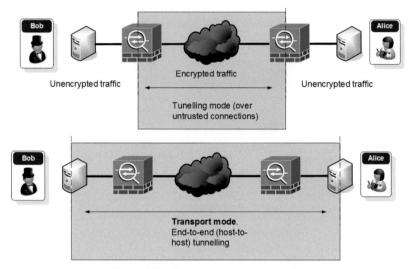

Figure 7.9 Tunneled or transport mode.

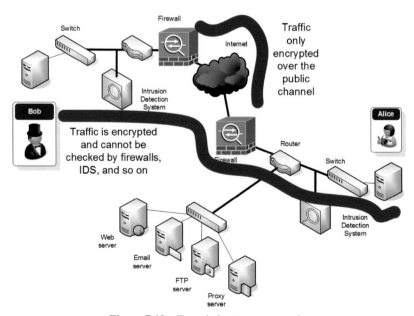

Figure 7.10 Tunneled or transport mode.

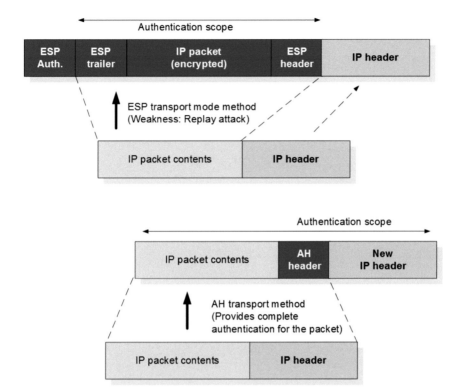

Figure 7.11 ESP and AH.

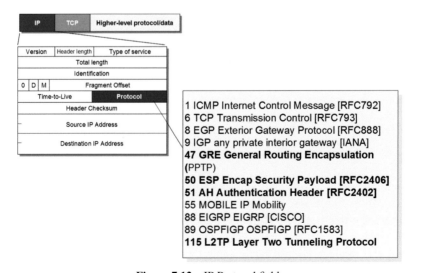

Figure 7.12 IP Protocol field.

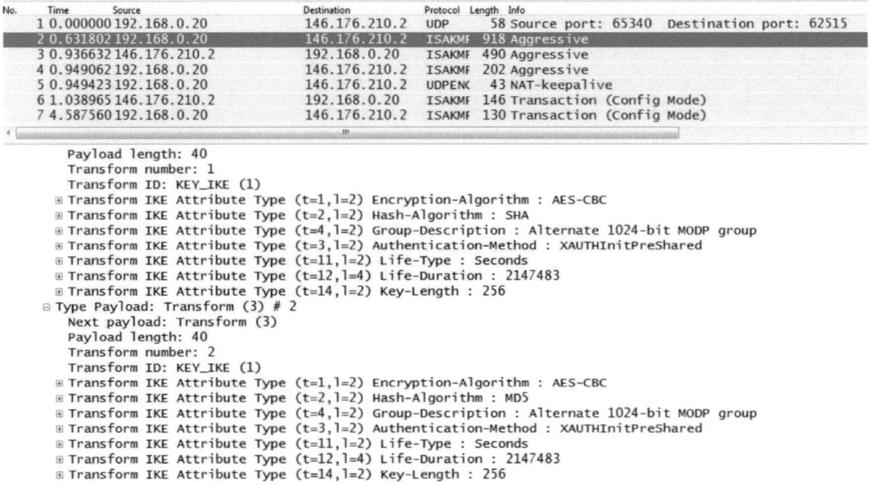

No.	Time	Source	Destination	Protocol	Length	Info
1	0.000000	192.168.0.20	146.176.210.2	UDP	58	Source port: 65340 Destination port: 62515
2	0.631802	192.168.0.20	146.176.210.2	ISAKMF	918	Aggressive
3	0.936632	146.176.210.2	192.168.0.20	ISAKMF	490	Aggressive
4	0.949062	192.168.0.20	146.176.210.2	ISAKMF	202	Aggressive
5	0.949423	192.168.0.20	146.176.210.2	UDPENC	43	NAT-keepalive
6	1.038965	146.176.210.2	192.168.0.20	ISAKMF	146	Transaction (Config Mode)
7	4.587560	192.168.0.20	146.176.210.2	ISAKMF	130	Transaction (Config Mode)

```
         Payload length: 40
         Transform number: 1
         Transform ID: KEY_IKE (1)
       ⊞ Transform IKE Attribute Type (t=1,l=2) Encryption-Algorithm : AES-CBC
       ⊞ Transform IKE Attribute Type (t=2,l=2) Hash-Algorithm : SHA
       ⊞ Transform IKE Attribute Type (t=4,l=2) Group-Description : Alternate 1024-bit MODP group
       ⊞ Transform IKE Attribute Type (t=3,l=2) Authentication-Method : XAUTHInitPreShared
       ⊞ Transform IKE Attribute Type (t=11,l=2) Life-Type : Seconds
       ⊞ Transform IKE Attribute Type (t=12,l=4) Life-Duration : 2147483
       ⊞ Transform IKE Attribute Type (t=14,l=2) Key-Length : 256
     ⊟ Type Payload: Transform (3) # 2
         Next payload: Transform (3)
         Payload length: 40
         Transform number: 2
         Transform ID: KEY_IKE (1)
       ⊞ Transform IKE Attribute Type (t=1,l=2) Encryption-Algorithm : AES-CBC
       ⊞ Transform IKE Attribute Type (t=2,l=2) Hash-Algorithm : MD5
       ⊞ Transform IKE Attribute Type (t=4,l=2) Group-Description : Alternate 1024-bit MODP group
       ⊞ Transform IKE Attribute Type (t=3,l=2) Authentication-Method : XAUTHInitPreShared
       ⊞ Transform IKE Attribute Type (t=11,l=2) Life-Type : Seconds
       ⊞ Transform IKE Attribute Type (t=12,l=4) Life-Duration : 2147483
       ⊞ Transform IKE Attribute Type (t=14,l=2) Key-Length : 256
```

Figure 7.13 ISAKMP first packet.

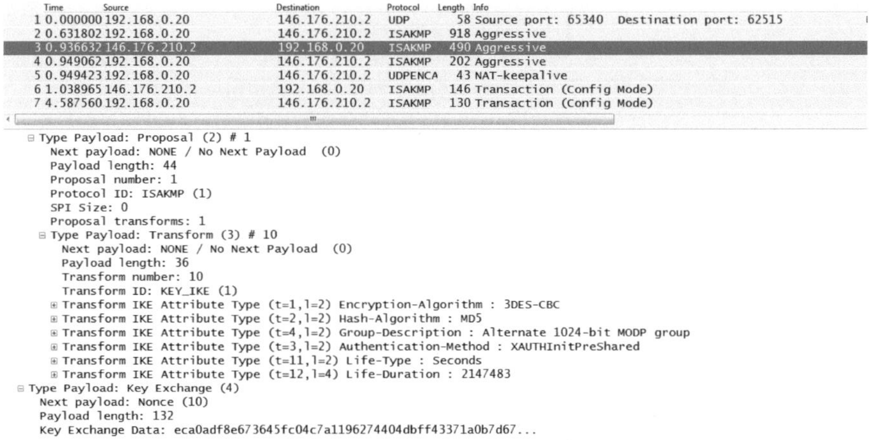

Time	Source	Destination	Protocol	Length	Info
1 0.000000	192.168.0.20	146.176.210.2	UDP	58	Source port: 65340 Destination port: 62515
2 0.631802	192.168.0.20	146.176.210.2	ISAKMP	918	Aggressive
3 0.936632	146.176.210.2	192.168.0.20	ISAKMP	490	Aggressive
4 0.949062	192.168.0.20	146.176.210.2	ISAKMP	202	Aggressive
5 0.949423	192.168.0.20	146.176.210.2	UDPENCA	43	NAT-keepalive
6 1.038965	146.176.210.2	192.168.0.20	ISAKMP	146	Transaction (Config Mode)
7 4.587560	192.168.0.20	146.176.210.2	ISAKMP	130	Transaction (Config Mode)

```
     ⊟ Type Payload: Proposal (2) # 1
         Next payload: NONE / No Next Payload  (0)
         Payload length: 44
         Proposal number: 1
         Protocol ID: ISAKMP (1)
         SPI Size: 0
         Proposal transforms: 1
       ⊟ Type Payload: Transform (3) # 10
           Next payload: NONE / No Next Payload  (0)
           Payload length: 36
           Transform number: 10
           Transform ID: KEY_IKE (1)
         ⊞ Transform IKE Attribute Type (t=1,l=2) Encryption-Algorithm : 3DES-CBC
         ⊞ Transform IKE Attribute Type (t=2,l=2) Hash-Algorithm : MD5
         ⊞ Transform IKE Attribute Type (t=4,l=2) Group-Description : Alternate 1024-bit MODP group
         ⊞ Transform IKE Attribute Type (t=3,l=2) Authentication-Method : XAUTHInitPreShared
         ⊞ Transform IKE Attribute Type (t=11,l=2) Life-Type : Seconds
         ⊞ Transform IKE Attribute Type (t=12,l=4) Life-Duration : 2147483
     ⊟ Type Payload: Key Exchange (4)
         Next payload: Nonce (10)
         Payload length: 132
         Key Exchange Data: eca0adf8e673645fc04c7a1196274404dbff43371a0b7d67...
```

Figure 7.14 ISAKMP return packet.

7.6 IKE

IPSec is a flexible framework, which supports many different types of encryption and hashing functions, and there thus has to be an initial negotiation phase, where the devices pass the encryption/hashing methods that they can support to each other. At the end of this phase, known as IKE (Internet Key Exchange) the devices should have agreed on a basic set of methods, for

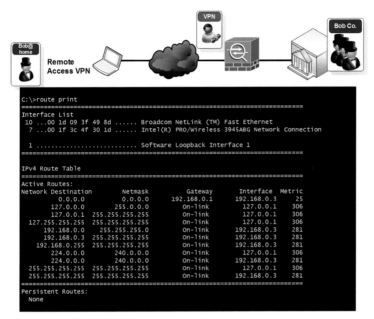

```
C:\>route print
=========================================================
Interface List
 10 ...00 1d 09 3f 49 8d ...... Broadcom NetLink (TM) Fast Ethernet
  7 ...00 1f 3c 4f 30 1d ...... Intel(R) PRO/Wireless 3945ABG Network Connection

  1 ........................ Software Loopback Interface 1
=========================================================

IPv4 Route Table
=========================================================
Active Routes:
Network Destination        Netmask          Gateway       Interface  Metric
          0.0.0.0          0.0.0.0      192.168.0.1     192.168.0.3     25
        127.0.0.0        255.0.0.0          On-link       127.0.0.1    306
        127.0.0.1  255.255.255.255          On-link       127.0.0.1    306
  127.255.255.255  255.255.255.255          On-link       127.0.0.1    306
      192.168.0.0    255.255.255.0          On-link     192.168.0.3    281
      192.168.0.3  255.255.255.255          On-link     192.168.0.3    281
    192.168.0.255  255.255.255.255          On-link     192.168.0.3    281
        224.0.0.0        240.0.0.0          On-link       127.0.0.1    306
        224.0.0.0        240.0.0.0          On-link     192.168.0.3    281
  255.255.255.255  255.255.255.255          On-link       127.0.0.1    306
  255.255.255.255  255.255.255.255          On-link     192.168.0.3    281
=========================================================
Persistent Routes:
  None
```

Figure 7.15 Sample routing table.

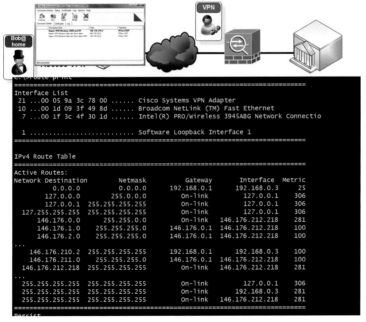

```
C:\>route print
=========================================================
Interface List
 21 ...00 05 9a 3c 78 00 ...... Cisco Systems VPN Adapter
 10 ...00 1d 09 3f 49 8d ...... Broadcom NetLink (TM) Fast Ethernet
  7 ...00 1f 3c 4f 30 1d ...... Intel(R) PRO/Wireless 3945ABG Network Connectio

  1 ........................ Software Loopback Interface 1
=========================================================

IPv4 Route Table
=========================================================
Active Routes:
Network Destination        Netmask          Gateway        Interface  Metric
          0.0.0.0          0.0.0.0      192.168.0.1      192.168.0.3     25
        127.0.0.0        255.0.0.0          On-link        127.0.0.1    306
        127.0.0.1  255.255.255.255          On-link        127.0.0.1    306
  127.255.255.255  255.255.255.255          On-link        127.0.0.1    306
      146.176.0.0    255.255.255.0          On-link  146.176.212.218    281
      146.176.1.0    255.255.255.0      146.176.0.1  146.176.212.218    100
      146.176.2.0    255.255.255.0      146.176.0.1  146.176.212.218    100
...
    146.176.210.2  255.255.255.255      192.168.0.1      192.168.0.3    100
    146.176.211.0    255.255.255.0      146.176.0.1  146.176.212.218    100
  146.176.212.218  255.255.255.255          On-link  146.176.212.218    281
...
  255.255.255.255  255.255.255.255          On-link        127.0.0.1    306
  255.255.255.255  255.255.255.255          On-link      192.168.0.3    281
  255.255.255.255  255.255.255.255          On-link  146.176.212.218    281
=========================================================
Persist...
```

Figure 7.16 Sample routing table (after VPN creation).

example a hand-held device might only be able to support basic encryption methods, so the negotiation phase is important in agreeing the foundation of the cryptography used in the connection. There are two main phases and are defined in the next sections.

7.6.1 Phase 1

In this phase the hosts negotiate a hash algorithm (SHA or MD5), and Diffie-Hellman group (Group 1, 2 or 5), encryption method (DES, 3DES or AES) and an authentication method (pre-share, RSA nonces or RSA signature). If this is successful, the hosts move onto the next phase. For example, on a PIX/ASA firewall the settings are:

```
firewall(config)# isakmp ?
Usage:   isakmp policy <priority> authen <pre-share|rsa-sig>
         isakmp policy <priority> encrypt <aes|aes-192|aes-256|des|3des>
         isakmp policy <priority> hash <md5|sha>
         isakmp policy <priority> group <1|2|5>
         isakmp policy <priority> lifetime <seconds>
```

The authentication they can use:

- **Pre-shared keys**. This uses pre-defined values for the authentication.
- **RSA encrypted nonces**. Peers encrypt with their private key and then the other side decrypts with the public key of the sending peer.
- **RSA signatures**. This uses asymetric public and private key pairs with a certificate authority (CA). Peers exchange their certificates, and contact the respected CA to validate them.

The configuration of the device on the left-hand side of Figure 7.17 becomes:

```
isakmp enable outside
isakmp key ABC&FDD address 176.16.0.2 netmask 255.255.255.255
isakmp identity address
isakmp policy 5 authen pre-share
isakmp policy 5 encrypt des
isakmp policy 5 hash sha
isakmp policy 5 group 1
isakmp policy 5 lifetime 86400
sysopt connection permit-ipsec

crypto ipsec transform-set MYIPSECFORMAT esp-des esp-sha-hmac
crypto map MYIPSEC 10 ipsec-isakmp
access-list 111 permit ip 10.0.0.0  255.255.255.0  176.16.0.0  255.255.255.0
crypto map MYIPSEC 10 match address 111
crypto map MYIPSEC 10 set peer 176.16.0.2
crypto map MYIPSEC 10 set transform-set MYIPSECFORMAT
crypto map MYIPSEC interface outside
```

The first line enabled on the outside interface. Next the Diffie-Hellman process requires a key-string, such as ABC&FDD (a pre-shared key which is defined on both devices), and which will be used with a peer at the address of 176.16.0.2 (which has a subnet mask of 255.255.255.255 so that it is only one host):

isakmp key <key-string> address <ip> [netmask <mask>] [no-xauth]
isakmp key ABC&FDD address 176.16.0.2 netmask 255.255.255.255

Next, if RSA encryption is being used for the public-key encryption, the hostname, or its IP address, are used to generate the key for RSA encryption. Otherwise with **pre-share** the identity is used to identify the peer. This is achieved using an address with:

isakmp identity <address\hostname\key-id> [<key-id-string>]
isakmp identity address

The address is normally used, but hostname is used where the IP address changes often. Each IKE has a policy number, where a 1 is the highest priority. Thus a higher value is typically used so that higher priorities can inserted at a future time. The following defines a policy number of 5 and that a pre-shared key is used (otherwise rsa-sig can be defined):

isakmp policy <priority> authen <pre-share\rsa-sig>
isakmp policy 5 authen pre-share

Then the encryption type can be defined, such as for the DES encryption algorithm (others include aes, aes-192, aes-256, and 3des):

isakmp policy <priority> encrypt <aes\aes-192\aes-256\des\3des>
isakmp policy 5 encrypt des

Next the hashing technique needs to be defined, as this will be used in the authentication process. The methods available, in this case, are MD5 and SHA. As available SHA has a larger hash code, and there is thus has less chance of creating the same signature for different unhashed values, and is typically used for enhanced security. Thus to define SHA:

isakmp policy <priority> hash <md5|sha>
isakmp policy 5 hash sha

Next the Diffie-Hellman method type is defined. For 768-bit Diffie-Hellman a Group 1 is used, while 1024-bit Diffie-Hellman uses Group 2, and 1582-bit Diffie-Hellman uses Group 5. Thus to setup Group 1 settings:

isakmp policy <priority> group <1|2|5>
isakmp policy 5 group 1

Finally the default lifetime is defined in terms of seconds. Thus to setup a period of 1 day (86,400 seconds) the following can be defined:

isakmp policy <priority> lifetime <seconds>
isakmp policy 5 lifetime 86400

Once the IKE is setup, the **IPSec parameters** can be defined. First we must allow the IPsec packets to pass through the device. Normally these would be interrupted by ACLs, which must be bypassed. This includes protocols: 50 (ESP), 51 (AH) and 500 (IKE). To do this the following is used:

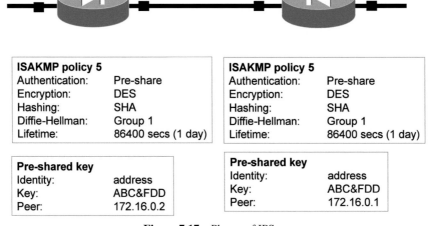

Figure 7.17   Phases of IPSec.

sysopt   connection   {permit-ipsec I permit-l2tp I permit-pptp I timewait I
{tcpmss [minimum] <bytes>} }
sysopt connection permit-ipsec

7.6.2 Phase 2

Once the secure connection has been created (in Phase 1), the second phase
is much faster (Figure 7.18), in which crypto maps are exchanged for:

- AH, ESP (or both).
- Encryption (DES, 3DES).
- ESP (tunnel or transport).
- Authentication (SHA/MD5).
- Security association (SA) lifetimes.

The first configuration defines the security protocol defined between the
peers, and the following defines a transform set named MYIPSECFORMAT

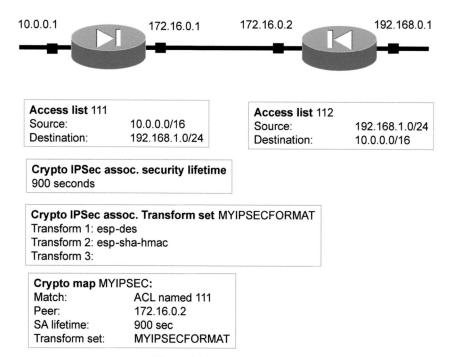

Figure 7.18   Phases of IPSec.

which uses DES for encapsulating security payload (ESP) and SHA-HMAC for the message authentication:

usage: crypto ipsec transform-set <trans-name> [ah-md5-hmac|ah-sha-hmac]
      [esp-aes|esp-aes-192|esp-aes-256|esp-des|esp-3des|esp-null]
      [esp-md5-hmac|esp-sha-hmac]
   crypto ipsec transform-set <trans-name> mode transport
crypto ipsec transform-set MYIPSECFORMAT esp-des esp-sha-hmac

Next a crypto map can be defined, where MYIPSEC defines the name associated with the map and 10 is a sequence number. These sequence numbers allow for different crypto combinations to be set for different peers which make connections on the interface that has the crypto map applied. There can only be one crypto map on each interface, thus sequence number blocks can apply different policies to a specific crypto map:

usage: crypto map <map-name> { (interface <if-name>)) |
      (client configuration address initiate|respond) |
      (<seqno> ipsec-manual|ipsec-isakmp|match|set ...)}
crypto map MYIPSEC 10 ipsec-isakmp

Next the access control list (number 111) can be defined to specify the traffic which will be encrypted. In the following, traffic from 10.0.0.0/24 to 176.16.0.0/24 will be encrypted:

access-list 111 permit ip 10.0.0.0 255.255.255.0 176.16.0.0 255.255.255.0

All other traffic is allow to pass, but not through the tunnel. After this, an access list number can be defined (in this case it is 111), where anything matching this list will either be encrypted (for outgoing data) or decrypted (for incoming data) as defined by the crypto map block (which is sequence number 10). Thus we can have different security settings depending on the sequence number:

usage: crypto map <map-name> { (interface <if-name>)) |
      (client configuration address initiate|respond) |
      (<seqno> ipsec-manual|ipsec-isakmp|**match**|set ...)}
crypto map MYIPSEC 10 match address 111

Next the peer which is associated with the crypto map security policy is defined:

usage: crypto map <map-name> { (interface <if-name>)) |
    (client configuration address initiate|respond) |
    (<seqno> ipsec-manual|ipsec-isakmp|match|**set** ...)}
crypto map MYIPSEC 10 set peer 176.16.0.2

Next the type of hashing and/or encoding is defined using the transform mapping:

usage: crypto map <map-name> { (interface <if-name>)) |
    (client configuration address initiate|respond) |
    (<seqno> ipsec-manual|ipsec-isakmp|match|**set** ...)}
crypto map MYIPSEC 10 set transform-set MYIPSECFORMAT

Next the crypto map can be applied onto an interface (only one is allowed on each interface):

usage: crypto map <map-name> { (**interface <if-name>**)) |
    (client configuration address initiate|respond) |
    (<seqno> ipsec-manual|ipsec-isakmp|match|set ...)}
crypto map MYIPSEC interface outside

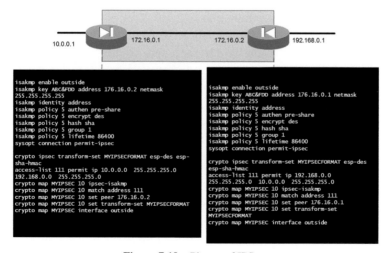

Figure 7.19   Phases of IPSec.

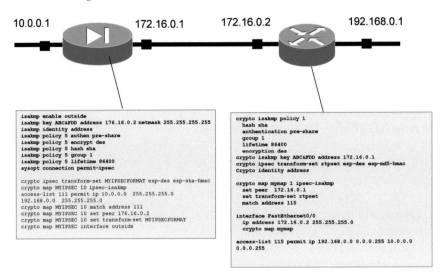

Figure 7.20 Phases of IPSec.

The completed IPSec setup is given in Figure 7.19 for two PIX/ASA firewalls, and Figure 7.20 shows the equivalent configuration for a PIX and a Cisco firewall.

7.7 Tor

As we move into an Information Age, there is a continual battle on the Internet between those who would like to track user activities, and those who believe in anonymity. The recent right to be forgotten debate has shown that very little can be hidden on the Internet, and deleting traces can often be difficult. The Internet, too, is be a place where crime can thrive through anonymity, so there is a continual tension between the two sides of the argument.

To law enforcement agencies the access to Internet-based information can provide a rich source of data for the detection and investigation of crime, but they have struggled to find evidence within the Tor (The Onion Routing) network. Its usage has been highlighted over the years, including in June 2013 when Edward Snowden used it to send information on PRISM to the Washington Post and The Guardian. The trace of a user's access to Web servers is thus confused with non-traceable accesses. This has caused a range of defence agencies to invest methods of compromising the infrastructure, especially to uncover the dark web. Its development received funding from Electronic Frontier Foundation, and further developed by The Tor Project – a non-profit making organisation. Many government agencies around the World now target its cracking, such as with the Russian government offering a bounty of $111,000. A strange feature in the history of Tor is that it was originally sponsored by the U.S. Naval Research Laboratory (which had been involved in onion routing), with its first version appeared in 2002. The original demonstrator was created by Roger Dingledine, Nick Mathewson, and Paul Syverson, and who have since been named, in 2012, as one of the Top 100 Global Thinkers.

Web traces contain a wide range of information, including user details from cookies, IP addresses, and even user behaviour (with the usage of user behaviour fingerprints). This information can then be used to target marketing to users, and also is a rich seam of information for the detection and investigating crime. The Tor network has long been a target of defence and law enforcement agencies, as it protects user identity, the contents of the accesses, and the source and destination locations. Connections within the Tor network can either just use it to route over public network networks (in a similar way to tunnelled VPN connection which are tunnelled through a gateway) and connect using HTTP or HTTPS (and where an exit gateway will define the host which is making the accesses), or can route directly from the browser to a Tor service (these connect to .onion sites).

Web sites which exist in the Tor network and which have a binding from the Tor browser to the end service are often known as residing in the dark web, and are not accessible to most search engines such as Google. Tor can thus be used to bind to a server, so that the server will only talk to a client which has been routed through the Tor network. This is the closed model of creating a Web infrastructure and which cannot be accessed by users on the Internet using non-Tor enabled browser.

With the Tor network, the routing is done using computers of volunteers around the world to route the traffic around the Internet, and within each hop the chances to trace the original source significantly reduces. In fact, it is rather like a pass-the-parcel game, where game players randomly pass the parcel to others, but where eventually the destination receiver will receive the parcel. As no-one has marked the parcel on its route, it's almost impossible to find out the route that the parcel took.

The encryption involves each of the routing nodes (the relay nodes) having a symmetric encryption key, and the data is encrypted with each of the keys (Figure 7.21). In this case the purple key is the encryption key of the first

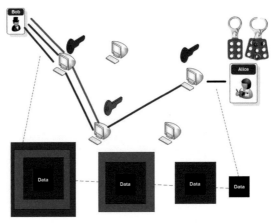

Figure 7.21   Onion routing.

node, and is the last to be encrypted. As the data goes through the network, each node decrypts with their key. The last part of the communication, out of the gateway, will thus be non-encrypted, but a protocol such as HTTPS can be used to protect the last part of the communication.

At the core of Tor is Onion Routing, which uses subscriber computers to route data packets – known as cells – over the Internet, rather than using publically available routers. One thing that must be said is that Tor aims to tunnel data through public networks, and keep the transmission of the data packets safe, which is a similar method that Google uses when you search for information (as it uses the HTTPS protocol for the search). So for a Tor network, let's ask a few questions:

- Can a remote Web site determine my IP address? For Tor, the answer is: No (it will contain the gateway address of the exit node, as we have in a VPN tunnel). For HTTPS, the answer is: Yes (the source IP address of the access will show up the source address of the HTTPS request, unless a proxy or VPN connection is used).
- Can the remote Web site determine my computer type? For Tor, the answer is: No (it hides the connection details). For HTTPS, the answer: Yes (the details of an HTTPS request will be passed, including details of the browser and computer type, unless some obfuscation is used on the browser or where a proxy is used).
- Can someone view the details of the data contained within the network packets? For Tor, the answer is: No (there are multiple keys used, and

an intruder would have to gain every key on the route to crack the communications). For HTTPS, the answer is: No (the key should be secret between the client and the server).

7.7.1 Tor Encryption

Tor traffic uses fix length cells which flow across the network, and create a circuit, and where each relay node only knows its predecessor and its successor. Each relay node then stores its own symmetric key for the connection and uses this to decrypt cells that are routed through it. For a connection between a client and a server we thus have:

- A stream cipher. This is symmetric key encryption (typically 128-bit AES in counter mode with an IV) for encrypted traffic, and where each relay node has negotiated its key with the server (and which the client will also know).
- Public-key key encryption. This uses either 1,024/2,048 bit RSA or Elliptic Curve for identity provision of the relay node.
- Elliptic-Curve Diffie-Hellman method for key negotiation. With this only the client and the server will know the keys involved for each of the relay nodes. It uses Curve25519 for the negotiation of the key and which uses the high-speed Elliptic-curve Diffie-Hellman method [9].
- A hash function. This is typically SHA-1, and is used to check the integrity of the data cells.

The client or server will initially send a cell and which is encrypted with each of the symmetric keys used in the Tor relay nodes, and encrypted in the sequence where the first relay node has the last encryption key applied (the outer layer of the onion), and the first one used has the key of the last relay node (the inner most layer of the onion). The negotiation of the key that each relay node uses is done on the creation of the circuit with the *ntor* handshake. This uses the Curve25519 method [9] and where the network shares the G and N value.

Along the route, each Tor relay node takes the cell from its predecessor and unencrypts with the negotiated symmetric key. It then passes the cell onto its successor. With Curve25519 we generate a 32-byte secret key (256 bits) and a 32-byte public key. A hash of the shared secret is then created as a 32-byte secret key (Curve25519(a,Curve25519(b,9)), as illustrated in Figure 7.22. This is used for a 128-bit or 256-bit AES key.

With the Diffie-Hellman parameters, the global defined parameter for G in the Tor network is 2 and N is:

244 *Tunneling*

```
FFFFFFFFFFFFFFFFC90FDAA22168C234C4C6628B80DC1CD129024E088A67CC74020
BBEA63B139B22514A08798E3404DDEF9519B3CD3A431B302B0A6DF25F14374FE135
6D6D51C245E485B576625E7EC6F44C42E9A637ED6B0BFF5CB6F406B7EDEE386BFB5
A899FA5AE9F24117C4B1FE649286651ECE65381FFFFFFFFFFFFFFFF
```

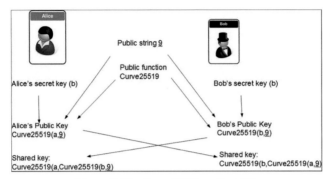

Figure 7.22   Curve25519 method [9].

The following provides an outline of the handshake:

```
from os import urandom
from eccsnacks.curve25519 import scalarmult, scalarmult_base
import binascii

a = urandom(32)
a\_pub = scalarmult_base(a)

b = urandom(32)
b_pub = scalarmult_base(b)

k_ab = scalarmult(a, b_pub)
k_ba = scalarmult(b, a_pub)

print "Bob private: ",binascii.hexlify(b_pub)
print "Alice private: ",binascii.hexlify(a_pub)
print "Bob shared: ",binascii.hexlify(k_ba)
print "Alice shared: ",binascii.hexlify(k_ab)
```

and a sample runs gives:

```
Bob public   : 2f2a9b6b82a1a696a4622923f1fe2ec7689383183ae32d1c011b9f
               ce02b2225d
Alice public: a44ae94deccf2313a4cffda0b93c1031b49902e7a34e86e73d30c4
               0eeac49f14
Bob shared   : 4cce4255bc41ff7c2b4a2d153c499dbaaf665d26426bf3f10f156c
               0849c19c49
Alice shared: 4cce4255bc41ff7c2b4a2d153c499dbaaf665d26426bf3f10f156c
               0849c19c49
```

📖 **Web link (Curve25519):** http://asecuritysite.com/encryption/curve25519

Each connection which is made between Tor relay nodes uses TLS/SSLv3 (and as a minimum must support TLS_DHE_RSA_WITH_AES_128_CBC _SHA).

Most current methods of cracking Tor traffic now are only able to analyse the network profile of the traffic, and from this determine the sites which are possibly being accessed. This works fairly well within a controlled environment, but it struggles within a real-life network. Juarez et al [10] carried out a critical approach on Website Fingerprinting (WF), and outlined that user's browsing habits and differences in location and version of Tor Browser Bundle, were normally omitted from attack models. Wang [11] outlines that while researchers have managed to fingerprint Tor traffic under lab conditions, there are significant differences with a realistic environment. In particular, they highlight that it may not be possible for an attacker to continually update their training set, and also where trained sequences in the lab will relate to single Web pages and trained on the first packet sent, and where noise is often added to the captured traffic due to network traffic from other pages. They present several new methods which are able to address the problems they highlight, and where packets can be collected in the wild.

He et al [12] also outline the problems related to the overlapping of Web accesses, and propose a method where the attacker delays HTTP requests that originated from users, in order to isolate their requests. For this they managed to identify access to the Alexa Top 100 top websites with a greater success than previous work.

7.7.2 Examining Tor Traffic

First let's access the same Web server, with Firefox and with the Tor browser (Figure 7.23). It can be see that the IP address differs for the same access. For a Tor browser when accessing the page: http://asecuritysite.com/ip/ we get:

```
Your IP Address  171.25.193.20
Your Hostname  tor-exit0-readme.dfri.se
Location  Long: 59.33
Lat: Europe/Stockholm
Country Code: SE
Country Name: Sweden
Region Name:
City:
Zip Code:
HTTP_USER_AGENT  Mozilla/5.0 (Windows NT 6.1; rv:31.0) Gecko/
20100101 Firefox/31.0
```

and for a non-Tor browser (using Firefox):

```
Your IP Address  82.41.182.60
Your Hostname  zzzz2-zgyl27-2-0-custzzz.sgyl.cable.virginm.net
Location  Long: 55.94
Lat: Europe/London
Country Code: GB
Country Name: United Kingdom
Region Name: Scotland
City: Edinburgh
Zip Code: EH11
HTTP_USER_AGENT  Mozilla/5.0 (Windows NT 6.1; WOW64; rv:34.0) Gecko/
20100101 Firefox/34.0
```

We can see that the IP address seen on the Web server is related to one of the routing elements of the Tor network. Then if we look at the IP address which appears in the log on the server, we see the Tor network address:

```
2014-12-23 21:24:59 212.227.84.95 GET /ip/details - 80 - 171.25.193.20
Mozilla/5.0+(Windows+NT+6.1;+rv:31.0)+Gecko/20100101+Firefox/31.0 200
0 0 2301
```

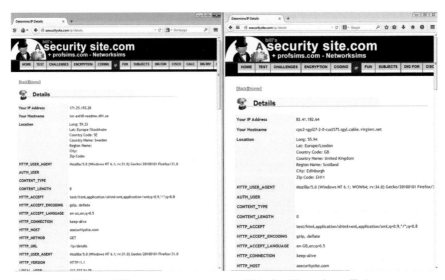

Figure 7.23 Accesses to a remote Web site (Tor and non-Tor).

If we quit the browser, and open another session on a Mac OS X browser, we see the IP address has changed and that the operating system and browser type has been hidden from the remote site.

```
Your IP Address 37.157.195.222
Your Hostname 12.transminn.cz
Location Long: 50.08
Lat: Europe/Prague
Country Code: CZ
Country Name: Czech Republic
HTTP_USER_AGENT Mozilla/5.0 (Windows NT 6.1; rv:31.0) Gecko/
20100101 Firefox/31.0
```

The log from the Web site now contains:

```
2014-12-23 21:55:47 212.227.84.95 GET /ip - 80 - 37.157.195.222
Mozilla/5.0+(Windows+NT+6.1;+rv:31.0)+Gecko/20100101+Firefox/31.0
200 0 0 41
```

With the Tor network, data packets are tunnelled through a number of routing elements. If we look at the IP address of the local machine we get:

```
Ethernet adapter Local Area Connection:

Connection-specific DNS Suffix  . : localdomain
Link-local IPv6 Address . . . . . : fe80::98e2:a1b:dc21:5bfc%10
IPv4 Address. . . . . . . . . . . : 172.16.121.169
Subnet Mask . . . . . . . . . . . : 255.255.255.0
Default Gateway . . . . . . . . . : 172.16.121.2
```

Now if we capture traffic from the Tor connection, we see that it communicates with a node at 5.39.76.36 on TCP port 9001.

```
   No. Time Source Destination Protocol Length Info
    5 11.546007000 172.16.121.169 5.39.76.36 TCP 597 1113 > 9001
[PSH, ACK] Seq=1 Ack=1 Win=64240 Len=543

   No. Time Source Destination Protocol Length Info
    6 11.546397000 5.39.76.36 172.16.121.169 TCP 60 9001 > 1113
[ACK] Seq=1 Ack=544 Win=64240 Len=0
```

An example PCAP file can be viewed on the link defined below. It can be seen from the trace that an analysis of the data packets does not contain any information that is useful, as a tunnel is used to transmit the data (Figure 7.24). There is thus no information related to DNS look-ups, or the standard signs of SYN/SYN-ACK and ACK (the three-way handshake). Notice that the length of each of the packets remains the same for each one send (597 bytes). Figure 7.25 outlines the usage of a Tor browser.

📖 **Web link (Workload):** http://asecuritysite.com/log/tor.zip

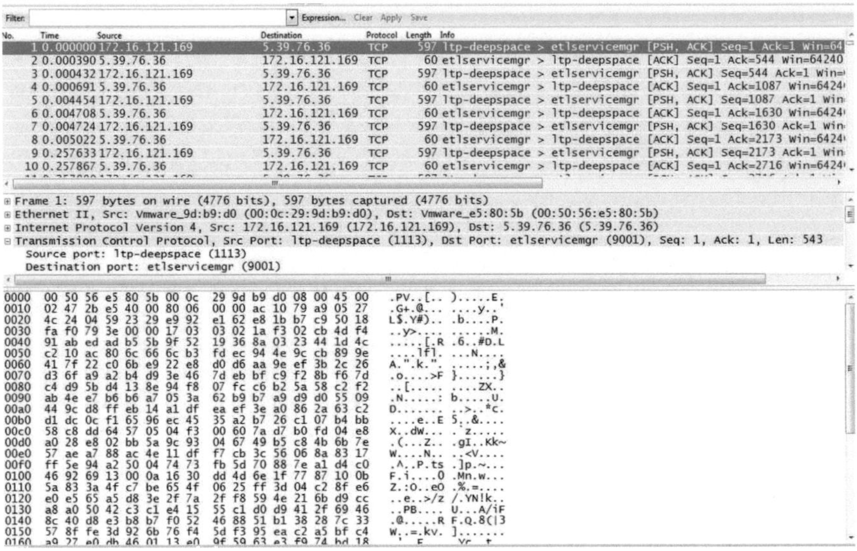

Figure 7.24 Sample network capture from Tor.

Figure 7.25 Using the Tor browser.

7.8 Lab/Tutorial

The lab and tutorial related to this chapter is available on-line at:

http://asecuritysite.com/crypto07

References

[1] Red Hat, "DROWN – Cross-protocol attack on TLS using SSLv2 (CVE-2016-0800)," *Red Hat*, 2016. [Online]. Available: https://access.redhat.com/security/vulnerabilities/drown.

[2] Mircrosoft, "Taming the Beast (Browser Exploit Against SSL/TLS) – Unleashed." [Online]. Available: https://blogs.msdn.microsoft.com/kaushal/2011/10/03/taming-the-beast-browser-exploit-against-ssltls/.

[3] B. Möller, T. Duong, and K. Kotowicz, "This POODLE Bites: Exploiting The SSL 3.0 Fallback," *Secur. Advis.*, pp. 1–6, 2014.

[4] J. Rizzo, "Practical Padding Oracle Attacks," *Proc. 4th USENIX Conf. Offensive Technol.*, vol. 10, pp. 1–9, 2010.

[5] US-CERT, "FREAK SSL/TLS Vulnerability." [Online]. Available: https://www.us-cert.gov/ncas/current-activity/2015/03/06/FREAK-SSLTLS-Vulnerability.

[6] "Weak Diffie-Hellman and the Logjam Attack." [Online]. Available: https://weakdh.org/.

[7] D. Adrian, K. Bhargavan, Z. Durumeric, P. Gaudry, M. Green, J. A. Halderman, N. Heninger, D. Springall, E. Thomé, L. Valenta, B. Vandersloot, E. Wustrow, and S. Z. Paul, "Imperfect Forward Secrecy: How Diffie-Hellman Fails in Practice," *Ccs*, pp. 5–17, 2015.

[8] I. Mantin and A. Shamir, "A practical attack on broadcast RC4," in *Lecture Notes in Computer Science (including subseries Lecture Notes in Artificial Intelligence and Lecture Notes in Bioinformatics)*, 2002, vol. 2355, pp. 152–164.

[9] D. J. Bernstein, "Curve25519: New Diffie-Hellman Speed Records," Springer, Berlin, Heidelberg, pp. 207–228, 2006.

[10] M. Juarez, S. Afroz, G. Acar, C. Diaz, and R. Greenstadt, "A Critical Evaluation of Website Fingerprinting Attacks," *Proc. 2014 ACM SIGSAC Conf. Comput. Commun. Secur. – CCS '14*, pp. 263–274, 2014.

[11] T. Wang and I. Goldberg, "On Realistically Attacking Tor with Website Fingerprinting," *Proc. Priv. Enhancing Technol.*, vol. 2016, no. 4, pp. 21–36, 2016.

[12] G. He, M. Yang, X. Gu, J. Luo, and Y. Ma, "A novel active website fingerprinting attack against Tor anonymous system," in *Proceedings of the 2014 IEEE 18th International Conference on Computer Supported Cooperative Work in Design, CSCWD 2014*, pp. 112–117, 2014.

8

Crypto Cracking

8.1 Introduction

Encryption provides a significant challenge for law enforcement agencies, especially in detecting threats against society. There is thus a drive to crack cryptography and also to investigate backdoors which allow investigators to break secret communications. Overall there are a number of methods that can be used to break encryption, such as when there is a flaw in the implementation of encryption method, or in the usage of weak encryption keys. For example, if we have a 128-bit encryption, but use a simple password to protect access to the encryption key, we significantly weaken the security of the key. In general, if all else fails, an investigator may have to rely on an exhaustive search of the keys to find the right one. In terms of a backdoor in cryptography, the two main methods which could be used are:

- **Key escrow**. This is where a copy of the encryption key is kept in escrow so that it can be used by a government agent.
- **A NOBUS** ('nobody but us') **backdoor**. This is where it is mathematically possible for government agents to crack the encryption, but no-one else can.

The attack on cryptography often centers around the cracking of hashes or on the discovery of the key used to encrypt. In symmetric encryption, the cracking processes searches for the key which encrypted the data, whereas with public key encryption, the search is often to find the private key which is associated with a given public key.

8.2 Key Escrow

One method that could be used is to crack everything that is encrypted, is to have a copy of the key which law enforcement would use if they required access to the data – an escrow key (Figure 8.1). This escrow key is a bit like

251

leaving your key under the doormat. The classic use case of used is with the Clipper Chip, where anyone who wanted to encrypt data would need a licence from law enforcement, and gain a chip to perform the encryption. A copy of the chip was than kept in case access was required – Government key escrow. Eventually, in 1997, the Clipper Chip project was abandoned as it was too difficult to enforce and would have been costly, and so was applied to a narrower set of applications, such as in regulated telecommunications systems. Along with the enforcement issue, there was also great risks of the keys becoming exposed (such as from an insider attack).

Of particular worry to the many is the insider (or trusted employee) threat, where the keys used either by an escrow system or for third-party encryption, could be breached, and cause a large-scale data loss. Thus the complexity of creating an escrow system which would scale across all the different agencies and data infrastructures involved would be well beyond current technology. Fraud and extortion could also result, along with the complexity of the coding involved from software vendors.

Figure 8.1   Government escrow.

The problems around key escrow are higlighted using two possible scenarious [1].

Scenario 1: Secure Tunnels and Escrow

The authors in [1] present the scenario of law enforcement being able to view encrypted data. Normally, with secure communications both public and private key are used. The encryption that happens in the secure tunnel is normally achieved with symmetric encryption (such as with AES or 3DES) and the associated key that will be used for the communication is protected using public key encryption. With SSL/TLS, the server sends its public key to the client (normally in the form of a digital certificate), and the client creates a new symmetric key and encrypts it with the server's public key and sends it back. The server then decrypts the encrypted key, and reveals the session key to be used. Once this has happened both sides have the same symmetric encryption key.

Within the paper the authors outline an approach where the symmetric key is encrypted a second time with a special escrow public key. Then we now have a single encryption process on the data, but both the server and law enforcement can read the stream. As we see in Figure 8.2, the public key from law enforcement is added to encrypt the session key and both are sent back to the server. Then a law enforcement agent can listen to the handshaking information and use their private key (which is secret) to reveal

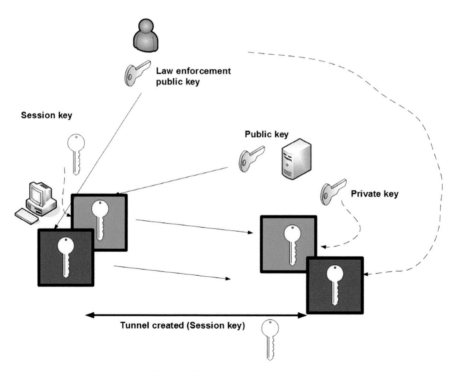

Figure 8.2 Double key creation.

the session key – which can then be kept in escrow (or used to decrypt the communications).

The authors thus outline that the double encryption of the session key is possible, but there are risks in the loss of the private key, and also in storing the session key, and where all the data that was encrypted by that key would be compromised, if lost. Their main issue with this type of system is who would actually control additional encryption. In the US, it may be the FBI, but what happens when you have cross-border communications? They speculate of the communications between the US and China, where both countries would have to agree to a single escrow agent.

Scenario 2: Encryption-by-default
Apart from secure tunnels, the other area that worries law enforcement is encryption-by-default, typical on mobile devices. On most systems, the encryption key are kept in escrow (typically on a domain server), so it is not too difficult to determine the key. With a mobile device the encryption key is stored in the TPM (Trusted Platform Module) chip, and can only be revealed with a password or fingerprint. Normally there is a lock-out time, or even a slow-down time, when brute-force is applied to the pass-phrase, which make it difficult to crack. In this scenario, again, the authors propose that the solution is to provide keys which are either provided by law enforcement or are stored in escrow. Again both methods are at risk of a breach of the escrow keys and from insider threats. The complexity of dealing with different nations states also would make it extremely complex for vendors.

8.3 Cracking the Code

A cryptosystem normally converts plaintext into ciphertext, using a key. There are several methods that an intruder can use to crack the cipher, including:

- **Exhaustive search**. Where the intruder uses brute force to decrypt the ciphertext and tries every possible key (Figure 8.3).
- **Known plaintext attack**. Where the intruder knows part of the cipher-text and the corresponding plaintext. The known ciphertext and plaintext can then be used to decrypt the rest of the ciphertext (Figure 8.4).
- **Man-in-the-middle**. Where the intruder is hidden between two parties and impersonates each of them to the other (Figure 8.5).

- **Chosen-ciphertext**. Where the intruder sends a message to the target, this is then encrypted with the target's private-key and the intruder then analyses the encrypted message. For example, an intruder may send an e-mail to the encryption file server and the intruder spies on the delivered message.
- **Active attack**. Where the intruder inserts or modifies messages (Figure 8.6).
- **The replay system**. Where the intruder takes a legitimate message and sends it into the network at some future time (Figure 8.7).
- **Cut-and-paste**. Where the intruder mixes parts of two different encrypted messages and is able to create a new message. This message is likely to make no sense, but may trick the receiver into doing something that helps the intruder.
- **Time resetting**. Some encryption schemes use the time of the computer to create the key. Resetting this time or determining the time that the message was created can give some useful information to the intruder.
- **Time attack**. This involves determining the amount of time that a user takes to decrypt the message; from this the key could be found.

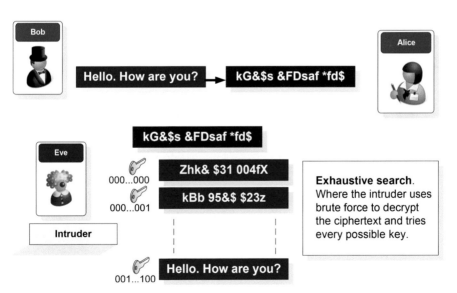

Figure 8.3   Exhaustive search.

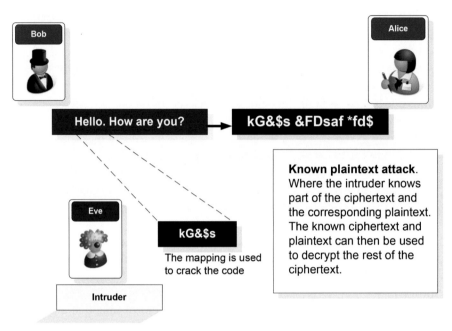

Figure 8.4   Known plaintext attack.

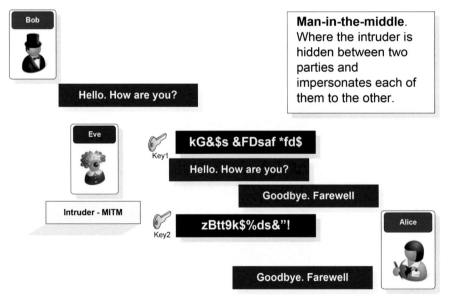

Figure 8.5   Man-in-the-middle.

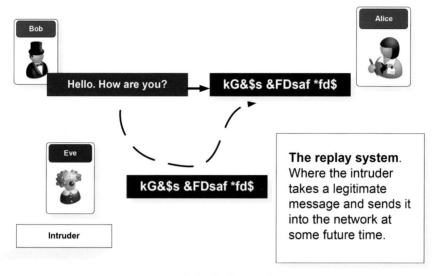

Figure 8.6 Replay attack.

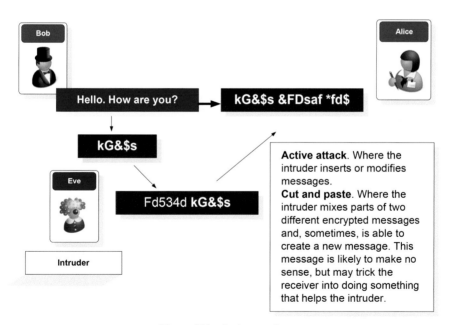

Figure 8.7 Active attack.

8.4 RSA Cracking

The RSA method suffers from several weaknesses. At the current time the limit of cracking RSA is for 768-bit keys and is attacked using the factorization of the modulus (N), but other methods use channel attacks, such as observing the current flows on a processor and which have managed to crack 1024-bit keys (in less than 100 hours) [2].

Within [3] RSA is attacked with a local FLUSH+RELOAD side-channel attack, where the "left-to-right sliding window" method leaks information about the exponent bits, and where the full key can be recovered. It involves a Level 3 Cache Side-Channel Attack where the cache memory stores the private RSA key. The attacker observes the memory utilisation of the cache (or from the electromagnetic radiation emitted in the decryption process). While it may be difficult on physical machines, the researchers outline that it is possible to extract the key from one VM (Virtual Machine) onto another. It is also likely that 2,048 bit RSA could be cracked with the same method, but would require more computing resource to crack. In the following sections we look at some fundamental problems with the RSA method.

8.4.1 RSA Crack with Different e Value

RSA can be cracked if we slightly modify the *e* key and get the same message ciphered. Bob selects an *e* of 5 and *N* of 15 ($p = 3$, $q = 5$), with a message of 7, so the cipher is (Figure 8.8):

$$C_1 = 7^5 \ (\text{mod } 15) = 7$$

Now we pass Bob a slightly different encryption key ($e_2 = 6$), and then get him to encrypt the same message:

$$C_2 = 7^6 \ (\text{mod } 15) = 4$$

Now we have done a few simple calculations and create an equation of $x.e_1 + y.e_2 = 1$, so if we raise C_1 and C_2 to the power of x and y, respectively, we generate:

$$(M^{e_1})^x \times (M^{e_2})^y$$

which then becomes:

$$M^{e_1.x+e_2.y}$$

which we can make equal to:

$$M^1$$

and where we can recover the message. So all we have to is to find the values of x and y which make this true:

$$x.e_1 + y.e_2 = 1$$
$$5.x + 6.y = 1$$

which is simple, as $x = -1$, and $y = 1$. So:

$$C_2 = M^{e1+1} \text{ and } C_1 = M^{e1}$$

If we divide the two we get:

$$\frac{C_2}{C_1} = \frac{M^{e1+1}(\text{mod } N)}{M^{e1}(\text{mod } N)}$$
$$(C_1)^{-1}.(C_2)^1 = M \ (\text{mod } N)$$
$$(7)^{-1}.(4)^1 = M \ (\text{mod } 15)$$
$$4 = 7 \times M \ (\text{mod } 15)$$

and thus $M = 7$ works (as 7×7 mod 15 equals 4). So the crack is that we solve for:

$$C_2 = C_1 \times M \ (\text{mod } N)$$

The following provides some outline code:

```
e1 = 5
n = 7*11
m= 7

def calcM(c1,c2,n):
        m=0
        r=[]
        while True:
                m = m + 1
                res = (c1 * m)   % n
                if (res==c2):
                r.append(m)
                if (m>60):
                        return(r)
        return r

e2 = e1 + 1
```

```
c1 = (m**e1) % n
c2 = (m**e2) % n

print 'M is\t\t',m
print 'e1 is\t\t',e1
print 'e2 is\t\t',e2
print 'N is\t\t',n
print 'Cipher 1 is\t',c1
print 'Cipher 2 is\t',c2

print '\n==== Eve then cracks by solving C2 = C1 x M (mod N)==='
crack = calcM(c1,c2,n)

print "Eve calculates message as",crack
print '\n========='
if (m in crack):
        print 'Message has been cracked'
else:
        print 'Message has not been cracked'
```

and a sample run gives:

```
M is            7
e1 is           5
e2 is           6
N is            15
Cipher 1 is     7
Cipher 2 is     4

==== Eve then cracks by solving C2 = C1 x M (mod N)===
Eve calculates message as   [7]

=========
Message has been cracked
```

📖 **Web link (RSA crack):** http://asecuritysite.com/encryption/crackrsa4

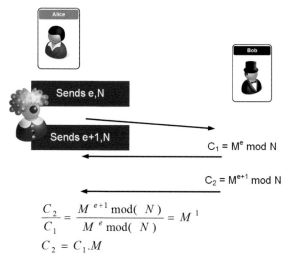

Figure 8.8   Active attack.

8.4.2 Cracking RSA by Factorizing N

One of the core methods in cracking RSA is to factorize N into the two core prime numbers: P and Q. If these can be determined, then PHI, which is equal to $(P-1)\times(Q-1)$, can be found, and where we can solved for:

$$(d \times e) \bmod (\text{PHI}) = 1$$

Example code to determine the decryption key (d) if we know P and Q is:

```
import gmpy2
p=7
q=5
e=0x010001

d = long(gmpy2.invert(e,(long(p)-1)*(long(q)-1)))
print '\nUsing e= ',int(e),' decryption key (d)=',d
```

A sample run is:

```
Using e=  65537  decryption key (d)= 17
```

We can check this with $P = 7$ and $Q = 5$, then $(65{,}537 \times 17) \bmod 24 = 1$. If we now check with a large value of N we get:

```
N=109,900,792,032,202,627,333,798,411,253,014,693,031,220,035,953,028,864,621
p= 8,842,485,744,298,254,476,968,088,873
q= 12,428,721,426,332,865,063,808,851,877

Using e=65537
  Decryption key (d)=
71,931,801,489,560,579,512,081,463,139,686,540,824,114,035,880,549,098,593
```

📖 **Web link (Factorization):** http://asecuritysite.com/encryption/crackrsa4

8.4.3 When M^e Is less than N

In RSA, we can crack the method if the value of m^e is less than N (where N is $P \times Q$). What we must make sure is the message is short, and e is relatively small. In the RSA encryption method we calculate our cipher (C) as (Figure 8.9):

$$C = M^e \ (\mathrm{mod} \ N)$$

where N is the modulus. But if (M^e) is less than (N), we get:

$$C = M^e$$

and:

$$M^e = C$$

So can take the log of each side:

$$\log_{10}(M^e) = \log_{10}(C)$$

and using logarithm rules:

$$e.\log_{10}(M) = \log_{10}(C)$$

we can now determine M from:

$$M = 10^{\log_{10}\left(\frac{C}{e}\right)}$$

Note that e is not the natural log in this case, but is the encryption key (e). So let's take an example. If we select:

$$P = 14{,}222{,}331{,}744{,}261{,}730{,}109 \text{ and}$$
$$Q = 6{,}549{,}179{,}332{,}223{,}292{,}769$$

then:

$$N \text{ is } 93{,}144{,}601{,}115{,}542{,}176{,}265{,}237{,}708{,}764{,}769{,}281{,}821.$$

Next we select a message (M) of 65, and e as 7. We can calculate the cipher as:

$$C = 65^7 \pmod{93, 144, 601, 115, 542, 176, 265, 237, 708, 764,}$$
$$769, 281, 821)$$

$$C = 49,022,278,90,625$$

$$M = 10^{log\left(\frac{49,022,278,90,625}{7}\right)} = 65$$

📖 **Web link (RSA crack):** http://asecuritysite.com/encryption/crackrsa2

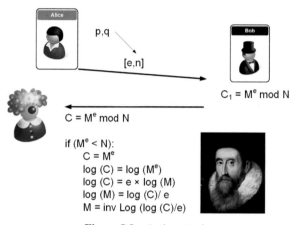

Figure 8.9 Active attack.

8.4.4 RSA Crack with Chinese Remainder Theory (CRT)

RSA can be cracked if the intruder records enough cipher text messages which use the same e value. Eve can take three N values (modulus) and a message, and create three cipher messages, and then use the Chinese Remainer Theorem (CRT) to solve the value of M^e, and simply use logarithms after that. For this she needs three cipher values which use the same e value and a different N value (which is similar to passing three public key values for Bob to encrypt). This gives:

$$C_1 = M^e \pmod{N_1}$$
$$C_2 = M^e \pmod{N_2}$$
$$C_3 = M^e \pmod{N_3}$$

which we can solve for M^e with CRT.

With $P_1 = 2$, $Q_1 = 3$ ($N_1 = 6$) $P_2 = 5$, $Q_2 = 7$ ($N_2 = 35$), and $P_3 = 11$, $Q_3 = 13$ ($N_3 = 143$), and with $e = 3$, and a message value of 5, we get:

$$5 = M^e \pmod 6$$
$$20 = M^e \pmod{35}$$
$$125 = M^e \pmod{143}$$

This can be solved with Chinese remainer theorem to give a value of 125 for M^e. This can then be solved by logs to give $M = 5$. The following shows values for a message of 1,500,000 (M) and $P_1 = 4519$, $Q_1 = 4523$, $P_2 = 4547$, $Q_2 = 4549$, $P_3 = 4561$, $Q_3 = 4567$:

```
N1= 20439437 N2= 20684303 N3= 20830087
Message= 1500000 e= 3
Cipher1= 6509102 Cipher2= 9683741 Cipher3= 3214286

=======Equations to solve=======
M^e mod 20439437=6509102
M^e mod 20684303=9683741
M^e mod 20830087=3214286

======Chinese Remainder Theorm Calc========
Result (M^e) is:   3375000000000000000

Calculated value of m is   1500000  Using 10^(log10(M^e)/e)
```

📖 **Web link (RSA crack):** http://asecuritysite.com/encryption/crackrsa3

8.4.5 Chosen Cipher Attack

In this attack, Eve gets Bob to cipher a chosen ciphertext. First Eve captures some cipher text, and then sends this back (with a random value raised to the power of Bob's encryption key (e)) and if Eve can determine the decrypted value, she can crack the message. First Eve listens for a cipher that she wants to crack (Figure 8.10):

$$C = M^e \pmod N$$

Next she takes this cipher and gets Bob to decrypt it (and also multiplying by a random value (r) to the power of Bob's e value):

$$C' = C \times r^e \pmod N$$

If Eve can determine the decrypted value for this cipher, she can determine the message as:

$$\left(C'\right)^{d} = (C \times r^{e})^{d} = M^{e \times d} \times r^{e \times d} = M \times r$$

as $(M^{e})^{d} \pmod N$ must equal $M^{1} \pmod N$. So Eve just takes the original cipher, and divides it by the random value (r). Some sample code is:

```
e=79
d=1019
N=3337
r=3
M=8

cipher=M**e % N
print 'Initial cipher:\t',cipher

cipher_dash = cipher * (r**e) % N
print 'Eve gets Bob to decipher:\t',cipher_dash

decipher = cipher_dash **d % N

print 'Bob says that the result is wrong:',decipher

print 'Eve determines as:',decipher/r
```

And a sample run gives:

```
==Initial values ====
e= 79 d= 1019 N= 3337
message= 8 r= 3

=============
Initial cipher: 2807
Eve gets Bob to decipher:   3022 (Cipher * r^e mod N)
Bob says that the result is: 24

=============
Eve determines the message as: 8
Eve has cracked message, as result is same as message
```

📖 **Web link (RSA crack):** http://asecuritysite.com/encryption/c_c

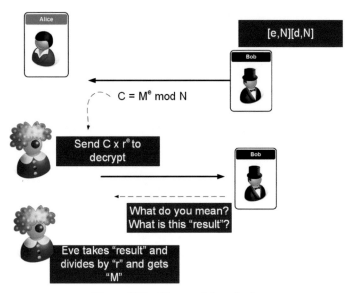

Figure 8.10   Chosen cipher attack.

8.4.6 Blinding Attack

In the blinding attack, we get Bob to sign for a message that is garbled. Eve has the message (M – "Pay Eve $1 million") and creates another message (Figure 8.11):

$$M' = r^e \, M \pmod{N}$$

where e is Bob's encryption key exponent and r is a random number. Eve gets Bob to sign for this. The signature is then:

$$S' = \left(M'\right)^d M \pmod{N}$$

Bob gives S' to Eve, and she just divides by r to get the signature for the original message:

$$\frac{S'}{r} = \frac{(M \times r^e)^d}{r} = \frac{M^d \times r^{ed}}{r} = \frac{M^d \times r^1}{r} = M^d \pmod{N}$$

So Eve takes Bob signature and adds it to the original message that Bob would not sign, and she can prove that Bob signed it. If she is sending to Alice the Banker, she would take the message:

```
"Pay Eve $1 million"
```

and add Bob signature for the message (S'/r), and then encrypt everything with Alice the Banker's public key. Alice will get the encrypted message and decrypts with her private key, and reads the message:

"Pay Eve $1 million"

and she then looks at the signature, and gets Bob's public key and checks the signature. It will match, so she will pay Eve one million dollars from Bob's account. Figure 8.11 shows the basic steps, and some simple code to demonstrate the principle is:

```
import sys
import os
import hashlib

e=79
d=1019
N=3337
r=2
Message='Pay Eve $1 million'

print '==Initial values ===='
print 'e=',e,'d=',d,'N=',N
print 'message=',Message,'r=',r
print '\n============='

array = os.urandom(1 << 20)
md5 = hashlib.md5()
md5.update(array)
digest = md5.hexdigest()
M = int(digest, 16) % N

print 'MD5 hash (mod N): ',M

signed=M**d % N
print 'Signed:\t',signed

val_sent_by_eve = M * (r**e) % N

signed_dash =val_sent_by_eve**d % N

print 'Bob sends Eve signature: ',signed_dash

result= signed_dash/r
print 'Eve send signature of:',result
```

```
print '\n=== Check =='
unsigned = result**e % N

print 'Unsigned value is:',unsigned
if (unsigned==M):
        print 'Success. Bob has signed it'
else:
        print 'Signatures do not compute'
```

A sample run is:

```
==Initial values ====
e= 79 d= 1019 N= 3337
message= Pay Eve $1 million r= 2

=============
MD5 hash (mod N):  3223
Signed: 914
Bob sends Eve signature:  1828
Eve send signature of: 914

=== Check ==
Unsigned value is: 3223
Success. Bob has signed it
```

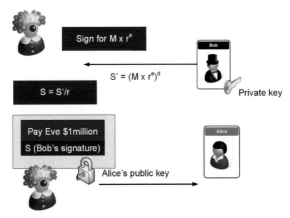

Figure 8.11   Chosen cipher attack.

📖 **Web link (RSA crack):** http://asecuritysite.com/encryption/c_c2

8.4.7 Bleichenbacher's Attack

In the Bleichenbacher's attack the intruder captures the cipher for the pre-shared key, and then re-ciphers with the additional of a value s. It has been the core of many attacks on SSL. Let's say that Eve is attacking the server. In the message she sends, there's a padding of the pre-shared key (as it is much smaller than the public modulus – N). In PKCS#1 v1.5 padding, we then have two bytes at the start (Figure 8.12):

$$0x00\ 0x02$$

Eve then captures the cipher in the handshake and which contains the SSL pre-shared key (M):

$$C = M^{e} \ (\mathrm{mod} \ N)$$

She then plays it back to the server, but adds an 's' value (where she multiplies the cipher (C) by s to the power of e (mod N)):

$$C' = C \times (s^{e}) \ (\mathrm{mod} \ N)$$

where e and N are the known public key elements. The server decrypts and gets:

$$M' = (C \times (s^{e}))^{d} \ (\mathrm{mod} \ N) = C^{d} \times s^{ed} \ (\mathrm{mod} \ N)$$
$$= M \times s \ (\mathrm{mod} \ N)$$
$$M = \frac{C'}{s}$$

When the server reads this, the first two bytes are likely to be incorrect, so it responds to say "Bad Crypto!". Eve then keeps trying with different s values, until the server gives her a positive response, and she's then on her way to finding the key. As we have 16 bits at the start, it will take us between 30,000 (1 in 2^{15} which is 1-in-32,728) and 130,000 attempts (1 in 2^{17} which is 1-in-131,073) to get a successful access. We use padding to make sure that M (the pre-shared key) is the same length as the modulus (N). As M is only 48 bytes, we need to pad to give a length equal to N (256 bytes for a 2048-bit key). Some sample code is:

```
import sys

e=79
d=1019
N=3337

def int_to_bytes(val, num_bytes):
    return [(val & (0xff << pos*8)) >> pos*8 for pos in range(num_bytes)]

print '==Initial values ===='
print 'e=',e,'d=',d,'N=',N
print '\n============='

pad = '\x00\x02\x55\x55'

val = int(pad.encode('hex'), 16)
print 'Padding is:',pad,' Int:',val

cipher = val**e % N

print 'Cipher is: ',cipher

for s in range(0,255):
        cipher_dash = (cipher*(s**e)) % N
        decode = cipher_dash **d % N
        result = int_to_bytes(decode,2)
        print s,
        if (result[0]==0 and result[1]==2):
                print '\n\\x00\\x02 Found it at s=',s
                break
```

and a sample run:

```
==Initial values ====
e= 79 d= 1019 N= 3337

=============
Padding is: UU  Int: 152917
Cipher is:  2652
   0   1   2   3   4   5   6   7   8   9  10  11  12  13  14  15  16  17  18
  19  20  21  22  23  24  25  26  27  28  29  30  31  32  33  34  35  36  37
  38  39  40  41  42  43  44  45  46  47  48  49  50  51  52  53  54  55  56
  57  58  59  60  61  62  63  64  65  66  67  68  69  70  71  72  73  74  75
  76  77  78  79  80  81  82  83  84  85  86  87  88  89  90  91  92  93  94
  95  96  97  98  99 100 101 102 103 104 105 106 107 108 109 110 111 112 113
 114 115 116 117 118 119 120 121 122 123 124 125 126 127 128 129 130 131 132
 133 134 135 136 137 138 139 140 141 142 143 144 145 146 147 148 149 150 151
 152 153 154 155 156 157 158 159 160 161 162 163 164 165 166 167 168 169 170
 171 172 173 174 175 176 177 178 179 180 181 182 183 184 185 186 187 188 189
 190 191 192 193 194 195 196 197 198 199 200 201 202 203 204 205 206 207 208
 209 210 211 212 213 214 215 216 217 218 219 220 221 222 223 224 225 226 227
 228 229 230 231 232 233
\x00\x02 Found it at s= 233
```

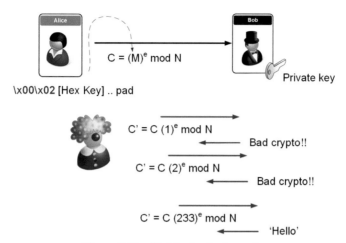

Figure 8.12 Bleichenbacher's attack.

📖 **Web link (RSA crack):** http://asecuritysite.com/encryption/c_c3

8.5 AES Cracking

AES has proven to be free from major vulnerabilities, but poor implementation of the encryption method often causes problems. These vulnerabilities include the usage of ECB, and where the cipher text does not change for the same input plain text. This allows for the copy-and-pasting of the cipher text, and also for it to be mapped from cipher text to plaintext. AES, like RSA, is also vulnerable to side channel attacks where the operation of the S-box can be observed using current flows in the processor [4]. One of the greatest weaknesses, though, is in using an encryption key which is derived from a simple password, as this considerably reduces the range of keys used.

8.5.1 AES Copy-and-Paste

AES can be susceptible to a copy-and-paste attack if ECB (Electronic Code Book) is used. With ECB the output will always be the same whenever we encrypt the same input block. For the following we will cipher each character for a given word ("napier) and then we will copy and paste each one and build a cipher stream. When we decrypt, we should be able to rebuild the original characters. In this way, Eve could determine the mapping for certain words and then copy-and-paste these back to Bob and rebuild valid messages:

```
from Crypto.Cipher import AES
import hashlib
import sys
import binascii
import Padding

word='edinburgh'
password='napier'

plaintext=''

def encrypt(plaintext,key, mode):
        encobj = AES.new(key,mode)
        return(encobj.encrypt(plaintext))

def decrypt(ciphertext,key, mode):
        encobj = AES.new(key,mode)
        return(encobj.decrypt(ciphertext))

ciphertext=''

key = hashlib.sha256(password).digest()

for ch in word:
        plaintext = Padding.appendPadding(ch,blocksize=Padding.AES_blocksize,
                    mode='CMS')

        ciphertext = ciphertext+ encrypt(plaintext,key,AES.MODE_ECB)
        print ""+binascii.hexlify(encrypt(plaintext,key,AES.MODE_ECB))

plaintext = decrypt(ciphertext,key,AES.MODE_ECB)
plaintext = Padding.removePadding(plaintext,mode='CMS')
print "  decrypt: "+plaintext
```

A sample run is:

```
Word:  napier
Password:  edinburgh
Key   a449d7560d2869d387374274863ccd268c3e8780b3efbad2a698683a48add29c

=========Ciphers====================
444604bbd56a89ab025827ed28feae90
555e423575ed1849dffc7a6479f82083
04d17aea8da89f35a14630976b07be06
2d16935b68db7636a40385065e2b16b9
c9f4268be5bad146e2e20ae7fda27815
54b05bc3f1b0d2a9a6d8dc3aedac6ebc
```

```
=========Concat Ciphers===================
Cipher:
444604bbd56a89ab025827ed28feae90555e423575ed1849dffc7a6479f8208304d17aea8da89
f35a14630976b07be062d16935b68db7636a40385065e2b16b9c9f4268be5bad146e2e20ae7fd
a2781554b05bc3f1b0d2a9a6d8dc3aedac6ebc

=========Decrypted===================
Decrypt: napier
```

📖 **Web link (RSA crack):** http://asecuritysite.com/encryption/crackaes

8.5.2 AES (Brute Force)

AES can be susceptible to brute force attacks when the encryption keys are generated by a password. In the code below we cause an exception for an incorrect encryption key, or where we create one which does not have plain text in the decrypted value. Although we use a 256-bit AES key, we are generating it from a password, so the number of keys possible is limited. In the following code we generate the keys for 'napier', 'test', 'password', 'foxtrot', '123456' and 'qwerty', and try these. If the decryption process creates an exception or is unprintable, we ignore it:

```
from Crypto.Cipher import AES
import hashlib
import sys
import binascii
import Padding

word='apple'
passwords=['napier','test','password','foxtrot','123456','qwerty']
password='napier'
passw=''

if (len(sys.argv)>1):
   word=str(sys.argv[1])

if (len(sys.argv)>2):
   password=str(sys.argv[2])

plaintext=''

def isprintable(s, codec='utf8'):
    try: s.decode(codec)
    except UnicodeDecodeError: return False
    else: return True

def encrypt(plaintext,key, mode):
    encobj = AES.new(key,mode)
    return(encobj.encrypt(plaintext))
```

```
def decrypt(ciphertext,key, mode):
    encobj = AES.new(key,mode)
    return(encobj.decrypt(ciphertext))

ciphertext=''

key = hashlib.sha256(password).digest()

print '\n=========Calculation===================='

print 'Word: ',word
print 'Password: ',password
print 'Key ',binascii.hexlify(key)

plaintext = Padding.appendPadding(word,blocksize=Padding.AES_blocksize,
            mode='CMS')

ciphertext = ciphertext+ encrypt(plaintext,key,AES.MODE_ECB)
print 'Cipher: ',binascii.hexlify(ciphertext)

print '\n=========Bruce Forece===================='

for passw in passwords:
    try:
            key = hashlib.sha256(passw).digest()
            plaintext = decrypt(ciphertext,key,AES.MODE_ECB)
            if (isprintable(plaintext)):
                print 'Plain text is ',plaintext,' and password is ', passw
except:
            print '',
```

The following uses a password of 'napier' and a secret word of 'edinburgh':

```
=========Calculation====================
Word:   edinburgh
Password:  napier
Key  5a6381be5d2e9bfdce1ea1a2801ae34685c908d14a729f5be23dfeddecef042d
Cipher:  16b5d75e609c7b8fd100ef933bb15542

 =========Calculation====================
Plain text is  edinburgh  and password is  napier
```

📖 **Web link (AES crack):** http://asecuritysite.com/encryption/crackaes2

8.5.3 AES Cracking with Non-Random Numbers

The generation of encryption keys can be guessed if the values are non-randomised and/or deterministic. In this case we will create a set of random numbers which are determistic and not pseudo-random:

```
import numpy as np
import hashlib
import sys
import binascii
import Padding

from Crypto.Cipher import AES
from Crypto import Random

i=4
msg="elephant_poo"

def encrypt(word,key, mode):
        plaintext=Padding.appendPadding(word,blocksize=Padding.AES_blocksize,
                mode='CMS')
        encobj = AES.new(key,mode)
        return(encobj.encrypt(plaintext))

def decrypt(ciphertext,key, mode):
        encobj = AES.new(key,mode)

        return(encobj.decrypt(ciphertext))

def isprintable(s, codec='utf8'):
        try: s.decode(codec)
        except UnicodeDecodeError: return False
        else: return True

rng = np.random.RandomState(0)

val=rng.randn(52)

print 'Value selected: ',val[i]

key= hashlib.sha256(str(val[i])).digest()
print 'Key=',hashlib.sha256(str(val[i])).hexdigest()

cipher = encrypt(msg,key,AES.MODE_ECB)
print binascii.hexlify(cipher)

for x in val:
        key= hashlib.sha256(str(x)).digest()
        decipher = decrypt(cipher,key,AES.MODE_ECB)
        if isprintable(decipher):
                print decipher
```

📖 **Web link (AES crack):** http://asecuritysite.com/encryption/aes_crack

8.6 **Digital Certificate Cracking**

One method of cracking encryption is to break the digital certificate which contains the private key which protects the symmetric encryption key. This type of system is used in disk encryption, where a symmetric key is used to encrypt the disk contents, and a public key, from a key pair, is used to encrypt this key. This key pair is then stored on the system as a digital certificate, and protected with a password. The following defines the method for creating the certificates. First we create a key pair (PVK) and a digital certificate (CER):

```
makecert.exe -n "CN=Test" -r -pe -a sha512 -len 4096 -cy authority -sv
bill.pvk bill.cer
```

Where -pe defines that the private key is exportable, and -n defines the certificate subject. In this case the key length is 4,096 bits (using the -len option), and -a defines the hashing method of SHA-512 (md5|sha1|sha256|sha384|sha512). Next we add the key pair (PVK) to the certificate (CER) to produce a digital certificate (PFX) and add a password (using the -po option):

```
pvk2pfx.exe -pvk bill.pvk -spc bill.cer -pfx bill01.pfx -po orange
```

We can then use the following C# code to try a range of passwords, and read the certificate. If there's an exception we ignore, otherwise we may have found a match and thus open up the certificate:

```
string path = Server.MapPath("/") + "bill.pfx";

string pass =
"apple,apricot,avocado,banana,bilberry,blackberry,blackcurrant,blueberry,
cantaloupe,nectarine,olive,orange,clementine,mandarine,tangerine,papaya,
pear,redcurrant,satsuma,strawberry,squash,tamarillo,tomato";
string[] pass1 = pass.Split(',');
string message = "";
foreach (string ss in pass1)
{
    message += h.showCer2(path, ss.Trim())+"\n";
}
public string showCer2(string f,string password)
{
   try
   {
      X509Certificate2Collection collection = new X509Certificate2Collection();
       collection.Import(f, password, X509KeyStorageFlags.PersistKeySet);

          foreach (X509Certificate2 cer in collection)
          {
```

```
                          try
                          {
                              hash1 = cer.SerialNumber;
                              hash2 = cer.GetEffectiveDateString();
                              hash3 = cer.Subject;
                              hash4 = cer.GetPublicKeyString();
                              hash5 = cer.GetKeyAlgorithm();
                              hash6 = cer.Issuer;
                              hash7 = cer.GetRawCertDataString();
                              // Import the certificate into an X509Store object
                          }
                          catch (Exception ex)
                          {
                              return ("Trying: "+password+ " - Exception");
                          }
                      }
              }
      catch (Exception ex)
      {
          return ("Trying: "+password+ " - Exception");
      }
      return ("Trying:" + password + " - Able to read");
      }
  }
```

A sample run gives:

```
Certificate: bill01.pfx
Trying: apple - Exception
Trying: apricot - Exception
Trying: avocado - Exception
..
Trying: olive - Exception
Trying: orange - Able to read. Serial: BD2BB5F6FB99ED8145AAC3441D6B285F
Trying: clementine - Exception
..
Trying: tamarillo - Exception
Trying: tomato - Exception
```

📖 **Web link (RC4):** http://asecuritysite.com/encryption/certcrack

8.7 Lab/Tutorial

The lab and tutorial related to this chapter is available on-line at:

http://asecuritysite.com/crypto08

References

[1] H. "Hal" Abelson, P. G. Neumann, R. L. Rivest, J. I. Schiller, B. Schneier, M. A. Specter, D. J. Weitzner, R. Anderson, S. M. Bellovin, J. Benaloh, M. Blaze, W. "Whit" Diffie, J. Gilmore, M. Green, and S. Landau, "Keys under doormats," *Commun. ACM*, vol. 58, no. 10, pp. 24–26, 2015.

[2] A. Pellegrini, V. Bertacco, and T. Austin, "Fault-Based Attack of RSA Authentication."

[3] D. J. Bernstein, J. Breitner, D. Genkin, L. Groot Bruinderink, N. Heninger, T. Lange, C. Van Vredendaal, and Y. Yarom, "Sliding right into disaster: Left-to-right sliding windows leak."

[4] O. Lo, W. J. Buchanan, and D. Carson, "Power analysis attacks on the AES-128 S-box using differential power analysis (DPA) and correlation power analysis (CPA)," *J. Cyber Secur. Technol.*, pp. 1–20, Sep. 2016.

9

Light-weight Cryptography
and Other Methods

9.1 Introduction

While AES and SHA work well together within computer systems, they
struggle in an IoT/embedded world as they often take up: too much processing
power; too much physical space; and consume too much battery power. So
NIST outlines a number of methods which can be used for light-weight
cryptography, and which could be useful in IoT and RFID devices [1]. They
thus define the device spectrum as:

- **Conventional cryptography**. Servers and Desktops. Tablets and smart
  phones.
- **Light-weight cryptography**. Embedded Systems. RFID and Sensor
  Networks.

With embedded systems, we commonly see 8-bit, 16-bit and 32-bit micro-
controllers, and which might struggle to cope with the real-time demands for
conventional cryptography methods. And, in the 40+ years since the first 4-
bit processor, there is even a strong market for 4-bit processors. RFID and
sensor network devices, especially, have limited numbers of gates available
for security, and are often highly constrained with the power drain on the
device.

So AES is typically a non-starter for many embedded devices. In light-
weight cryptography, we often thus see: smaller block sizes (typically 64 bits
or 80 bits); smaller keys (often less than 90 bits); and less complex rounds
(and where the S-boxes often just have 4-bits).

For light-weight cryptography the main constraints that we have are
typically related to power requirements, gate equivalents (GEs), and timing.
With passive RFID devices, we do not have an associated battery for the
power supply, and where the chip must power itself from energy coupled

from the radio wave. An RFID device is thus likely to be severely constrained in the power drain associated with any cryptography functions, along with being constrained for the timing requirements and for the number of gates used. Even if an RFID device has an associated battery (active RFID), it may be difficult to recharge the battery, so the drain on power must often be minimised.

There is thus often a compromise between the cryptography method used and the overall security of the method. Thus often light-weight cryptography methods balance performance (throughput) against power drain and GE, and do not perform as well as main-stream cryptography standards (such as AES and SHA-256). Along with this the methods must also have a low requirement for RAM (where the method requires the usage of running memory to perform its operation) and ROM (where the method is stored on the device). In order to assess the strengths of various methods we also define the physical area that the cryptography function will use on the device – and which is defined in μm^2.

9.2 Light-Weight Symmetric Methods

One of the first methods to show promise for a replacement for AES for light-weight cryptography was PRESENT [2]. It uses smaller block sizes and has the potential for smaller keys (such as for an 80-bit key). PRESENT uses either an 80-bit (10 hex characters) or a 128-bit encryption key (16 hex characters). It operates on 64 bit block and uses an SPN (substitution-permutation network) method. With SPN, as with AES (Rijndael), we operate on blocks of plaintext and apply a key and then use a number of rounds which we use substitution boxes (S-boxes) and permutation boxes (P-boxes). The operations used are typically achieved through XOR/bitwise rotation, and parts of the key are introduced though the rounds of operation. The decryption process is then the reverse of the encryption rounds, and the S-boxes/P-boxes are reversed in their operation.

Within Figure 9.1 we see an example of a single round, and where 8-bits of data is entered, and then EX-ORed with the first eight bits of the key. Next the output from this operation is fed into an S-box which maps in the inputs to the output (for example 0x0 will be mapped to 0xC). After this we feed the output into a P-box which will scramble the bits in a defined way. The output of this is then fed into the next round, and which will follow the same process, but this time our input is from the previous round, and from the next eight bits of the key.

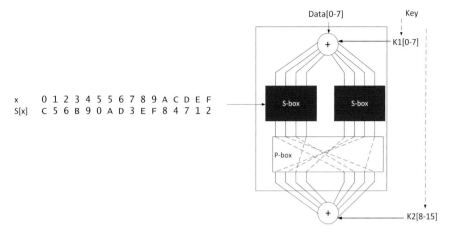

x   0 1 2 3 4 5 5 6 7 8 9 A C D E F
S[x]   C 5 6 B 9 0 A D 3 E F 8 4 7 1 2

Figure 9.1   SPN method.

An S-box substitutes a small block of bits (the input of the S-box) by another block of bits (the output of the S-box). This substitution should be one-to-one, to ensure invertability (hence decryption). In particular, the length of the output should be the same as the length of the input (in Figure 9.1 it has S-boxes with 4 input and 4 output bits), which is different from S-boxes in general that could also change the length, as in DES (Data Encryption Standard), for example. An S-box is usually not simply a permutation of the bits. Rather, a good S-box will have the property that changing one input bit will change about half of the output bits (or an avalanche effect). It will also have the property that each output bit will depend on every input bit.

📖 **Web link (PRESENT):** http://asecuritysite.com/encryption/present

Within PRESENT, we take a block of 64 bits and apply an 80-bit or a 128-bit key. Overall it has 32 rounds (Figure 9.2), which is made up of: a round key operation; an S-box layer; and a P-box layer. The key round operation takes part of the key and EX-ORs it with the data input into the round. It then operates on 4x4 bit S-boxes, and which considerably cuts down on processing power (Figure 9.3). In AES we map for 16 bit inputs to 16-bit outputs (0x00 to 0xFF) but for PRESENT we have 4-bit values and which map onto 16 output values (0x0 to 0xF). In Figure 9.4 we see the permutation of the bits for inputs of 64 bits, so that Bit 1 is mapped to Bit 16. The output from the layer provides the input to the next layer.

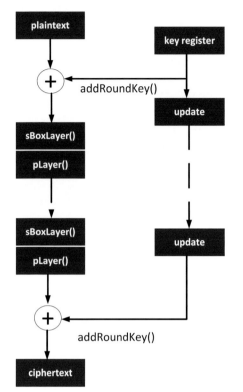

```
generateRoundKeys()
for i=0 to 31 do
    addRoundKey(STATE,Kᵢ)
    sBoxLayer(STATE)
    pLayer(STATE)
end for
addRoundKey(STATE,K₃₂)
```

Figure 9.2 PRESENT method [2].

x	0 1 2 3 4 5 5 6 7 8 9 A C D E F
S[x]	C 5 6 B 9 0 A D 3 E F 8 4 7 1 2

Figure 9.3 sboxlayer mapping [2].

i	0	1	2	3	4	5	6	7	8	9	10	11	12	13	14	15
$P(i)$	0	16	32	48	1	17	33	49	2	18	34	50	3	19	35	51
i	16	17	18	19	20	21	22	23	24	25	26	27	28	29	30	31
$P(i)$	4	20	36	52	5	21	37	53	6	22	38	54	7	23	39	55
i	32	33	34	35	36	37	38	39	40	41	42	43	44	45	46	47
$P(i)$	8	24	40	56	9	25	41	57	10	26	42	58	11	27	43	59
i	48	49	50	51	52	53	54	55	56	57	58	59	60	61	62	63
$P(i)$	12	28	44	60	13	29	45	61	14	30	46	62	15	31	47	63

Figure 9.4 pLayer mapping [2].

Another contender for light-weight cryptography is the super-fast XTEA method. XTEA (eXtended TEA) is a block cipher which uses a 64-bit block size and a 64-bit key. It was designed by David Wheeler and Roger Needham at the Cambridge Computer Laboratory, and part of an unpublished technical report in 1997. The amazing thing about XTEA is that it does its operations with just a few lines of code:

```
#include <stdint.h>

/* take 64 bits of data in v[0] and v[1] and 128 bits of key[0] - key[3] */

void encipher(unsigned int num_rounds, uint32_t v[2], uint32_t const key[4]) {
    unsigned int i;
    uint32_t v0=v[0], v1=v[1], sum=0, delta=0x9E3779B9;
    for (i=0; i < num\_rounds; i++) {
        v0 += (((v1 << 4) ^ (v1 >> 5)) + v1) ^ (sum + key[sum & 3]);
        sum += delta;
        v1 += (((v0 << 4) ^ (v0 >> 5)) + v0) ^ (sum + key[(sum>>11) & 3]);
    }
    v[0]=v0; v[1]=v1;
}
void decipher(unsigned int num_rounds, uint32_t v[2], uint32_t const key[4]) {
    unsigned int i;
    uint32_t v0=v[0], v1=v[1], delta=0x9E3779B9, sum=delta*num_rounds;
    for (i=0; i < num_rounds; i++) {
        v1 -= (((v0 << 4) ^ (v0 >> 5)) + v0) ^ (sum + key[(sum>>11) & 3]);
        sum -= delta;
        v0 -= (((v1 << 4) ^ (v1 >> 5)) + v1) ^ (sum + key[sum & 3]);
    }
    v[0]=v0; v[1]=v1;
}
```

📖 **Web link (XTEA):** http://asecuritysite.com/encryption/xtea

Other block ciphers, too, are now being called back from retirement, including RC5, as they have proven to be fairly simple in their operation, but relatively secure. The great thing about RC5 is that it has a variable block size (32, 64 or 128 bits), and has key sizes from 0 to 2,040 bits. Along with this, it can have from 0 to 255 rounds. When it was first created, the recommended implementation was a block size of 64 bits, a 128-bit key and 12 rounds, but, in an IoT world, this can be optimised to the device.

For light-weight crypto, the NSA released SIMON in 2013, and which was optimized for hardware implementations. It has key sizes of 64, 72, 96, 128, 144, 192 or 256 bits, and block sizes of 32, 48, 64, 96 or 128 bits:

Web link (SIMON): http://asecuritysite.com/encryption/simon

and for SPECK (which is optimized for software implementations):

📖 **Web link (SPECK):** http://asecuritysite.com/encryption/speck

9.3 Light-Weight Hashing

While we will have 32-bit or 64-bit processors in our mobile phones and desktops, and have much more the 1GB of memory, in an IoT world we often measure memory capacity in just a few KiloBytes, and where 8-bit processors are common. The cost of a simple 8-bit processor can be defined in 10s of cents, compared with hundreds of dollars for our complex processors. And so our crypto hash functions for MD5 and SHA-1, and most of our other modern hash methods, are just not efficient for IoT devices. NIST have thus recommended new hashing methods such SPONGENT, PHOTON, Quark and Lesamnta-LW. These methods produce a much smaller memory footprint, and have a target an input of just 256 characters (whereas typical hash functions support up to 2^{64} bits).

SPONGENT uses the sponge function (Figure 9.5) [3]. With the sponge construction, we use a fixed-length permutation (or transformation) and a padding rule. This construction thus takes a variable length input and maps it to a variable-length output. The input is $(Z2)^*$ of any length and then converts it into $(Z2)^n$, where n is defined as part of the process. Overall the method uses a finite-state machine process, and iterates through the states with the addition

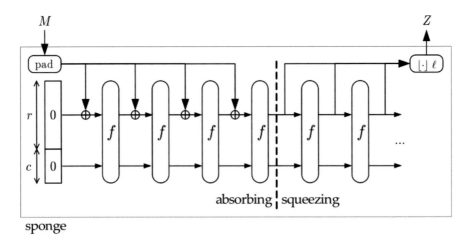

Figure 9.5   Sponge function [3].

of the input data. The concept of sponge function was created Bertoni, who created Keccak [4]. They can use either use a publicly known unkeyed permutation (P-Sponge) or with a random function (T-Sponge). Along with their usage in hashing, they can also be used in creating stream ciphers.

The sponge construction uses a function f which has a variable-length input and a defined output length. It operates on a fixed number of bits (b) – the width. The sponge construction then operates on a state of $b=r+c$ bits. r is defined as the bitrate and c as the capacity (Figure 9.5). Initially an input string is padded using a reversible padding rule (such as adding NULL characters), and then segmented into blocks of r bits. Next the b bits of the state are set to zero, and the sponge construction next defines:

- **Absorbing phase**. This is where the r-bit input blocks are X-ORed into the first r bits of the state, interleaved with application of the function f. After all the input blocks have been processed, we then move to a squeezing phase.
- **Squeezing phase**. This is where the first r bits of the state are outputted as blocks and, interleaved with the function f. The number of bits of the output are defined as part of the process.

Overall the last c bits of a state are never changed by the input blocks and never outputted within the squeezing phase. For an 88-bit hash we have (SPONGENT-088-080-00 - Spongent-88/80/8: n=88 bits, b=88 bits, c=80 bits, r=8 bits, R=45) and for 128-bit (SPONGENT-128-128-008 - Spongent-128/128/8: n=128 bits, b=136 bits, c=128 bits, r=8 bits, R=70).

 📖 **Web link (SPONGENT):** http://asecuritysite.com/encryption/spongent

Lesamnta-LW which uses AES methods as its core. One thing to notice about Lesamnta-LW is that the S-box structure is the same as you would find in AES. The authors think that it only requires 8.24 kGates, and has a throughput of 125Mbit/sec (which is five times faster than SHA-256, which also gives a 256-bit hash): For the RAM requirements on an 8-bit processor, the authors estimate that Lesamnta only requires 50 bytes of RAM.

 Web link (Lesamnta-LW): http://asecuritysite.com/encryption/lw

Quark is defined in three main methods: u-Quark, d-Quark, and s-Quark, and uses a sponge function. It can be used for hashing and in stream encryption. u-Quark has the lowest footprint and provides 64-bit security on 1,379 digital gate, whereas s-Quark provides 112-bit security.

📖 **Web link (Quark):** http://asecuritysite.com/encryption/quark

PHOTON is light-weight cryptography method for hashing and is based on an AES-type approach. It can create 80-bit, 128-bit, 160-bit, 224-bit and 256-bit hashes [5]. It takes an arbitrary-length input and produces a variable-length output. The method is defined as PHOTON-n-r-r', where *n* is the hash size, *r* is the input bit rate, and r' is the output bit rate. A sample list of hashed values for "abc" are:

```
Photon 80 signature (PHOTON-80/20/16)
("abc") = 3151cb8f09f5a4908531
------------
Photon 128 signature (PHOTON-128/16/16)
("abc") = e1bb314c7c9ace3ea0ed6fd1d762d216
------------
Photon 160 signature (PHOTON-160/36/36)
("abc") = c11d4cd3da84bc245430ba7cf696d0092941ba58
------------
Photon 224 signature (PHOTON-224/32/32)
("abc") = 7798abbae697af77eaa56f358ec9845ee947c6d3c7daca9e7ae476ec
------------
Photon 256 signature (PHOTON-256/32/32)
("abc") = c412435e329f6f4837a5e55eda83d66d8a8eae5d9744931f9c7cbb7e55584df6
------------
```

Web link (Photon): http://asecuritysite.com/encryption/photon

The internal state is defined as *t* (bits), and is calculated as $t=c+r$ (where *c* is the capacity). With PHOTON we use a sponge function and where we take input bits and XOR with bits taken from the current state. Overall there are three main phases:

- **Initialisation**. This phase takes the input bit stream and breaks into *r* bits (and pads if required).
- **Absorbing**. In this phase, for all the message blocks, we take *r*-input bits and XOR with *r* bits of the state, and interleave with a *t*-bit permutation function.
- **Squeezing**. In this phase, we extract *r* bits from the current internal state, and apply a permutation function (P) to it. This will continue until the number of output bits is equal to the required hash size.

The internal permutation function (P) is similar to AES with 12 rounds, and which each round has the functions of (Figure 9.6):

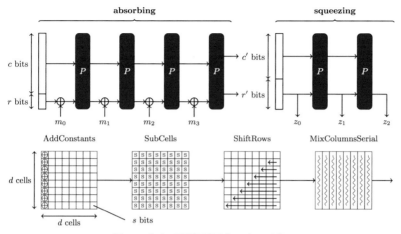

Figure 9.6 PHOTON functions [5].

- **AddConstants**. In this function, the first column with the internal state is XOR-ed with round (r) and internal constants.
- **SubCells**. In this function, the internal state is fed through the PRESENT S-box (Figure 9.3).
- **ShiftRows**. In this function, the internal state cell row [i] is cyclically shifted by i positions to the left.
- **MixColumnsSerial**. In this function, the internal state cell column is multiplied by the MDS (Maximum Distance Separable) matrix.

9.4 Other Light Weight Ciphers

Chaskey Cipher is light-weight cryptography method for signing messages (MAC) using a 128-bit key. The hardware implementation only requires 3,334.33 gate equivalent with an operating clock frequency of 1 MHz. With SHA-256 we need around 15,000 gates, while Keccak (SHA-3) requires 4,658 gates.

```
Message:                  hello
Key (128 bits - 32 hex):  BD63710BAF4753D0367DBF6A875ACAAB

Signature:                db6a554716651bc3a818e0c1d01d582d
Encrypt (CBC):            18c381d3811319c24af6cd71af70f97f
```

📖 **Web link (Chaskey):** http://asecuritysite.com/encryption/chas

Rabbit is a light-weight stream cipher and was written by Martin Boesgaard, Mette Vesterager, Thomas Christensen and Erik Zenner. It creates a key stream from a 128-bit key and a 64-bit initialization vector.

📖 **Web link (Rabbit):** http://asecuritysite.com/encryption/chas

Mickey V2 is a light-weight stream cipher and was written by Steve Babbage and Matthew Dodd. It creates a key stream from an 80-bit key and a variable length initialization vector (of up to 80 bits). The keystream has a maximum length of 2^{40} bits.

Web link (Mickey): http://asecuritysite.com/encryption/mickey

Trivium is a light-weight stream cipher and was created Christophe De Cannière and Bart Preneel, and has a low footprint for hardware. It uses an 80-bit key, and generates up to 2^{64} bits of output, with an 80-bit IV.

Web link (Trivium): http://asecuritysite.com/encryption/trivium

Grain is a light-weight stream cipher and was written by Martin Hell, Thomas Johansson and Willi Meier. It has a relatively low gate count, power consumption and memory. It has an 80-bit key, and has two shift registers and a non-linear output function.

Web link (Grain): http://asecuritysite.com/encryption/grain

CLEFIA is a light-weight block cipher and was written by Taizo Shirai, Kyoji Shibutani, Toru Akishita, Shiho Moriai, and Tetsu Iwata, and can be implemented with 6K gates. It was defined by Sony, and has 128, 192 and 256 bit keys, and 128-bit block sizes. Along with this it is included in ISO/IEC 29192 International Standard for a lightweight block cipher method (ISO/IEC 29192-2:2012).

Web link (CLEFIA): http://asecuritysite.com/encryption/clefia

Enocoro is a light-weight stream cipher and was defined by Hitachi. It has a 128-bit key and a 64-bit IV value. Along with this it is included in ISO/IEC 29192 International Standard for a lightweight stream cipher method (ISO/IEC 29192-3:2012).

Web link (Enocoro): http://asecuritysite.com/encryption/enocoro

RC5 ("Ron's Cipher 5"), created in 1994 by Ron L. Rivest, also shows great potential for a light-weight cryptography method. It is a block cipher which has a variable block size (32, 64 or 128 bits), a variable number of rounds, and a variable key size (0 to 2,048 bits). It can thus be used to match the encryption to the capabilities of the device. If it is a low-powered device with a limited memory and a relatively small physical footprint, we could use a 32-bit block size and an 80-bit key, with just a few rounds. But we can ramp up the security if the device can cope with it, and use 128-bit block sizes and a 128-bit key. It can also be flexible, where a single change on either side can improve or reduce the requirements.

The flexibility around the key size, block size and rounds, supports a range of design choices, in a way that AES struggles with. AES, for example, uses relatively large key sizes of 128 bits, 192 bits and 256-bits, with 128-bit block sizes. It also has a fixed number of rounds depending on the key size, such as 10 rounds for 128-bit encryption, 12 rounds for 192-bit encryption, and 14 rounds for 256-bit encryption. These requirements, for an IoT device, often consume considerable amounts of memory and processing resource, and could have a significant effect on the power consumption, draining the battery resource. The following uses RC5/32/12/16 (32-bit blocks, 12 rounds and 16-byte key – 128 bits):

 📖 **Web link (RC5):** http://asecuritysite.com/encryption/rc5

For light-weight cryptography PHOTON, SPONGENT and Lesamanta-LW are defined as standards for hashing methods within ISO/IEC 29192-5:2016, PRESENT and CLEFIA for block methods within ISO/IEC 29192-2:2012, and Enocoro and Trivium for stream methods within ISO/IEC 29192-3:2012.

Our normal public key methods do not quite work on RFID devices, so let's look at the proposed method for proving that a RFID device is real. The method proposed by the ISO/IEC is ISO/IEC 29192-4:2013/Amd 1 is named ELLI (Elliptic Light). It uses Elliptic Curves along with a Diffie-Hellman related handshake between the RFID tag and the RFID reader [6].

Within Elliptic Curve we start with a point on a curve (P) which is known. Then we multiply this point with a large number (ε) to produce another point (A) on the curve:

$$A = \varepsilon\, P$$

and where A will be the public key, and ε is the private key. If ε is large enough, it is then difficult to compute ε even though we have A and P. Now let's look at the basics of ELLI. For this RFID tag contains a random value of ε (the private key), and the RFID reader generates a random value of λ. On creating the RFID tag we calculate (Figure 9.7):

$$B = \varepsilon\,P$$

along with the signature of B which has been signed by a key that the RFID reader can validate. Thus the tag contains: [ε, B, PublicKeySign(B)]. Each time the RFID reader wants to validate the tag it takes its random value (λ) and computes:

$$A = \lambda\,P$$

Next the RFID reader sends A to the RFID tag. The RFID tag then multiplies the value of A by its private key (ε) to get C:

$$C = \varepsilon\,A$$

It then sends back its public key (B), the value of C and the signature of the public key which the reader can verify. The reader then computes D:

$$D = \lambda\,B$$

and compares C and D. If they are the same we have verified the private key. This is true as:

$$C = \varepsilon\,(A) = \varepsilon\,(\lambda\,P)$$
$$D = \lambda\,(B) = \lambda\,(\varepsilon\,P)$$

It is secure as it uses the Elliptic Curve Diffie Hellman Problem (ECDHP). If Eve wants to produce a fake RFID tag she receives the challenge of:

$$A = \lambda\,P$$

and now must return a valid response (C), along with a public key which has been signed by an authority. Since Eve only has A and B, so she cannot compute a valid response for C as she does not know λ and ε, in order to compute:

$$\lambda.\varepsilon.P$$

📖 **Web link (ELLI):** http://asecuritysite.com/encryption/elli

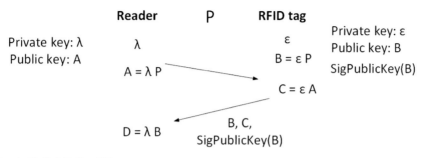

Figure 9.7   Abstraction of the ELLI method.

9.5 Secret Shares

While most of our cryptography uses key-based cryptography, there is also a move toward keyless systems, where we can store data without the usage of keys. So, let's say we have we have two pirates and they need to define the route to get to the buried treasure. Now they think back to Pirate School and remember that they should each put down a marker, along the route, so that both markers are in a line from the start point to the treasure. They can then get together and put their marker down in each place, and then they follow the route and find the treasure (Figure 9.8). This works with more pirates, but now only two pirates are needed to get together to reveal the treasure.

As we know being a pirate is a risky business, and they might not make it back to the treasure island. So, let's say we now set up a Pirate's Co-op, where the pirates share their booty. In their first meeting we now have three pirates and we need to define a route to the treasure. If we want any two from three pirates, we basically just add another point on the straight line, so that two pirates can get together. That works fine, but not very scalable.

So, if they don't trust each other, and are willing to take a risk of one pirate not returning to the island, they must now define a 2nd order equation to show the way. Let's say it is $4x^2+6x+1$, and each pirate takes a point on the route: (9,331), (6,151) and (3,43). So only when the three pirates get together, they put their markers down, and then can find the route to the treasure (Figure 9.9).

This is the basics of an amazing little method which allows shares to be distributed, without revealing the original data. It provides us with a way to

Figure 9.8   Treasure map.

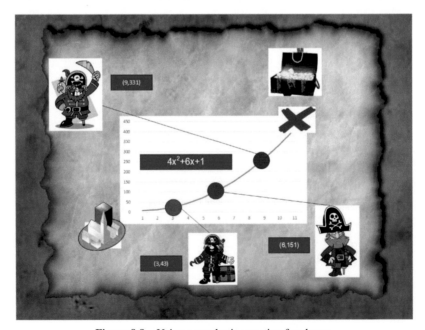

Figure 9.9   Using a quadratic equation for shares.

create keyless encryption. In 1979, Adi Shamir (who represents the "S" in RSA) created a secret sharing algorithm which allows a secret to be split into parts, and only when a number of them are added together will the original message be created. In these times when we need to integrate trust, his algorithm has many application areas.

So let's take an example ... let's say that there are six generals who have control over firing a missile, and there are three bases, with two generals on each base. Unfortunately we are worried that one of the generals might make a rash decision, so we agree that the generals will not get the secret password to fire the missile. We are also worried that a base could be taken over by a malicious force, so we agree that no two generals will be able to gain the password. So to overcome these problems we decide that a least three generals must agree together to generate the correct password to fire the missile.

Obviously, if this was a real scenario, we would create a password which would change over time, where we generate a one-time password, which cannot be used again (just in case!). For this we can use either a time-based mechanism, where the password is only relevant for a certain time window, such as with Time-based One Time Password (TOTP). Otherwise we can use an original seed password, which then changes each time we access it, where only those who know the original seed will be able to generate the next password. For this we can use a counter-based system such as for the Hash-based One Time Password (HOTP). With these schemes the same password would not be generated for each consecutive access, so that even when they had generated their password, they could only use it for a certain amount of time (with TOTP) or would differ for the next access (with HOTP).

In Shamir's secret sharing scheme, the number of generals can be represented by the number of shares, and the number generals who are required to generate the secret is represented by the threshold. Thus, in this example, we have a share value of six, and a threshold value of three. So we give out one of the shares to each of the generals, so that none of them have the same one. We will then require three generals to put together their shares in order for them to generate the password for the missile.

📖 **Web link (Shamir):** http://asecuritysite.com/encryption/shamir

So for a password of "Fire The Missile", with a share of six, and threshold of three we get:

```
000if30kGfSjyDNilwU0Tyz5awf2uk=
001PoYDi8eMEoMI6Zkbkn2n1GABoas=
002HxjSfFEEakXLqUnfGolr27XG3Ws=
003qGMlZ/Fa9+YOyozQWch/6nnYpik=
004RHyZDtPQP9/yXgv4cojC6Zg6Ets=
0058wduFXOOonw3Pc73McnW2FQkaZk=
```

for which each of these can be distributed to each of the generals. When three of them combine their codes, the result will be:

```
Trying a share of 1: %ýôꞁ gÒꞁ ÍŠ \Ñ<³å¬Úé
Trying a share of 2: ?D·CümQFH¤Ú¼A7™r¹Ì
Trying a share of 3: Fire The Missile
```

The basic theory relates to the number of points within a mathematical equation, that are required to reveal what the equation is (and thus determine the secret). For example if we have a secret of 15, and can only be revealed when two people combine their information. For this we thus need a linear equation, such as:

$$y = 2x + 15$$

The two pieces of information that could be generated to reveal the equation would thus be two points on this line, such as:

$$(0,15) \text{ and } (1,17)$$

If we only know one point, such as (1,17), we cannot determine the equation of the line, but knowing two points we can determine the gradient:

$$m = \frac{17 - 15}{1 - 0} = 2$$

and from here we can determine the point it cuts the y-axis (c):

$$c = y - mx = 17 - 2 \times 0 = 15$$

and thus we have the equation ($y = 2x + 15$), where the secret is 15. In this example we have only two shares, but if we require three shares we need a parabola, such as:

$$y = 2x^2 + 5x + 15$$

and share three points on the equation to generate the secret. If we require four shares then a cubic curve is required, and so on. In most situations, it would be too processor intense to encrypt data with the secret share method, especially if we had a complete equation (such as using an any 8 from 10 share), so we often use symmetric encryption (such as for AES or ChaCha) to encrypt the data, and then create a share of the key. For example, if we had an 8 from 10 secret share policy, we would generate a 256-bit encryption key and then encrypt the data. Next we would then take the 256-bit AES key and create a secret share where 8 of the 10 shares can come together to recreate the key.

📖 **Web link (Shamir decode):** http://asecuritysite.com/encryption/shamir_decode

9.6 Post Quantum Cryptography

The race for a true supercomputer is on, but there's one contender which could trump them all, and leave them all behind. With the Cloud increasingly providing massively parallel systems, there is one type of computer which will cause shock waves around the world, and change everything that we know ... the quantum computer. In quantum computing, a *qubit* (or quantum bit) is equivalent to a bit in classical computer systems, and relates to the polarization of a single photon (vertical or horizontal polarization – equivalent to a 1 or a 0). In quantum mechanics, a qubit can be a superposition of both states at the same time, which allows for ultra-fast computing.

Many companies, including Google and IBM, are advancing the quantum computers, and could operate be over 100 million times faster than existing processors. The core of providing identity is focused on public key, where one key (the private key) is used sign an entity, and the other key (the public key) is used to verify the signature. One of the most popular methods used is RSA, and which generates its keys based on the multiplication of two prime numbers.

At present, it is fairly difficult for computers to determine the two prime numbers that were used to create the modulus (N). But an important application of quantum computing will be in factorizing numbers for primes number. Also with this quantum computers would be able to crack discrete logarithm methods, such Elliptic Curve and ElGamal. For RSA cracking, a key focus is likely to be on Shor's algorithm which is a quantum algorithm to determine the prime number factors of a number (N). Overall the time taken is logarithmically related to the value of N, which significantly reduces the processing time.

Quantum computers have fast multiplication circuits, and thus can be used to perform multiplications and search a range of prime numbers at a speed which would break most existing RSA implementations. If the modulus (N) is cracked for prime numbers, it is not too difficult to determine the decryption key that is associated with the public key.

NIST has now started to move on this with a paper outlining the requirement for new cryptography methods [7], as a large-scale quantum computer will break most of the currently available public key systems, including RSA, Elliptic Curve and ElGamal. Along with providing signatures, public key is often used to protect symmetric keys using in symmetric encryption (such as with AES or ChaCha). A large-scale breach of public key methods would thus lead to a complete lack of trust on the Internet.

The main contenders for quantum robust methods are:

- Lattice-based cryptography – This classification shows great potential and is leading to new cryptography, such as for fully homomorphic encryption, and code obfuscation.
- Code-based cryptography – This method was created in 1978 with the McEliece cryptosystem but has barely been using in real applications. The McEliece method uses linear codes that are used in error correcting codes, and involves matrix-vector multiplication. An example of a linear code is the Hamming code.
- Multivariate polynomial cryptography – These focus on the difficulty of solving systems of multivariate polynomials over finite fields. Unfortunately, many of the methods that have been proposed have already been broken.
- Hash-based signatures – This would involve created digital signatures using hashing methods. The drawback is that a signer needs to keep a track of all the messages that have been signed, and that there is a limit to the number of signatures that can be produced.

McEliece Method

In just two pages, Robert McEliece, in 1978, outlined a public key encryption method based on Algebraic Coding – now known as the McEliece Cryptography method. It was an asymmetric encryption method (with a public key and a private key), and, at the time, looked to be a serious contender for a trapdoor method. Unfortunately for the method, RSA became the King of the Hill, and the McEliece method was pushed to the end of the queue for designers. It has basically drifted in the 38 years since. But, as an era of quantum computers is dawning, it is now being reconsidered, as it is seen to be immune to attacks using Shor's algorithm.

The McEliece method uses code-based cryptography. Its foundation is based on the difficulty in decoding a general linear code, and is generally faster than RSA for encryption and decryption. Within it we have a proba-bilistic key generation method, which is then used to produce the public and the private key. The keys are generated with the parameters of *n*, *k* and *t*. With this we create an [*n,k*] matrix of codes, and which is able to correct *t* errors.

In his paper, McEliece defines *n*=1024, *k*=524, *t*=50, and which gives a public key size of 524 × (1024-524) = 262,000 bits. In example given next we use *k*=1608, *N*=2048 and *t*=40, which gives a public key size of 1608 × (2048-1608) – 707,520 bits (88,440 bytes – 86 kB). We thus end up with a public key which is around 86 kB long. For quantum robustness, it is recommended that *N* is 6960, *k* is 5,412 and *t* is 119, and which would give a key size of 8,373,911 bits – around 1,046,738 bytes (approx. 1MB). The original work did not provide support for a signing using one of the keys, but this has now been addressed with the method.

📔 **Web link (McEliece):** http://asecuritysite.com/encryption/mce

A sample run is for a message of "To Be Or Not To Be" is:

```
Public key: K: 1608, N: 2048, T:40, Name:MPKCPublicKey
Private key: K: 1608, N: 2048, T:40, Name:MPKCPrivateKey
Message: To Be Or Not To Be.
Cipher: 434553df9ad6441afa012ba209b3489a236044f0b9f2af653641ac6a7b75
7a447288ad2f5904dd90c10069cf474f49b763f908b525b3d4e5c3145e4eebabc2a2
7f04f0739f005e7e84b6f96b2258dd73d2625507d4f32cfc39ca69a77c340fa77e57
ffb82d42e8dca9b6b2b93cc2fd9592266105f35903b30f7767c27e3b60e2b49be812
fee642a012c587abe11d09703423805625e55437c68ef579860ef8713ddea9927c66
e62487a18f2ba3e8cc34a33e389785b3e435d0443330572d9c5c6f10a010944a10c5
ae2dc97802b3a37155cbf4fd0b79c377730468256515ec442752737b30c8e283c02c
fdb3e36ca7f8c3543eb95926337b8a31385251875304ab37190c740ab7814d0cd36a
facd624013a6d9
Decrypt: To Be Or Not To Be.

Signing (only first 50 hex chars shown):
583ec5766fc3ca22a0b67851997593833405c74ac14cd2985a
- Sign and verify test worked.
- Private key comparison works.
```

Lattice methods

Lattice-based cryptography uses asymmetric cryptographic primitives based on lattices. It has been known about for several decades, and is now being investigated because of its quantum robustness, whereas many of the existing public key methods such as RSA and Diffie-Hellman cryptosystems can be broken with quantum computers. The following uses the NTRU (Nth degree TRUncated polynomial ring) open source public-key cryptosystem and where we generate the public and private key from N, p and q:

📖 **Web link (Lattice):** http://www.asecuritysite.com/encryption/lattice

Bob and Alice agree to share N, p and q, and then Bob generates two polynomials (f and g), and generates his key pair from this. Alice receives this, and she generates a random polynomial, and encrypts some data for Bob. Bob then decrypts the message with his private key.

Overall Lattice methods are faster than RSA, and can compete with symmetric key encryption. It also has a low memory footprint and are well matched to mobile devices and smartcard applications, and can support different levels of security:

- Moderate Security: n=167, p=3, q=128.
- Standard Security n=251, p=3, q=128.
- Standard Security n=347, p=3, q=128.
- Highest Security n=503, p=3, q=128.

A sample run is:

```
==== Bob generates public key =====
Bob picks two polynomials (g and f):
f(x)=  [1, 1, -1, 0, -1, 1]
g(x)=  [-1, 0, 1, 1, 0, 0, -1]
d   = 2

Bob's Public Key: [21L, 118L, 11L, 29L, 66L, 60L, 26L, 52L, 11L, 35L,
70L, 119L,  38L, 105L, 96L, 58L, 85L, 40L, 36L, 94L, 97L, 80L, 13L,
125L, 19L, 89L, 119L, 82L, 59L, 93L, 124L, 60L, 41L, 53L, 19L, 98L,
30L, 80L, 44L, 115L, 117L, 48L, 33L, 86L, 77L, 68L, 122L, 127L, 114L,
4L, 36L, 101L, 50L, 69L, 13L, 121L, 97L, 43L, 126L, 25L, 77L, 22L,
10L, 39L, 23L, 89L, 50L, 68L, 94L, 40L, 15L, 43L, 126L, 119L, 110L,
37L, 28L, 2L, 17L, 40L, 96L, 46L, 65L, 4L, 117L, 93L, 43L, 117L, 39L,
54L, 63L, 65L, 48L, 32L, 25L, 9L, 127L, 122L, 126L, 108L, 8L, 95L,
45L, 32L, 41L, 78L, 41L, 24L, 26L, 35L, 82L, 64L, 20L, 53L, 14L, 21L,
77L, 105L, 61L, 51L, 66L, 13L, 9L, 122L, 30L, 39L, 115L, 37L, 114L,
60L, 2L, 108L, 99L, 83L, 86L, 103L, 102L, 113L, 120L, 10L, 109L, 40L,
```

```
76L, 110L,  65L, 104L, 125L,  13L,  67L, 113L, 103L,  26L,   3L,  69L,  52L,
68L,  89L,  45L,  27L,  34L,  14L, 104L,  20L,  91L,  37L,  16L,  65L]
```

```
==== Alice generates public key =====
Alice's Original Message    :  [1, 0, 1, 0, 1, 1, 1]
Alice's Random Polynomial   :  [-1, -1, 1, 1]
Encrypted Message           :  [30L,  82L,  0L,  41L, 103L, 127L,  28L,
16L,  69L,  96L,   2L,  83L, 100L,  10L, 124L,  95L,  46L,  87L,  73L, 113L,
39L, 115L,  38L, 117L, 103L,  90L,  64L, 105L,  73L,  19L,  28L,  32L,  92L,
14L,  87L,  59L,  88L,  21L,  12L, 109L,  60L, 110L,  69L,  10L,  10L,  50L,
47L,  72L, 103L,   9L,  91L,  71L,  51L,  54L,  79L,  83L, 104L, 110L,  19L,
95L,  73L,  28L,  82L,  22L,  38L,  67L,  25L, 110L,  59L,  80L,  65L, 100L,
42L,  79L,  76L,  38L, 108L,  95L,  10L,  47L,  33L,   1L,  75L,  91L,  98L,  89L,
83L,  22L,  68L,  73L, 117L,  23L,  12L,  16L,  40L,  10L,  19L, 123L,  48L,
45L,  12L,   9L,  56L,  78L,  73L,   2L, 118L,  34L,  79L,  12L,  55L,   1L,  99L,
91L,  51L, 114L,  35L,  71L,  52L,  82L,  19L,  99L,  29L, 100L, 122L,  58L,
122L,   7L,   9L,  62L,  11L,  64L,  77L,  40L, 114L, 107L,  20L,  50L,  44L,
127L,  86L,  71L,   9L,  17L,  79L,  51L,  94L,  93L,  63L,   2L, 104L,  25L,  49L,
43L, 108L, 112L,  20L,  86L, 127L,  91L,  72L,  85L,  28L,  21L, 116L,
46L,  13L]
```

```
==== Bob decrypts =====
Decrypted Message           :  [1L, 0L, 1L, 0L, 1L, 1L, 1L]
```

Unbalanced Oil and Vinegar

The multivariate polynomial problem is now being applied in quantum robust cryptography, where we create a trap door to allow us to quickly solve for n variables with m equations (which are multivariate polynomials). One scheme is the Unbalanced Oil and Vinegar (UOV) scheme and was created by J. Patarin et al [8]. The signature is created using a number of equations:

$$y_1 = f(x_1, x_2 ... x_n)$$
$$y_2 = f(x_2, x_2 ... x_n)$$
$$\cdots$$
$$y_m = f(x_1, x_2 ... x_n)$$

where $y_1, y_2, ... y_m$ is the message that is to be signed, and where $x_1, x_2 ... x_n$ is the signature for the message. So a simple example:

$$5x + 4y + 10w + 9z = 99$$
$$6x + 3y + 2w + 3z = 38$$

$$8x + 2y + 7w + z = 51$$
$$x + 9y + 4w + 6z = 57$$

In this case, the message is 99, 38, 51 and 57, and the signature is 5, 4, 10 ... 6. Some sample code and a sample run is given here:

📖 **Web link (UOV):** http://asecuritysite.com/encryption/rain

A sample run is:

```
Random:3o8s7d8kr34j9105p5gt2upumj
----Private Key----
Doc length:27
B1:
 92 203 100 225 86 248 29 40 185 146 73 116 217 117 163 220 253 114
 61 3 209 136 143 210 224 171 158
B2:
 238 24 25 71 64 134 35 12 59 239 175 154 150 55 255 249 121 245 239
 251 163 22 176 238 0 23 41 118 153 231 74 110 43
InvA1:
Dimension: [27][27] (Only displaying first 10)
 222 234 146 190 102 148 36 180 234 179 75
InvA1:
Dimension: [33][33] (Only displaying first 10)
 104 222 98 141 229 251 126 127 104 85 174
Vi: 6 12 17 22 33

----Public Key----
Doc length:27
Coeff Quadratic:
Dimension: [27][561] (Only displaying first 10)
 240 233 216 74 41 169 63 73 161 50 27
Coeff Scalar:
 51 48 139 43 78 22 75 33 87 178 149 98 196 153 135 240 85 76 51 10
 103 1 211 23 232 192 53
Coeff Singlar:
Dimension: [27][33] (Only displaying first 10)
 249 197 70 63 127 209 197 123 222 142 197

----Message and signing----
Message:hello
Signature:e074c938922deff14b4f01cff2e61dc4f1f881adecc21684ef30fe72ea
357d388b
Success in signature
```

The variables used include:

- Doc – Number of polynomials in Rainbow.
- Vi – Number of Vinegar-variables per layer ({6, 12, 17, 22, 33})

- B1 – Translation part of the private quadratic map L1.
- InvA1 – Inverse matrix of A1.
- B2 – Translation part of the private quadratic map L2.
- InvA2 – Inverse matrix of A2.

General Merkle signature scheme

A Merkle tree is a tree that defines each non-leaf node with a value or a label and contains a hash of its children. This builds a hash trees and is used to provide a verification of large-scale data structures.

 📖 **Web link (Merkle tree):** http://asecuritysite.com/encryption/merkle

The General Merkle signature scheme (GMSS) applies a binary tree to provide a signature which is signed by a private key and authenticated with a public key:

 📖 **Web link (GMSS):** http://www.asecuritysite.com/encryption/gmss

A sample run is:

```
Message:hello
Public key:3e0dbaf83d49191905726822dd1581c239f6ca571f13321fb62b4938
Signature Length (hex chars):4992
Signature Length (first 10 hex char):000000005447577a6c91
It Works!
```

9.7 Lab/Tutorial

The lab and tutorial related to this chapter is available on-line at:

http://asecuritysite.com/crypto09

References

[1] K. A. McKay, L. Bassham, M. S. Turan, and N. Mouha, "Report on lightweight cryptography," 2017.
[2] A. Bogdanov, L. R. Knudsen, G. Leander, C. Paar, A. Poschmann, M. J. B. Robshaw, Y. Seurin, and C. Vikkelsoe, "PRESENT: An Ultra-Lightweight Block Cipher," *Cryptogr. Hardw. Embed. Syst. – CHES 2007*, pp. 450–466.

[3] A. Bogdanov, M. Knežević, G. Leander, D. Toz, K. Varici, and I. Verbauwhede, "{SPONGENT}: The Design Space of Lightweight Cryptographic Hashing," 2011.

[4] G. Bertoni, J. Daemen, M. Peeters, and G. Van Assche, "Keccak," Springer, Berlin, Heidelberg, 2013, pp. 313–314.

[5] J. Guo, T. Peyrin, and A. Poschmann, "The PHOTON Lightweight Hash Functions Family," *Crypto*, pp. 222–239, 2000.

[6] M. Braun, E. Hess, and B. Meyer, "Using Elliptic Curves on RFID Tags," *IJCSNS Int. J. Comput. Sci. Netw. Secur.*, vol. 8, no. 2, 2008.

[7] L. Chen, S. Jordan, Y.-K. Liu, D. Moody, R. Peralta, R. Perlner, and D. Smith-Tone, "NISTIR 8105 Draft – Report on Post-Quantum Cryptography," 2016.

[8] A. Kipnis, J. Patarin, and L. Goubin, "Unbalanced Oil and Vinegar Signature Schemes," Springer, Berlin, Heidelberg, 1999, pp. 206–222.

10

Blockchain and Crypto-currency

10.1 Introduction

There is a general move towards the usage of cryptocurrencies, and where electronic coins are kept within an electronic wallet. Often this wallet does not actually contain electronic coins, but holds the public and private key used to support the transfer of the coins from one account to another. The coins, themselves, are often created by a mining process, where currency miners perform some work and are then are rewarded with coins when they succeed in their task. One the most popular crypto-currencies is Bitcoin (BTC). A key focus for the crypto-currency to protect against someone spending money that they do not have, so Bitcoin uses a publicly available ledger of transactions – known as a Blockchain. This allows the Bitcoin network to know the number of bitcoins that a given user has in their account. Someone who then tries to spend more than the number of bitcoins that they have, will not be allowed to complete the transaction.

Within Bitcoins, we have the genesis record, and which relates to the first transaction created. Figure 10.1 shows the very first mined block in Blockchain, and that it was rewarded with 50 BTC, and created on 3 January 2009 at 18:15. It had just one transaction for 50 BTC (Figure 10.2). The nounce value used relates to the value that is required to create a given format of the hash (in this case the genesis block required 10 preceding zeros).

One of the greatest contributions Bitcoins to computer science, though, has been in the creation of Blockchain technology and which allows for the creation of a trustworthy ledger of transactions within a trusted infrastructure. So, while the Blockchain method used with bitcoins is efficient for crypto currency transactions, there are other alternatives, including Ethereum, Ripple, Litecoin, Monero, Ethereum Classic, Dash, Steem, KiloCoin and Augur. With the Ethereum Blockchain, we can support both cryptocurrency transactions (Ether) and also peer-to-peer smart contracts (this concept will be covered later in the chapter).

Summary

Height	0 (Main chain)
Hash	000000000019d6689c085ae165831e934ff763ae46a2a6c172b3f1b60a8ce26f
Previous Block	0000000000000000000000000000000000000000000000000000000000000000
Next Blocks	00000000839a8e6886ab5951d76f411475428afc90947ee320161bbf18eb6048
Time	2009-01-03 18:15:05
Difficulty	1
Bits	486604799
Number Of Transactions	1
Output Total	50 BTC
Estimated Transaction Volume	0 BTC
Size	0.285 KB
Version	1
Merkle Root	4a5e1e4baab89f3a32518a88c31bc87f618f76673e2cc77ab2127b7afdeda33b
Nonce	2083236893
Block Reward	50 BTC
Transaction Fees	0 BTC

Figure 10.1   Genesis record for Bitcoin.

Figure 10.2   The first transaction within Bitcoins.

10.2 Bitcoins, Blockchain and Miners

Bitcoin was created in 2009 by someone known as Satoshi Nakamoto, and borrowed from a whole lot of research methods. Overall it does not require the support of a central government or any organisation to regulate it, nor a broker to manage payments. Conventional currencies usually have a central bank that creates money and then controls its supply. The Bitcoin currency is instead created when users *mine* for it, using their computers to perform complex calculations through special software. The algorithm behind Bitcoin is designed to limit the number of bitcoins that can ever be created. Each miner processes transactions then has a reward, and the reward reduces over time, and which should reduce the supply of the coins. In 2016, the reward for a successful mining process was reduced from 25 BTC to 12.5 BTC. This reward will continue to reduce until the currency is forked (and where new parameters are used), or when we reach a saturation level.

All Bitcoin transactions are thus recorded on a public database known as a Blockchain. Every time someone mines for Bitcoin, it is recorded with a new block that is transmitted to every Bitcoin application across the network, and which is similar to a bank updating its online records. This Blockchain can then be either public, where all the transactions are viewable, or can be built within a private space.

10.2.1 Bitcoin Transactions

Figure 10.3 shows an overview of the creation of the private key and the public key. From a randomly generated 256-bit private, we generate a 512-bit Elliptic Curve public key. In this case Bob's private key is (based on a 256-bit key):

```
5JQdwmJiEAEb3VxRN9oNAokCq7gGSt1JZeGycD4fxRxT2Z1FkiA
```

and his public key is (which has 34 Base-58 characters):

```
1GqdnsdQfMXWjEEdpkhqcL9DhU9aNqRHUH
```

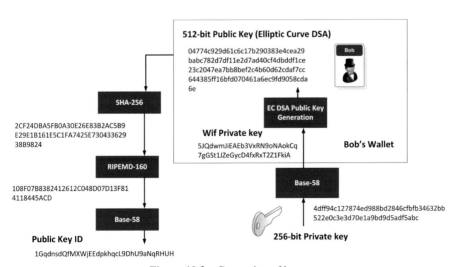

Figure 10.3 Generation of keys.

When we look at a wallet we can view the number of bitcoins in the account and the number of transactions related the account (Figure 10.4). A transaction is then defined as a transfer of bitcoins from one account ID (the public key ID) to another one (Figure 10.5). This transaction is confirmed on the Blockchain by the miners, which normally process all the transaction within a 10 minute time period.

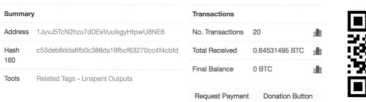

Figure 10.4   Bitcoin wallet.

Figure 10.5   Bitcoin transaction.

Bitcoins generate a 256-bit random key which is converted into a Wif (Wallet Interchange Format) key, and which has a 256-bit private key and a 512-bit public key. It then uses Elliptic Curve Ciphers (ECC) to sign for transactions:

```
Private key:
4c0333a50b7724c71b89df148d83f64d49d896e21701007eeb8cada52744aca2

Public key:
0489fc7b8c3f655a10840d35c76ebb5596694045e49e940fb1e7a759da4edf0fafc
45bbbea6f5a56abf14c145c529c8eda9d3ad606f3a0bf4ca01ce991d4987b97

Wif:    5JPmDetQXXvc5aT5efyrg7BxHbH4135owRzq9DD7n2eWQCta5MN

Address:    16RAf9CjnstWCfBJGfrzSSMfTeHJVt8QWw

Signed:
4830450220264c4dce5f1cf0dff8d32d21c5d5cf6baed428b12ae6f8594924246a61
1e9ee602210096ef8e7054ec7a39f0a35d8de3fd50090b1d125c0e795af8cf3d577b
676407ca01410489fc7b8c3f655a10840d35c76ebb5596694045e49e940fb1e7a759
da4edf0fafc45bbbea6f5a56abf14c145c529c8eda9d3ad606f3a0bf4ca01ce991d4
987b97
```

The WiF address is in an Base-58 format for the random key, and is stored in the Bitcoin Wallet. For example, a sample private key is:

```
Private key: 5c04990cf2fb95ca8749d4021100ee98b0744e81a5ec00a2177aeaf4b29c00d3
```

We then convert this into WiF format (Base-58) to give:

```
5JWp4FM7sfAAE88DW3yvGF5mQyrsEXeWzXZn79bg61Vg8YMfJjA
```

This can be stored in a Bitcoin wallet. Next, we can take the private key and a hash value, and convert it into a useable Bitcoin address, such as:

```
1A3CohNBuB6kFAMtp3KFEYwv3Eu58F2HyN
```

The format of the keys is defined in Figure 10.6, where we create a 256-bit private key and convert this to a WiF private key. Next, we generate a 512-bit public key, and then take a 160-bit RIPEM-160 hash and convert to a Bitcoin address. Bitcoins use Elliptic Curve Ciphers (ECC) with a 256-bit private key (and a 512-bit public key).

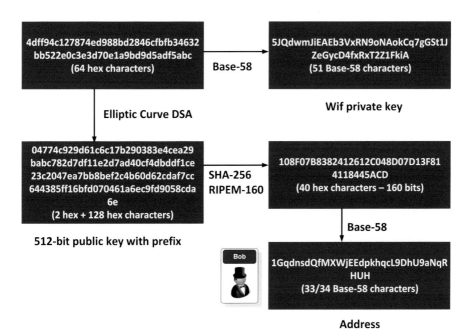

Figure 10.6 Key generation.

10.2.2 Mining Process

A transaction involves the spending of bitcoins, and basically transfers the ownership of the bitcoins from one person to a new address. The owner of the coins must create the transaction, and only requires the public ID of the recipient. In Figure 10.7 we see a transfer to bitcoins from Alice to Bob. First Bob sends his public key, and Alice then takes her private key and creates a signature for the transaction, and also adds her public key. This is part of the IN element of the transaction. The OUT part defines the number of bitcoins to be transferred and Bob's ID (his public key ID). The aim of this process is to sign the transaction with the private key of the person who is transferring the bitcoins (the sender). For this we take the transaction message and hash it twice. We then use Elliptic Curve DSA (SECP 256k1) to sign this hash with the private key of the sender. The output is then concatenated with the public key. This transaction will then be captured by miners who will compile a list of the latest transactions.

If valid, the transaction is then recorded within a mining process, where mining nodes gather new transactions and compute a hash of the new block, and which should also contain the hash of the previous block, and then build a transaction log. This happens at times period of around up to 10 minutes. The puzzle challenge to the miners is thus to compute a new

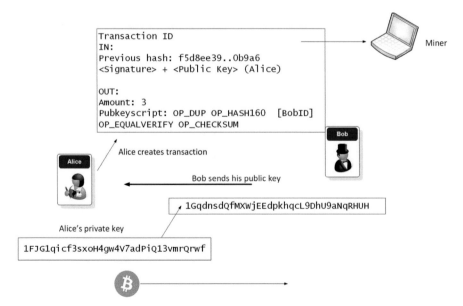

Figure 10.7   Transfer of BTC from Alice to Bob.

hash to a given format. Once complete, this becomes part of the official Blockchain in the network, and the miners reach a concensus on the current Blockchain. The node which has successfully completed the creation of the current Blockchain is then rewarded with some bitcoins (such as 12.5 BTCs). The high reward thus makes it a highly competitive process, and the difficulty of the cryptography challenge makes it a serious challenge. Only those with specialist hash generators will have the computing power required to create the required hash. Often GPUs (Graphical Processing Units) are used, as these have specialist hardware-driven methods (ASICs) and have multiple processors (often with several thousand processes on a single GPU). This then becomes a parallel processing task, and where the miners with the fastest resources are more likely to gain the reward.

There are no centralised servers with Bitcoin, and instead, there is a distributed peer-to-peer network where nodes exchange transactions, blocks and addresses with the rest of the network. On a new transaction the node sends out the new transaction to a peer node, who will send it to others, until it spans the whole network. The mining nodes then pick up the transactions, and start mining, and then broadcast the mined block. After a while, the node will receive the mined block back and which will show the new block with the successful transaction.

Figure 10.8 outlines the mining process and where the Blockchain is built by taking new transactions and then adding on the previous block, and then finding the required hash for the next block. Whichever miner finds the required hash for the new block they will be rewarded and the other miners will agree on the current version of the Blockchain. This operation happens every 10 minutes or so, thus a transaction will not be verified until it is added onto the Blockchain. Figure 10.7 outlines the information contained within a block, and which includes the hash (and the previous hash), along with the next block number. In Blockchain, we ask workers to generate a SHA-256 hash with a certain number of leading zeros. For this there will be a continual hashing until a number of preceding zeros is found and using a nounce value to create the required hash value. In the case in Figure 10.7 we see that both of the hashes requires 18 preceding zeros:

```
Hash
000000000000000000d98e57b83834a2d1f4387a93d06861bcf3ea5fc498bd55
Previous Block
00000000000000000012138e05f0779765277a9d2ab7e4a2a70882790abf98a0c
```

In this case we will add a nonce to create a hash with one leading zero:

📖 **Web link (Workload):** http://asecuritysite.com/encryption/block

Within the block, we see the number of transactions and the number of bitcoins transferred. It can be seen, also, that the reward for finding the hash is 12.5 BTCs. The reward for mining the hash, though changes over time, and there are break-block times where the reward is halved. The size of the block used in Bitcoin was increased from 1MB to 2MB in 2017 as the number of

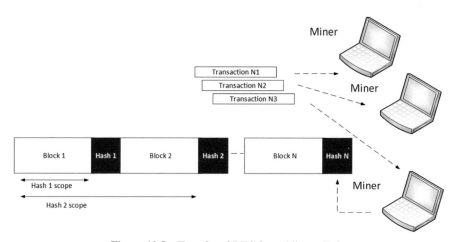

Figure 10.8   Transfer of BTC from Alice to Bob.

Block #475370

Summary	
Number Of Transactions	1937
Output Total	10,443.01703436 BTC
Estimated Transaction Volume	555.96160374 BTC
Transaction Fees	0.87013657 BTC
Height	475370 (Main Chain)
Timestamp	2017-07-11 21:44:58
Received Time	2017-07-11 21:44:58
Relayed By	AntPool
Difficulty	708,659,466,230.33
Bits	402754864
Size	998.17 KB
Version	0x20000000
Nonce	1203121562
Block Reward	12.5 BTC

Hashes	
Hash	00000000000000000d98e57b83834a2d1f4387a93d06861bcf3ea5fc498bd55
Previous Block	00000000000000000012138e05f0779765277a9d2ab7e4a2a70882790abf98a0c
Next Block(s)	00000000000000000010e3117695c04d66c31cfa8489b70579dcc2f12c5a2daae
Merkle Root	140d91abab9501d50ace079ba12c80125f48c2b5fe7d9da685ea3ee8ea767e82

Figure 10.9   Information on block.

transactions often exceed the 1MB block size, and some transactions thus had to wait until the next block to be processed. The 2MB increase thus overcame a significant bottle neck. As the miners agree, as a consensus, to the change, this is known as a soft fork. With a hard fork, we would end up with two versions of the Blockchain, and thus two different currencies.

The owner of a Bitcoin address has a private key, and this is used to sign the transaction (proving their ownership). When spent, their public key is then used to verify the digital signature on the transaction. This works in the same way that PKI works. The signature is a 256-bit cryptographic hash of the contents.

10.3 Ethereum

Ethereum was created by Vitalik Buterinin in 2015 and was built on the Bitcoin/Blockchain concept but included the concept of smart contracts. After a hack, in 2016, the Ethereum currency split into two: Ethereum (ETH) and Ethereum Classic (ETC). The focus of the attack was on the DAO (Decentralised Autonomous Organisation), which is an investment fund and is an offshoot from the Bitcoin crypto-currency. DAO – created by Ethereum – was developed as an investment fund where users purchase tokens – named Ether – for real cash, and then could spend these to support start-ups and other investment opportunities. Within the DAO hack, hackers took control of virtual cash worth $60m, and where they used an opening in a system which allowed developers to integrate their code – known as a debug hack – and where developers have a higher level of trust on the system, or where developers forget to turn off a trusted feature on a production instance.

As there are no rules about how to cope with this type of scenario, there was great debate about how the attack could be nullified, including rolling-back the clocks on the computer systems to a time before the attack. This would cause many in the industry to thus question the robustness of crypto currencies where one agency could reset the whole system, and then move it back to a time that was preferable to them. Vitalik Buterin proposed two solutions:

- Reset the whole system back to the time before the hack.
- Define a 27-day lock down on all non-trusted IP addresses to give time for the system to be fixed, and for no transfers to happen in that period.

Both of these were possible, as the scope of the system was still under the control of Ethereum. Overall they went ahead with a hard code fork. So, on

20 July 2015, the currency it split itself into two: Ethereum Classic (the original version) and a new version which fixed the bug. It is this ability to fork the currency which will worry some people, as it can be used to control its operation. Up to now, though Bitcoins have been relatively free from hard forks.

10.3.1 Gas

Within Ethereum applications we define the concept of *gas*. This is basically the unit that is used to measure the amount of work that is required to perform a single Keccak-256 hash, and where 30 gas are consumed for a single hash and 6 more gas for each 256 bits of data hashed. In this way there is a motivation to keep contracts small, as they will be less costly. Gas thus provides a way to define the fee that miners receive in performing operations on the blockchain. This differs from Bitcoin which only charge for the number of kiloBytes in a transaction. When it comes to the actual payment of the transaction fees, there is a payment of ether to the miners who create the blocks.

Ethereum transactions thus have a fee associated with them. If the fee is too low, then the miners will not process the transaction. When gas is consumed it is paid to the miner, and cannot be recovered back. If the transaction fee is set too high, there are likely to be many eager miners who are keen to profit from the high fee, and the transaction is likely to be prioritized. Overall, though, miners only charge for the work they have done, and they will return back any excess gas which they have not used. A miner can decide whether it needs to change the use of gas according to the price of gas varying. This overcomes the changes in transaction fees that happen in Bitcoin. In Ethereum, just like Bitcoin, there is a block limit, so you'll end up paying more if you overspill into another block (which means you should be efficient with your code and data). The gas price per transaction aims to overcome denial of service and infinite loops, and where 0.00001 Ether or 1 Gas is used to execute a line of code. If there is not enough Ether, no transaction will be performed. This also aims to make code designers efficient and not use waste bandwidth and CPU utilization.

10.3.2 Practical Implementation of Ethereum

The application named geth is a command line interface for the Ethereum blockchain and implemented in Go. Ethereum can thus create a public blockchain, or a private one. Initially we can create a Blockchain and which is stored in the d:\eth6 folder:

```
C:\Program Files\Geth> \geth –datadir=d:eth6 init customg.json
INFO [06-26 | 21:42:43] Allocated cache and file handles    database=d:\\eth6
\\geth\\chaindata cache=16 handles=16
INFO [06-26 | 21:42:43] Writing custom genesis block
INFO [06-26 | 21:42:43] Successfully wrote genesis state    database=chaindat
a              hash=10367b.67437b
INFO [06-26 | 21:42:43] Allocated cache and file handles    database=d:\\eth6
\\geth\\lightchaindata cache=16 handles=16
```

The genesis block is created with a configured file of:

```
{
   "config": {
     "chainId": 15,
     "homesteadBlock": 0,
     "eip155Block": 0,
     "eip158Block": 0
   },
   "difficulty": "200000000",
   "gasLimit": "0x3d0900",
   "alloc": {
     "228041751ddb7365cc4bc75c4985d14d5db2432f": { "balance": "30000000" },
     "cdfc92d1b5dd1c9ee1c9e2368abc86a193ae35a5": { "balance": "40000000" },
     "c9c425ae15a0e66500ecf5b7a1c10c6ed35600b9": { "balance": "0x400000000000000" }

   }
}
```

The difficulty value defines the level at which the miners will have to operate in order to mine the blockchain. In this case we have given the account "0xc9c425ae15a0..c6ed35600b9" a balance of 0x400000000000000. Next we will start Geth:

```
C:\Program Files\Geth>geth –datadir=d:\eth6
INFO [06-26 | 21:42:53] Starting peer-to-peer node        instance=Geth/v1.
6.6-stable-10a45cb5/windows-amd64/go1.8.3
INFO [06-26 | 21:42:53] Allocated cache and file handles    database=d:\\eth6
\\geth\\chaindata cache=128 handles=1024
WARN [06-26 | 21:42:53] Upgrading chain database to use sequential keys
INFO [06-26 | 21:42:53] Database conversion successful
INFO [06-26 | 21:42:53] Initialised chain configuration      config="{ChainID:
 15 Homestead: 0 DAO:  DAOSupport: false EIP150:  EIP155: 0 EIP158: 0
Metropolis:  Engine: unknown}"
INFO [06-26 | 21:42:53] Disk storage enabled for ethash caches  dir=d:\\eth6\\get
h\\ethash count=3
INFO [06-26 | 21:42:53] Disk storage enabled for ethash DAGs   dir=C:\\Users\\Ad
ministrator\\AppData\\Ethash count=2
WARN [06-26 | 21:42:53] Upgrading db log bloom bins
INFO [06-26 | 21:42:53] Bloom-bin upgrade completed        elapsed=10.000ms
INFO [06-26 | 21:42:53] Initialising Ethereum protocol      versions="[63 62]
" network=1
INFO [06-26 | 21:42:53] Loaded most recent local header      number=0 hash=103
67b.67437b td=200000000
```

```
INFO [06-26 | 21:42:53] Loaded most recent local full block      number=0 hash=103
67b.67437b td=200000000
INFO [06-26 | 21:42:53] Loaded most recent local fast block      number=0 hash=103
67b.67437b td=200000000
INFO [06-26 | 21:42:53] Starting P2P networking
```

Next we will connect to the Geth and create a new account:

```
C:\Program Files\Geth> geth attach
Welcome to the Geth JavaScript console!

instance: Geth/v1.6.6-stable-10a45cb5/windows-amd64/go1.8.3
coinbase: 0xc9c425ae15a0e66500ecf5b7a1c10c6ed35600b9
at block: 0 (Thu, 01 Jan 1970 00:00:00 GMT)
 datadir: d:\eth6
 modules: admin:1.0 debug:1.0 eth:1.0 miner:1.0 net:1.0 personal:1.0 rpc:1.0 txp
ool:1.0 web3:1.0

> personal.newAccount("Qwerty")
"0xce1373ddfa2232dc9ca82d98420be7a2e11962b5"

> web3.eth.accounts
["0xc9c425ae15a0e66500ecf5b7a1c10c6ed35600b9", "0xbb4fcfac2efd3dbc35117dc979ce5c
43ca5c615b", "0xce1373ddfa2232dc9ca82d98420be7a2e11962b5"]
```

We can look at the initial balances in the accounts:

```
> eth.getBalance("0xce1373ddfa2232dc9ca82d98420be7a2e11962b5")
0
> eth.getBalance("0xc9c425ae15a0e66500ecf5b7a1c10c6ed35600b9")
288230376151711744
> personal.unlockAccount('0xc9c425ae15a0e66500ecf5b7a1c10c6ed35600b9','Qwerty')
 true
```

Next we can transfer some currency from one account to another:

```
> eth.sendTransaction({from: '0xc9c425ae15a0e66500ecf5b7a1c10c6ed35600b9', to: '
0xce1373ddfa2232dc9ca82d98420be7a2e11962b5',value:1000})
"0x4029e82ac13fd2a56078c2747f2ff55b42db12c8fa40dbde8c6350b128476243"
>
> eth.getTransaction('0x4029e82ac13fd2a56078c2747f2ff55b42db12c8fa40dbde8c6350b1
28476243')
{
  blockHash: "0x0000000000000000000000000000000000000000000000000000000000000000
0000000000",
  blockNumber: null,
  from: "0xc9c425ae15a0e66500ecf5b7a1c10c6ed35600b9",
  gas: 90000,
  gasPrice: 18000000000,
  hash: "0x4029e82ac13fd2a56078c2747f2ff55b42db12c8fa40dbde8c6350b128476243",
  input: "0x",
  nonce: 0,
  r: "0xedbbbe21778eab7a3b3f82198854e6354abff4348dc9668ec337a786749a4d3a",
```

```
   s: "0x27228d637ac06acf1ffdcd93ff5a2dbd59f23353d196b97ff2ee7e2a14527595",
   to: "0xce1373ddfa2232dc9ca82d98420be7a2e11962b5",
   transactionIndex: 0,
   v: "0x41",
   value: 1000
}
```

If we look at the balances there have not been any transfers:

```
> eth.getBalance("0xce1373ddfa2232dc9ca82d98420be7a2e11962b5")
0
> eth.getBalance("0xc9c425ae15a0e66500ecf5b7a1c10c6ed35600b9")

288230376151711744
```

We can now start the miner:

```
> miner.start()
null

> eth.getBalance("0xc9c425ae15a0e66500ecf5b7a1c10c6ed35600b9")

288230376151711744
> eth.getBalance("0xce1373ddfa2232dc9ca82d98420be7a2e11962b5")

0
```

We can transfer again:

```
> eth.sendTransaction({from: '0xc9c425ae15a0e66500ecf5b7a1c10c6ed35600b9', to: '
0xce1373ddfa2232dc9ca82d98420be7a2e11962b5',value:100000})
"0x2e25093e25cbf511c2892cb38b45a5c9f6f9b2785774cd5830cf5bd978839165"
> eth.getBalance("0xce1373ddfa2232dc9ca82d98420be7a2e11962b5")

0
> eth.getBalance("0xc9c425ae15a0e66500ecf5b7a1c10c6ed35600b9")

288230376151711744
```

The mining process then adds some credits to the inital account:

```
> eth.getBalance("0xc9c425ae15a0e66500ecf5b7a1c10c6ed35600b9")
5288230376151711744

> eth.getBalance("0xce1373ddfa2232dc9ca82d98420be7a2e11962b5")
0
```

After a few minutes the mining process complete and shows:

```
> eth.getBalance("0xce1373ddfa2232dc9ca82d98420be7a2e11962b5")
200000
```

If we look at the blockchain we see there are two blocks that have been created:

```
> eth.blockNumber
2
```

Within Geth we see:

```
INFO [06-26 | 22:26:08] Commit new mining work          number=2 txs=2 un
cles=0 elapsed=2.000ms
INFO [06-26 | 22:26:55] Successfully sealed new block    number=2 hash=783
ace.91c41f
INFO [06-26 | 22:26:55] ?? mined potential block         number=2 hash=78
3ace.91c41f
INFO [06-26 | 22:26:55] Commit new mining work           number=3 txs=0 un
cles=0 elapsed=0s
```

10.3.3 Smart Contracts

Along with creating a new currency (Ether), the main contribution of Ethereum is to create the concept of peer-to-peer smart contracts which enables users to create their own contracts, and which will be strictly abided to. For example, a concert organiser could create a contract which defined the range of seats to be sold, and the cost of these seat, and then define that the cost of the seats will increase by 10% as the ticket sales reach certain limits. The contact could also define transaction and cancelation fees and where suppliers could be paid directly on each sale.

Now let's create a contract. For this we need to create a compiled version of the contract, and use JavaScript and compile using the Solidity compiler. A code sample is:

```
pragma solidity ^0.4.0;
contract test2{
    uint a ;
    function test2() {
        a = 1;
    }
    function val() returns(uint){
        return a;
    }
}
```

```
contract test3 is test2{
    uint b = a++;
    function show() returns(uint){
        return b;
    }
}
```

The Ethereum site of https://ethereum.github.io/browser-solidity/ provides an online compiler (Figure 10.10) and where we can create a compiled version of the code for the Ethereum blockchain.

Figure 10.10   Compiling with Solidity.

Now we copy from Web3 deploy and place in a JavaScript file, and then load it from Geth with:

```
>loadScript('sayhello2.js')
```

and next define the account to run the script:

```
> web3.eth.defaultAccount = '0x821eacc2a570c1aeb9b5aa64b5b915d4c1e1f3ee'
```

We can now start our miners:

```
> miner.start()
null
> null [object Object]
Contract mined! address: 0x8d487f4a719b5a1cf47c61cc83e757b8d269f877 transactionH
```

```
ash: 0xf4bb0fa6ddc1d9e1921a55d576d68acf5b715d00cd89cc7268ece3653c50de50
null [object Object]
Contract mined! address: 0xf3872dc9ced78283ad3a511e970891807dd38590 transactionH
ash: 0xab90aa5169f4ebfcbc139874208cabb29416feb3f12c296c93466d7d8090f805
null [object Object]
Contract mined! address: 0x7a74b5da4168f0a06a752301a3711c8991acaf88 transactionH
ash: 0x6ce2a63c59d124d5ecd4681a368243ba7de8aeacc735d41583f834789cba0b16
```

Finally we can view the contract as:

```
> test_sol_test2
{
  abi: [{
      constant: false,
      inputs: [],
      name: "val",
      outputs: [{...}],
      payable: false,
      type: "function"
  }, {
      inputs: [],
      payable: false,
      type: "constructor"
  }],
  address: "0x7a74b5da4168f0a06a752301a3711c8991acaf88",
  transactionHash: "0x6ce2a63c59d124d5ecd4681a368243ba7de8aeacc735d41583f834789c
ba0b16",
  allEvents: function(),
  val: function()
}
> test_sol_test3
{
  abi: [{
      constant: false,
      inputs: [],
      name: "val",
      outputs: [{...}],
      payable: false,
      type: "function"
  }, {
      constant: false,
      inputs: [],
      name: "show",
      outputs: [{...}],
      payable: false,
      type: "function"
  }],
  address: "0xbd570c2f87b8af945146177377276901fd82b12d",
  transactionHash: "0xc028384b4d8ea0e283c9cd3a6a747ab3efff859bb591d55f710ca20b09
665808",
  allEvents: function(),
  show: function(),
  val: function()
}
```

And then test:

```
> test_sol_test2.val()
"0xd69b536cd4055a45e209f3274d9b9370f33c88b474c0dca294b665efa2ac5d2d"
> test_sol_test3.val()
"0x4a5fa248e8f6c2223082518106c3e784d54e4ff70793c9d4f65c9ef931cd667c"
```

Now we will create a contract to do a bit of maths. Let's say we want to calculate the square root of a value:

```solidity
pragma solidity ^0.4.0;

contract mymath {
   function sqrt(uint x) constant returns (uint y) {
   uint z = (x + 1) / 2;
   y = x;
   while (z < y) {
       y = z;
       z = (x / z + z) / 2;
   }
}

}
```

When we create the JavaScript for the compiled version, and we load and run we get:

```
> personal.unlockAccount('0xc7552f45deb093cafb47286a0bc9415845ca3735','Qwerty')
true
> loadScript('mycontract.js')
null [object Object]
true
Contract mined! address: 0xc706a04b759a32dbec85702dd3864584e737aa77 transactionH
ash: 0xece670dcb578a78dec4d2338755ecade084a517310daacf37fd46fe336341563
null [object Object]
Contract mined! address: 0xfafb5f4d0db2c545592ac9134292162b03088295 transactionH
ash: 0x46204af57db69df078e1ae637b50fa76d8415ee1c1e3bd7e1c2990f328dc85ce
null [object Object]
Contract mined! address: 0x83e0bbb8abe2f0976fde9cf5db05333de067b0df transactionH
ash: 0xabea9606989bcc1bf93513213d298c84d47c7e8e1b397eaf536ebffb793d9304

> test_sol_mymath.sqrt(9)
3
> test_sol_mymath.sqrt(12)
3
> test_sol_mymath.sqrt(81)
9
```

Web link (Demo): http://asecuritysite.com/subjects/chapter91

Let's expand with more maths functions:

```
pragma solidity ^0.4.0;

contract mymath {
   function sqrt(uint x) constant returns (uint y) {
     uint z = (x + 1) / 2;
     y = x;
     while (z < y) {
         y = z;
         z = (x / z + z) / 2;
     }
}
function sqr(uint a) constant returns (uint) {
     uint c = a * a;
     return c;
   }

function mul(uint a, uint b) constant returns (uint) {
     uint c = a * b;
     return c;
   }

   function sub(uint a, uint b) constant returns (uint) {
     return a - b;
   }

   function add(uint a, uint b) constant returns (uint) {
     uint c = a + b;
     return c;
   }
}
```

We then compile this with the Solidity compiler to give:

```
var test_sol_mymathContract = web3.eth.contract([{"constant":true,"inputs":
[{"name":"x","type":"uint256"}],"name":"sqrt","outputs":[{"name":"y","type":
"uint256"}],"payable":false,"type":"function"},{"constant":true,"inputs":
[{"name":"a","type":"uint256"},{"name":"b","type":"uint256"}],"name":"add",
"outputs":[{"name":"","type":"uint256"}],"payable":false,"type":"function"},
{"constant":true,"inputs":[{"name":"a","type":"uint256"}],"name":"sqr",
"outputs":[{"name":"","type":"uint256"}],"payable":false,"type":"function"},
{"constant":true,"inputs":[{"name":"a","type":"uint256"},{"name":"b","type":
"uint256"}],"name":"sub","outputs":[{"name":"","type":"uint256"}],"payable":
false,"type":"function"},{"constant":true,"inputs":[{"name":"a","type":
"uint256"},{"name":"b","type":"uint256"}],"name":"mul","outputs":[{"name":
"","type":"uint256"}],"payable":false,"type":"function"}]);
var test_sol_mymath = test_sol_mymathContract.new(
    {
      from: web3.eth.accounts[0],
      data: '0x6060604052341561000c57fe5b5b...d5217a13a07400029',
      gas: '4700000'
    }, function (e, contract){
```

```
    console.log(e, contract);
    if (typeof contract.address !== 'undefined') {
        console.log('Contract mined! address: ' + contract.address +
' transactionHash: ' + contract.transactionHash);
    }
})
```

We can then add this onto the Blockchain with:

```
> web3.eth.accounts
["0xc7552f45deb093cafb47286a0bc9415845ca3735", "0x0851db3e133a15cd1c32531ffff96b
4526e3cbcd"]
> personal.unlockAccount('0xc7552f45deb093cafb47286a0bc9415845ca3735','Qwerty')
true
> loadScript('mymath.js')
null [object Object]
true
> web3.eth.defaultAccount = '0xc7552f45deb093cafb47286a0bc9415845ca3735'
"0xc7552f45deb093cafb47286a0bc9415845ca3735"
> miner.start()
null
> null [object Object]
Contract mined! address: 0xb7d8bcde9849896b9887dc31863c64875945fce5 transactionH
ash: 0xd5bd0ffed4b1d8ab199b93815c44ee9bec635c69a7ab8bcd21de21b0e732ed5f

> miner.stop()
true
> test_sol_mymath
{
  abi: [{
      constant: true,
      inputs: [{...}],
      name: "sqrt",
      outputs: [{...}],
      payable: false,
      type: "function"
  }, {
      constant: true,
      inputs: [{...}, {...}],
      name: "add",
      outputs: [{...}],
      payable: false,
      type: "function"
  }, {
      constant: true,
      inputs: [{...}],
      name: "sqr",
      outputs: [{...}],
      payable: false,
      type: "function"
  }, {
      constant: true,
      inputs: [{...}, {...}],
      name: "sub",
      outputs: [{...}],
```

```
      payable: false,
      type: "function"
  }, {
      constant: true,
      inputs: [{...}, {...}],
      name: "mul",
      outputs: [{...}],
      payable: false,
      type: "function"
  }],
  address: "0xb7d8bcde9849896b9887dc31863c64875945fce5",
  transactionHash: "0xd5bd0ffed4b1d8ab199b93815c44ee9bec635c69a7ab8bcd21de21b0e7
32ed5f",
  add: function(),
  allEvents: function(),
  mul: function(),
  sqr: function(),
  sqrt: function(),
  sub: function()
}
> test_sol_mymath.sqrt(9)
[3]
> test_sol_mymath.sqrt(12)
[3]
> test_sol_mymath.sqrt(16)
[4]
> test_sol_mymath.sqrt(81)
[9]
> test_sol_mymath.add(3,4)
[7]
> test_sol_mymath.add(4,2)
[6]
> test_sol_mymath.sqr(4)
[16]
> test_sol_mymath.mul(4,3)
[12]
```

📖 **Web link (Demo):** http://asecuritysite.com/subjects/chapter92

A basic "Hello World" contact can be created with:

```
pragma solidity ^0.4.0;

contract mycontract {
    /* Owner of the type address*/
    address owner;

    /* initialization and sets the owner of contract */
    function mycontract() { owner = msg.sender; }

    /* Recover the funds on the contract */
    function kill() { if (msg.sender == owner) selfdestruct(owner); }
}
```

```
contract showmessage is mycontract {
    /* Message in contract */
    string message;

    /* Initialise on contract */
    function showmessage(string _msg) public {
        message= _msg;
    }

    /* show function */
    function show() constant returns (string) {
        return message;
    }
}
```

We then compile this with the Solidity compiler to give (where the data part of the Json string defines the compiled code):

```
var browser_hello_sol_mycontractContract = web3.eth.contract([{"constant":
false,"inputs":[],"name":"kill","outputs":[],"payable":false,"type":
"function"},{"inputs":[],"payable":false,"type":"constructor"}]);
var browser_hello_sol_mycontract = browser_hello_sol_mycontractContract.new(
   {
     from: web3.eth.accounts[0],
     data: '0x606060405234....de13a65c0029',
     gas: '4700000'

   }, function (e, contract){
    console.log(e, contract);
    if (typeof contract.address !== 'undefined') {
        console.log('Contract mined! address: ' + contract.address +
' transactionHash: ' + contract.transactionHash);
    }
 })

var _msg = "Hello" ;
var browser_hello_sol_showmessageContract = web3.eth.contract([{"constant":
false,"inputs":[],"name":"kill","outputs":[],"payable":false,"type":
"function"},{"constant":true,"inputs":[],"name":"show","outputs":[{"name":
"","type":"string"}],"payable":false,"type":"function"},{"inputs":[{"name":
"_msg","type":"string"}],"payable":false,"type":"constructor"}]);
var browser_hello_sol_showmessage = browser_hello_sol_showmessageContract.new(
   _msg,
   {
     from: web3.eth.accounts[0],
     data: '0x606060405234156100....13970029',
     gas: '4700000'
   }, function (e, contract){
    console.log(e, contract);
    if (typeof contract.address !== 'undefined') {
        console.log('Contract mined! address: ' + contract.address +
' transactionHash: ' + contract.transactionHash);
    }
 })
```

We can then add this onto the Blockchain with:

```
> personal.unlockAccount(web3.eth.accounts[0],'Qwerty')
true
> loadScript('hello.js')
null [object Object]
null [object Object]
true
> null [object Object]
Contract mined! address: 0x3023606f3c8d9fe9f521aeed92a4500c9f026cd7 transactionH
ash: 0xdaa242a5d0c628ad339491176f3c8c08aff9bf87014c6259d5897424b5d1fccf
null [object Object]
Contract mined! address: 0x8422605c83fc69bb3569587857d164461bbe105c transactionH
ash: 0x669a43d64b210ee4c032694d34923e134f164987abba17bc86331cd6956c8abc
```

We can view the mined contract:

```
> browser_hello_sol_showmessage
{
  abi: [{
      constant: false,
      inputs: [],
      name: "kill",
      outputs: [],
      payable: false,
      type: "function"
  }, {
      constant: true,
      inputs: [],
      name: "show",
      outputs: [{...}],
      payable: false,
      type: "function"
  }, {
      inputs: [{...}],
      payable: false,
      type: "constructor"
  }],
  address: "0x8422605c83fc69bb3569587857d164461bbe105c",
  transactionHash: "0x669a43d64b210ee4c032694d34923e134f164987abba17bc86331cd695
6c8abc",
  allEvents: function(),
  kill: function(),
  show: function()
}
```

We now run with:

```
> browser_hello_sol_showmessage.show()
"Hello"
```

  Web link (Demo): http://asecuritysite.com/subjects/chapter93

Ethereum allows us to store transactions on a Blockchain and for these to be verifiable. So let's look at an example where we have a buyer (sender) and a seller (receiver). We first create a contract (market) and create a constructor (market) which will be initialised with a given number of tokens (supply). The findBalanceOf value will then be used to show the balance of an address (the user), and we'll implement a sendCoin() function to send a number of tokens from the buyer to the seller. First we create our code:

```
pragma solidity ^0.4.0;

contract market {
   mapping (address => uint) public findBalanceOf;
   event MoneyTransfer(address sender, address receiver, uint amount);

 /* Initializes contract with initial supply tokens to the
    creator of the contract */
 function market(uint supply) {
       findBalanceOf[msg.sender] = supply;
   }

 /* Very simple trade function */
   function sendCoin(address receiver, uint amount) returns(bool sufficient) {
       if (findBalanceOf[msg.sender] < amount) return false;
       findBalanceOf[msg.sender] -= amount;
       findBalanceOf[receiver] += amount;
       MoneyTransfer(msg.sender, receiver, amount);
       return true;
   }
}
```

We then compile this with the Solidity compiler to give:

```
var supply = /* var of type uint256 here */ ;
var browser_market_sol_marketContract = web3.eth.contract([{"constant":true,
"inputs":[{"name":"","type":"address"}],"name":"findBalanceOf","outputs":
[{"name":"","type":"uint256"}],"payable":false,"type":"function"},{"constant":
false,"inputs":[{"name":"receiver","type":"address"},{"name":"amount","type":
"uint256"}],"name":"sendCoin","outputs":[{"name":"sufficient","type":"bool"}],
"payable":false,"type":"function"},{"inputs":[{"name":"supply","type":
"uint256"}],"payable":false,"type":"constructor"},{"anonymous":false,"inputs":
[{"indexed":false,"name":"sender","type":"address"},{"indexed":false,"name":
"receiver","type":"address"},{"indexed":false,"name":"amount","type":
"uint256"}],"name":"MoneyTransfer","type":"event"}]);
var browser_market_sol_market = browser_market_sol_marketContract.new(
   supply,
   {
     from: web3.eth.accounts[0],
     data: '0x6060604052341561000c..4e067c3188f2b0029',
     gas: '4700000'
```

```
    }, function (e, contract){
      console.log(e, contract);
      if (typeof contract.address !== 'undefined') {
            console.log('Contract mined! address: ' + contract.address +
  ' transactionHash: ' + contract.transactionHash);
      }
})
```

First, we will view the accounts on the system, and log into one of them:

```
> web3.eth.accounts
["0xc7552f45deb093cafb47286a0bc9415845ca3735",
"0x0851db3e133a15cd1c32531ffff96b4526e3cbcd"]

> loadScript('market.js')
Error: authentication needed: password or unlock undefined
true

> personal.unlockAccount('0xc7552f45deb093cafb47286a0bc9415845ca3735',
'Qwerty')
true
```

Next, we'll load up the compiled contract and start the miner:

```
> loadScript('market.js')
null [object Object]
true
> miner.start()
null
Contract mined! address: 0xcddb2cda65a39a7cc7765a41b4d2be5fbe30e2fc transactionH
ash: 0x4516a1611d3bae6729563791ba882c8619d6ba0b9b0ecd2cecceb16887bd0e32
```

Let's see the initial credits on the accounts when the contract is created:

```
> browser_market_sol_market.findBalanceOf(eth.accounts[0]) + tokens"
"10000 tokens"
> browser_market_sol_market.findBalanceOf(eth.accounts[1]) + " tokens"
"0 tokens"
```

Now we see the contract being mined and we can transfer tokens from one account to another:

```
> browser_market_sol_market.sendCoin.sendTransaction(eth.accounts[1], 1000,
{from: eth.accounts[0]})
"0xf72e3e8ca69a62ce9380ae23d5a858e94213f2b5d8e96c7ea0a4da9c1ef33c77"

> browser_market_sol_market.findBalanceOf(eth.accounts[0]) + " tokens"
"10000 tokens"
> browser_market_sol_market.findBalanceOf(eth.accounts[1]) + " tokens"
```

```
"1000 tokens"
> browser_market_sol_market.findBalanceOf(eth.accounts[0]) + " tokens"
"9000 tokens"

> browser_market_sol_market.sendCoin.sendTransaction(eth.accounts[1], 1000,
{from: eth.accounts[0]})
"0xa26555a901fd994af362a5dd07e7da7c15c4e39cbfd85b98aab0b5c4c3c3b547"

> browser_market_sol_market.findBalanceOf(eth.accounts[0]) + " tokens"
"8000 tokens"
> browser_market_sol_market.findBalanceOf(eth.accounts[1]) + " tokens"
"2000 tokens"
```

📖 **Web link (Demo):** http://asecuritysite.com/subjects/chapter94

In Blockchain we create transactions and these are then captured by miners by
certain periods and then the miners compete to create a new hash, and which
creates a new block (Figure 10.11). With Ethereum we can add data onto the
Blockchain with a transaction. Let's stop the miners and add the string "hello"
to the Blockchain. For this we create a variable named myData, and then use
the sendTransaction() method:

```
> miner.stop()
true

> var myData="0x68656c6c6f";
undefined

>  eth.sendTransaction({from:eth.accounts[0], gas:3141592, data: myData})
Error: authentication needed: password or unlock
    at web3.js:3104:20
    at web3.js:6191:15
    at web3.js:5004:36
    at anonymous:1:2

> personal.unlockAccount(web3.eth.accounts[0],'Qwerty')
true

>  eth.sendTransaction({from:eth.accounts[0], gas:3141592, data: myData})
"0x88add08d4f88af17ba03f68d675d57be39f43e27bd96002bdcec23e396d0a369"
```

We now have a transaction number, so let's examine it:

```
> eth.getTransaction("0x88add08d4f88af17ba03f68d6..6002bdcec23e396d0a369")
{
  blockHash: "0x0000000000000000000000000000000000000000000000000000000000000000",
  blockNumber: null,
  from: "0xc7552f45deb093cafb47286a0bc9415845ca3735",
  gas: 3141592,
```

```
  gasPrice: 18000000000,
  hash: "0x88add08d4f88af17ba03f68d675d57be39f43e27bd96002bdcec23e396d0a369",
  input: "0x68656c6c6f",
  nonce: 74,
  r: "0x8b918d7e489450991f99db45717c38e0a53d4f96a9b328668f3efdf070be80df",
  s: "0x7554476d9abd3924c14835d592120b3b6212ed3b4b9eda475da8c563e6a24e3",
  to: null,
  transactionIndex: 0,
  v: "0x41",
  value: 0
}
```

We can see the input property defines the data ("hello"), but there is no block number yet as it hasn't been mined yet. So let's start the miner:

```
> miner.start()
null
```

The miner doesn't pick it up straight away, as the blockNumber is still null:

```
> eth.getTransaction("0x88add08d4f88af17ba03f68d6..6002bdcec23e396d0a369")
{
  blockHash: "0x0000000000000000000000000000000000000000000000000000000000000000",
  blockNumber: null,
  from: "0xc7552f45deb093cafb47286a0bc9415845ca3735",
  gas: 3141592,
  gasPrice: 18000000000,
  hash: "0x88add08d4f88af17ba03f68d675d57be39f43e27bd96002bdcec23e396d0a369",
  input: "0x68656c6c6f",
  nonce: 74,
  r: "0x8b918d7e489450991f99db45717c38e0a53d4f96a9b328668f3efdf070be80df",
  s: "0x7554476d9abd3924c14835d592120b3b6212ed3b4b9eda475da8c563e6a24e3",
  to: null,
  transactionIndex: 0,
  v: "0x41",
  value: 0
}
```

But eventually we see it:

```
>eth.getTransaction("0x88add08d4f88af17ba03f68d675d57be39f43e27bd96002bdcec
                    23e396d0a369")
{
  blockHash: "0x3709ba4981574413e4c5af775f586fa89d3a73060356c0f59ab5143524e75caf
",
  blockNumber: 5134,
  from: "0xc7552f45deb093cafb47286a0bc9415845ca3735",
  gas: 3141592,
  gasPrice: 18000000000,
  hash: "0x88add08d4f88af17ba03f68d675d57be39f43e27bd96002bdcec23e396d0a369",
  input: "0x68656c6c6f",
```

```
  nonce: 74,
  r: "0x8b918d7e489450991f99db45717c38e0a53d4f96a9b328668f3efdf070be80df",
  s: "0x7554476d9abd3924c14835d592120b3b6212ed3b4b9eda475da8c563e6a24e3",
  to: null,
  transactionIndex: 0,
  v: "0x41",
  value: 0
}
```

We see now that that the transaction is within Block 5134, so let's examine that block:

```
> web3.eth.getBlock(5134)

{
  difficulty: 958134,
  extraData: "0xd983010606846765746887676f312e382e338777696e646f7773",
  gasLimit: 4712388,
  gasUsed: 53343,
  hash: "0x3709ba4981574413e4c5af775f586fa89d3a73060356c0f59ab5143524e75caf",
  logsBloom: "0x000000000000000000000000000000000000000000000000000000000000
00000000000000000000000000000000000000000000000000000000000000000000000000000000
00000000000000000000000000000000000000000000000000000000000000000000000000000000
00000000000000000000000000000000000000000000000000000000000000000000000000000000
00000000000000000000000000000000000000000000000000000000000000000000000000000000
00000000000000000000000000000000000000000000000000000000000000000000000000000000
00000000000000000000000000000000000000000000000000000000000000000000000000000000
00000000000000000000000000000000000000000000",
  miner: "0xc7552f45deb093cafb47286a0bc9415845ca3735",
  mixHash: "0x2760004c46842240f6cf8cd888cf28d94dc1764178d82fa06c2d25361909b82d",

  nonce: "0x721e33400bb8c24b",
  number: 5134,
  parentHash: "0xdc6254b48f8379062f8a56943ed148649022c4adb5025ffb0b98b5f
              22206bb8a",
  receiptsRoot: "0xc251f3128f854445a2d16fb1936824836d137b951e43bb330dcc3
                272e6fda9ec",
  sha3Uncles: "0x1dcc4de8dec75d7aab85b567b6ccd41ad312451b948a7413f0a142f
               d40d49347",
  size: 630,
  stateRoot: "0x0667d97ebd8090ee546863505a2439adcaa3788ab6e851410c9b696a
              0060b39a",
  timestamp: 1498920608,
  totalDifficulty: 2428672816,
  transactions: ["0x88add08d4f88af17ba03f68d675d57be39f43e27bd96002bdcec2
                 3e396d0a369"],
  transactionsRoot: "0xea5082474a5a95aa28668a1ca465ffb5d1e93a2d16a2cb74b4
                     7d85ce22bfd742",
  uncles: []
}
```

We can see there is only one transaction there (0x88...), which is the one where we added "hello". Now let's view the transaction:

```
> web3.eth.getBlock(5134,true).transactions[0]
{
  blockHash: "0x3709ba4981574413e4c5af775f586fa89d3a73060356c0f59ab5143524e75caf",
  blockNumber: 5134,
  from: "0xc7552f45deb093cafb47286a0bc9415845ca3735",
  gas: 3141592,
  gasPrice: 18000000000,
  hash: "0x88add08d4f88af17ba03f68d675d57be39f43e27bd96002bdcec23e396d0a369",
  input: "0x68656c6c6f",
  nonce: 74,
  r: "0x8b918d7e489450991f99db45717c38e0a53d4f96a9b328668f3efdf070be80df",
  s: "0x7554476d9abd3924c14835d592120b3b6212ed3b4b9eda475da8c563e6a24e3",
  to: null,
  transactionIndex: 0,
  v: "0x41",
  value: 0
}
```

And finally view the data we added to the Blockchain:

```
> web3.eth.getBlock(5134,true).transactions[0].input
"0x68656c6c6f"
```

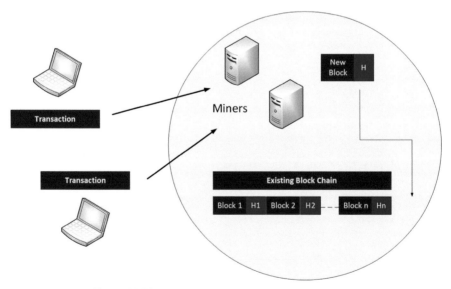

Figure 10.11   Transactions and blockchain in Ethereum.

📖 **Web link (Demo):** http://asecuritysite.com/subjects/chapter95

Figure 10.12 provides an overview of the operation of blocks and transactions and where we use Keccak-256 to create the hashing method for the blocks. Overall one Keccak-256 hash takes 30 gas, along with six more gas for every 256 bytes of data.

📖 **Web link (Keccak):** https://asecuritysite.com/encryption/s3

We can use this script to mine across given block numbers and for a specified account:

```
function getTransactionsByAccount(myaccount, startBlockNumber, endBlockNumber) {
  if (endBlockNumber == null) {
    endBlockNumber = eth.blockNumber;
    console.log("Using endBlockNumber: " + endBlockNumber);
  }
  if (startBlockNumber == null) {
    startBlockNumber = endBlockNumber - 1000;
    console.log("Using startBlockNumber: " + startBlockNumber);
  }
  console.log("Searching for transactions to/from account  \"" + myaccount + "\"
within blocks "  + startBlockNumber + " and " + endBlockNumber);

  for (var i = startBlockNumber; i <= endBlockNumber; i++) {
    if (i % 1000 == 0) {
      console.log("Searching block " + i);
    }
    var block = eth.getBlock(i, true);
    if (block != null && block.transactions != null) {
      block.transactions.forEach( function(e) {
        if (myaccount == "*" || myaccount == e.from || myaccount == e.to) {
          console.log("  tx hash          : " + e.hash + "\n"
          + "   nonce           : " + e.nonce + "\n"
          + "   blockHash       : " + e.blockHash + "\n"
          + "   blockNumber     : " + e.blockNumber + "\n"
          + "   transactionIndex: " + e.transactionIndex + "\n"
          + "   from            : " + e.from + "\n"
          + "   to              : " + e.to + "\n"
          + "   value           : " + e.value + "\n"
          + "   time            : " +block.timestamp +" "+new Date(block.
timestamp * 1000).toGMTString() + "\n"
          + "   gasPrice        : " + e.gasPrice + "\n"
          + "   gas             : " + e.gas + "\n"
          + "   input           : " + e.input);
        }
      })
    }
  }
}
```

We can now load the JavaScript file and determine transactions for a given user on blocks 1 to 20:

```
> loadScript('trans.js')
true
> getTransactionsByAccount(eth.accounts[0],1,20)
Searching for transactions to/from account "0xc7552f45deb093cafb47286a0bc941
5845ca3735" within blocks 1 and 20
  tx hash           : 0xcc943bcd0210882febc3bce1c2c118967976ea86449d3f9e5a023
                      3af8bf5e696
    nonce           : 0
    blockHash       : 0xde3fb2117453b32a4d9403a3268b2ba27398a06f62a1c86955d98
                      6a735f563b4
    blockNumber     : 1
    transactionIndex: 0
    from            : 0xc7552f45deb093cafb47286a0bc9415845ca3735
    to              : 0x0851db3e133a15cd1c32531ffff96b4526e3cbcd
    value           : 100000
    time            : 1498553568 Tue, 27 Jun 2017 08:52:48 GMT
    gasPrice        : 18000000000
    gas             : 90000
    input           : 0x
```

Now let's try blocks 1 to 40, and 1 to 60:

```
> getTransactionsByAccount(eth.accounts[0],1,40)
Searching for transactions to/from account "0xc7552f45deb093cafb47286a0bc94
15845ca3735" within blocks 1 and 40
  tx hash           : 0xcc943bcd0210882febc3bce1c2c118967976ea86449d3f9e5a02
                      33af8bf5e696
    nonce           : 0
    blockHash       : 0xde3fb2117453b32a4d9403a3268b2ba27398a06f62a1c86955d9
                      86a735f563b4
    blockNumber     : 1
    transactionIndex: 0
    from            : 0xc7552f45deb093cafb47286a0bc9415845ca3735
    to              : 0x0851db3e133a15cd1c32531ffff96b4526e3cbcd
    value           : 100000
    time            : 1498553568 Tue, 27 Jun 2017 08:52:48 GMT
    gasPrice        : 18000000000
    gas             : 90000
    input           : 0x
undefined
> getTransactionsByAccount(eth.accounts[0],1,60)
Searching for transactions to/from account "0xc7552f45deb093cafb47286a0bc9415845
ca3735" within blocks 1 and 60
  tx hash           : 0xcc943bcd0210882febc3bce1c2c118967976ea86449d3f9e5a0233af8
bf5e696
    nonce           : 0
    blockHash       : 0xde3fb2117453b32a4d9403a3268b2ba27398a06f62a1c86955d986
                      a735f563b4
    blockNumber     : 1
    transactionIndex: 0
    from            : 0xc7552f45deb093cafb47286a0bc9415845ca3735
    to              : 0x0851db3e133a15cd1c32531ffff96b4526e3cbcd
    value           : 100000
    time            : 1498553568 Tue, 27 Jun 2017 08:52:48 GMT
    gasPrice        : 18000000000
```

```
      gas               : 90000
      input             : 0x
    tx hash             : 0xbfe234697a506bfb7b2c19202bdeb9938e53eb9ae78104b22f3ff954
                          77547861
      nonce             : 1
      blockHash         : 0xcd416ab6a3fb87eb88ce5d830d78f888db48d31d5b3eef200241ba5a
                          eb46b377
      blockNumber       : 45
      transactionIndex: 0
      from              : 0xc7552f45deb093cafb47286a0bc9415845ca3735
      to                : 0x0851db3e133a15cd1c32531ffff96b4526e3cbcd
      value             : 100000
      time              : 1498553643 Tue, 27 Jun 2017 08:54:03 GMT
      gasPrice          : 18000000000
      gas               : 90000
      input             : 0x
undefined
```

We can now use some JavaScript to view the transaction based on its hash:

```javascript
function printTransaction(txHash) {
  var tx = eth.getTransaction(txHash);
  if (tx != null) {
    console.log("  tx hash          : " + tx.hash + "\n"
        + "    nonce          : " + tx.nonce + "\n"
        + "    blockHash      : " + tx.blockHash + "\n"
        + "    blockNumber    : " + tx.blockNumber + "\n"
        + "    transactionIndex: " + tx.transactionIndex + "\n"
        + "    from           : " + tx.from + "\n"
        + "    to             : " + tx.to + "\n"
        + "    value          : " + tx.value + "\n"
        + "    gasPrice       : " + tx.gasPrice + "\n"
        + "    gas            : " + tx.gas + "\n"
        + "    input          : " + tx.input);
  }
}
```

And now print the transaction:

```
> printTransaction("0xbfe234697a506bfb7b2c1...78104b22f3ff95477547861")
    tx hash             : 0xbfe234697a506bfb7b2c19202bdeb9938e53eb9ae78104b22f3ff9
                          5477547861
    nonce               : 1
    blockHash           : 0xcd416ab6a3fb87eb88ce5d830d78f888db48d31d5b3eef200241ba
                          5aeb46b377
    blockNumber         : 45
    transactionIndex: 0
    from                : 0xc7552f45deb093cafb47286a0bc9415845ca3735
    to                  : 0x0851db3e133a15cd1c32531ffff96b4526e3cbcd
    value               : 100000
    gasPrice            : 18000000000
    gas                 : 90000
    input               : 0x
```

Now we will find all the transactions in blocks, and can use the JavaScript of:

```
function checkTransactionCount(startBlockNumber, endBlockNumber) {
  console.log("Searching for non-zero transaction counts between blocks "  +
startBlockNumber + " and " + endBlockNumber);

  for (var i = startBlockNumber; i <= endBlockNumber; i++) {
    var block = eth.getBlock(i);
    if (block != null) {
      if (block.transactions != null && block.transactions.length != 0) {
        console.log("Block #" + i + " has " + block.transactions.length + "
transactions")
      }
    }
  }
}
```

Let's search for transactions between Block 0 and Block 1000:

```
> checkTransactionCount(0,1000)
Searching for non-zero transaction counts between blocks 0 and 1000
Block #1 has 1 transactions
Block #45 has 1 transactions
Block #68 has 2 transactions
Block #82 has 2 transactions
Block #120 has 10 transactions
Block #126 has 10 transactions
Block #155 has 1 transactions
Block #156 has 1 transactions
Block #872 has 2 transactions
Block #873 has 1 transactions
```

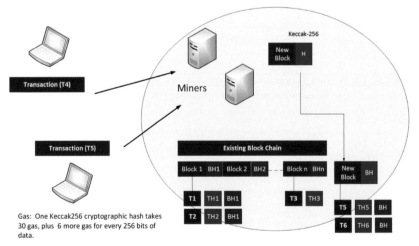

Figure 10.12   Ethereum hashing.

 📖 **Web link (Demo):** http://asecuritysite.com/subjects/chapter96

Once we have created an Ethereum contact on the blockchain, we can then allow others to use it, by giving them the address of the contract. Demos thus are defined here:

 📖 **Web link (Demo):** http://asecuritysite.com/subjects/chapter97
 📖 **Web link (Demo):** http://asecuritysite.com/subjects/chapter98

10.4 Lab/Tutorial

The lab and tutorial related to this chapter is available on-line at:

http://asecuritysite.com/crypto10

References

[1] K. A. McKay, L. Bassham, M. S. Turan, and N. Mouha, "Report on lightweight cryptography," 2017.
[2] A. Bogdanov, L. R. Knudsen, G. Leander, C. Paar, A. Poschmann, M. J. B. Robshaw, Y. Seurin, and C. Vikkelsoe, "PRESENT: An Ultra-Lightweight Block Cipher," *Cryptogr. Hardw. Embed. Syst. – CHES 2007*, pp. 450–466.
[3] A. Bogdanov, M. Knežević, G. Leander, D. Toz, K. Varici, and I. Verbauwhede, "{SPONGENT}: The Design Space of Lightweight Cryptographic Hashing," 2011.
[4] G. Bertoni, J. Daemen, M. Peeters, and G. Van Assche, "Keccak," Springer, Berlin, Heidelberg, 2013, pp. 313–314.
[5] J. Guo, T. Peyrin, and A. Poschmann, "The PHOTON Lightweight Hash Functions Family," *Crypto*, pp. 222–239, 2000.

11

Zero-knowledge Proof (ZKP) and Privacy Preserving

11.1 Introduction

Many systems are designed where the user must provide the plaintext version of their password. Unfortunately if Eve is listening, then she could determine the password. With ZKP we create a protocol where Alice can prove to Bob that she knows a secret, without actually revealing the secret. In this way Bob and Alice just need to negotiate their agreement of the secret, and from there on Alice will not have to release the secret to Bob. We can also use this method when we do not trust anyone, such as where Bob and Alice want to play a game of toss coin over a telephone link. If they do not trust each other, how can they toss a coin using cryptography, and then prove the result of the toss, so that neither Bob nor Alice can lie? There are thus many applications in proving things without actually revealing the information that you are proving. This could be to prove your identity, or even your date of birth. Overall these methods are defined as Privacy Enhancing Technologies (PET), and where we can reveal information without revealing our answer.

So ZKP is all about finding out something about a user or a device, without actually revealing the actual data involved. For example we might want to know the number of people who are using an App who are under 18 years old, and who are over 18 years old, without actually asking for their age (or in storing it). Some common methods are:

- Discrete logs.
- Graphs.
- Feige-Fiat-Shamir.
- Non-interactive random oracle access.
- Fair coin flip.
- Voting with Paillier crypto system.
- Oblivious transfer.

- Scrambled circuits.
- Millionaire's Problem.
- Secure Function Evaluation (SFE).
- Randomized Aggregable Privacy-Preserving. Ordinal Response (RAP-POR).
- Secure Remote Password (SRP) protocol.

11.2 ZKP: Discrete Logs

In cryptography, let's say that Alice wants to prove that Bob knows a value (x), such that:

$$Y = g^x \pmod P$$

where g is a pre-selected value, P is a prime and Y is a result. Both Bob and Alice know these values, and it's difficult to know the value of x, as there are many values of x that would fit. To prove that Bob knows the value of x, he creates a random number (r) and sends the result of this calculation to Alice:

$$C = g^r \pmod P$$

He then sends:

$$Cipher_1 = g^{(x+r) \bmod (P-1)} \pmod P$$

Alice then calculates:

$$Cipher_2 = C.Y \pmod P$$

If the values are the same ($Cipher_1$ equals $Cipher_2$), Bob has proven that he knows the secret (which is x). The following provides an example:

```
p= 6353   (the prime number)
g= 5436
x= 54643   (the secret)
r= 643215   (the random value)
==============
Y=g**x % p= 369
====Bob sends====
C=g**r % p= 925
Cipher1=g^(x+r)%(p-1) mod p= 4616
====Alice calculates====
Cipher2=C.Y mod P= 4616
==============
Well done ... have you proven that you know x
```

An outline of the code used is:

```
p=71
g=13
x=7
r=8

print 'p=',p
print 'g=',g
print 'x=',x
print 'r=',r
print '========'

y= g**x % p
print 'Y=',y

C = g**r % p
print 'C=',C

print '========'
val1=g**((x+r)%(p-1))  % p
print 'g^(x+r)%(p-1) mod p=',val1

val2=C*y %p
print 'C.y mod P=',val2

if (val1==val2):
        print 'Well done ... have you proven that you know x'
else:
        print 'Not proven'
```

  📖 **Web link (ZPF):** http://asecuritysite.com/encryption/z

11.3 Commutative Encryption

How can Bob generate a receipt of a purchase but not give away Bob's identity or the details of his booking? For this let's take an example of booking a seat in a theatre at a Festival, where we have 100 seats in a theatre, and Bob wants to book one of them. But he does not want the theatre company to know which seat that he has booked, or his identity. Bob also wants a receipt of purchase that he can verify his booking. One way would be to get a trusted agent to look after the bookings, but he don't trust them either. So how can we do this? One was is with commutative encryption.

Initially the theatre company generates 100 receipts for each of the seats, and then encrypts them with its public key. Next when Bob wants to make a booking they send him the encrypted receipts that they have left, and Bob select one at random, and then encrypts it with his public key. He sends them all back, including the one he's encrypted. The theatre checks to see which one has been changed, and then decrypts it with their private key, and sends it back to Bob.

Bob then decrypts with his private key, and he can now view the receipt for the booking, and the theatre company cannot determine which seat he has, but Bob will have the receipt of his booking. This is illustrated in Figure 11.1.

All this is made possible with commutative encryption and where we can encrypt and decrypt in any order. One method is to use the RSA method to generate the keys required, but for everyone to share a common P and Q value (and thus a shared N value).

📖 **Web link (ZPF):** http://asecuritysite.com/encryption/comm

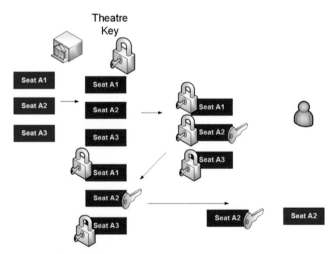

Figure 11.1   Commutative encryption example.

11.4 Graphs and Hamiltonian Cycles

Bob wants Victor to have his secret treasure and which is hidden in a maze, and which is guarded by a troll. Bob tells the troll (Peggy) that he will send someone to pick up the treasure, and that she can tell whom he sends because they will find their way through the maze. So Bob shows Victor the maze, and he sets off. When he gets there, there are a whole lot of people there who also say that Bob has also sent them, and want to get into the maze, including Eve the Beast. But how does the troll know that Victor is the one that Bob sent, as others are listening to what he is saying? This is a common problem in zero-knowledge proofs. In this case Peggy (the troll) wants to make sure that Victor knows the secret (the way through the maze), but does not want him to reveal it. For this we can have a Hamiltonian cycle – which is the route through the maze which returns to the same point, and where we visit each of the junctions (vertices) on the maze. If the maze is complex, and Peggy provides a map of the maze which has different labels for the junctions, it will be difficult for Eve to find the Hamiltonian cycle.

Figure 11.2 is an example of an extremely simple maze. We could walk from 1 to 3 to 5 and then back to 1, but we have missed out 2 and 4. So what is a solution for us to be able to visit each junction (vertice) just once? In this case the Hamiltonian cycle will be 1 -> 5 -> 4 -> 2 -> 3 -> 1, and which return us back to 1, and having visited all the junctions (vertices) in the maze.

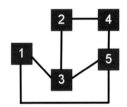

Figure 11.2   Simple maze.

In this case we have 10 **Hamiltonian cycle** routes, with two which will return to the starting point (1):

    1 -> 5 -> 4 -> 2 -> 3 -> 1
    1 -> 3 -> 2 -> 4 -> 5 -> 1
    3 -> 2 -> 4 -> 5 -> 1 -> 3
    3 -> 1 -> 5 -> 4 -> 2 -> 3
    2 -> 4 -> 5 -> 1 -> 3 -> 2
    2 -> 3 -> 1 -> 5 -> 4 -> 2
    5 -> 4 -> 2 -> 3 -> 1 -> 5
    5 -> 1 -> 3 -> 2 -> 4 -> 5
    4 -> 5 -> 1 -> 3 -> 2 -> 4
    4 -> 2 -> 3 -> 1 -> 5 -> 4

We have also have 12 **Hamiltonian paths** – which are routes where we go to every node, but do not have to end at the same one:

```
1 -> 5 -> 4 -> 2 -> 3
1 -> 3 -> 5 -> 4 -> 2
1 -> 3 -> 2 -> 4 -> 5
3 -> 2 -> 4 -> 5 -> 1
3 -> 1 -> 5 -> 4 -> 2
2 -> 4 -> 5 -> 1 -> 3
2 -> 3 -> 1 -> 5 -> 4
5 -> 4 -> 2 -> 3 -> 1
5 -> 1 -> 3 -> 2 -> 4
4 -> 2 -> 3 -> 5 -> 1
4 -> 5 -> 1 -> 3 -> 2
4 -> 2 -> 3 -> 1 -> 5
```

While this looks simple; for a complex graph, it is computationally infeasible to find the Hamiltonian cycle, as it is a known as an NP-complete. Victor just has to show to Peggy that he knows the Hamiltonian cycle, and then Peggy will know that he knows the secret.

Both Peggy and Victor know the original graph (G) of the maze, and Peggy knows the Hamiltonian cycle for it. Peggy then recreates a new graph (H) with the same connections, but with different names for the nodes (vertices). As the graphs are the same in their structure, Peggy will easily be able to translate between the two graphs. For example, as shown in Figure 11.3, she can change the labels to letters ('A', 'B', ...). As it is the same structure for the graph (G), Victor will be able to find the Hamiltonian cycles, but others, such as Eve the Beast, will have to search for them, and which is an extremely difficult problem for a complex graph.

Peggy (the troll) will then ask Victor some questions. She can either ask him to show how he converted from G to H, or to reveal a Hamiltonian cycle in H. If he reveals the Hamiltonian cycle, he will show the translation (the route) for it to Peggy. Eve will struggle to find a Hamiltonian cycle within reasonable time limits.

This solution is an example of the travelling salesman problem (TSP) where a salesman must visit each city only once, and return to the same city they started from. This is known to be a NP-complete problem, and is perfect for a puzzle which is easy to solve if we know the answer, but difficult if we do not. In this way we have a puzzle which is easy for Victor to solve, but Eve will find it difficult with a new graph each time.

📖 **Web link (ZPF):** http://asecuritysite.com/encryption/maze

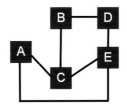

Figure 11.3 Simple maze.

11.5 Feige-Fiat-Shamir

With Feige-Fiat-Shamir, first Bob and Alice agree on two prime numbers (say 101 and 23), and calculate a value of n:

$$n = (101 \times 23) = 2,323.$$

Bob has three secret numbers of $s_1=5$; $s_2=7$; and $s_3=3$ (note that these numbers need to be co-prime to n, so that they do not share a factor with n). Now Alice generates three random numbers which are either 0 or a 1, such as $a_1=1$; $a_2=0$; and $a_3=1$, and sends these to Bob. Next Bob generates a random number (such as $r=13$). He calculates a value of x which is:

$$x = r^2 \pmod{n}$$

which gives $x = 169$. Bob calculates:

$$y = (r \times (s_1{}^{a_1}) \times (s_2{}^{a_2}) \times (s_3{}^{a_3})) \pmod{n}$$

and which gives 195, which he sends to Alice. Next Alice computes:

$$v_1 = (s_1{}^2) \pmod{n}$$
$$v_2 = (s_2{}^2) \pmod{n}$$
$$v_3 = (s_3{}^2) \pmod{n}$$

and then computes:

$$y = (x \times (v_1{}^{a_1}) \times (v_2{}^{a_2}) \times (v_3{}^{a_3})) \pmod{n}$$

If $(y^2 \bmod n)$ is equal to y that Bob has sent, Bob has proven that he knows the secret values. Here is the code:

```
n=101*23
r=13
s1=5
s2=7
s3=3
a1=1
a2=0
a3=1
print 'N=',n
x = (r**2) % n
print 'x=',x
print 's1=',s1,'s2=',s2,'s3=',s3
print 'a1=',a1,'a2=',a2,'a3=',a3

y = (r * ((s1**a1) * (s2**a2) * (s3**a3)) ) % n
print 'Y=',y, ' y^2 mod n = ',(y**2 % n)

v1=(s1**2) %n
v2=(s2**2) %n
v3=(s3**2) %n

y2 = (x * ((v1**a1) * (v2**a2) * (v3**a3)) ) % n

print 'Y=',(y**2) %n
```

and a sample run:

```
N= 2323

x= 169

s1= 5 s2= 7 s3= 3

a1= 1 a2= 0 a3= 1

Y= 195 y∧2 mod n = 857

Y∧2= 857
```

We can see the values are the same at the end. Alice can go through several rounds of providing a_1, a_2 ... a_n, in order to prove that Bob knows the information.

📖 **Web link (ZPF):** http://asecuritysite.com/encryption/z2

11.6 Non-interactive Random Oracle Access for Zero-knowledge Proof

Alice can prove her identity using *non-interactive random oracle access* for zero-knowledge proofs. In the case we will look at the use of a hash value for the non-interactive random oracle part. Normally Alice would pass a random value to whoever it is that wants to prove her identity, but she will use the hash value to randomise the puzzle and answer.

1. First everyone agrees on a puzzle and has a secret (x). The puzzle is:

$$y = g^x$$

   where we agree on g and x is the secret that Alice proves that she knows. Let's say that g is 13, and x is 11, so:

$$g^x = 1,792,160,394,037$$

2. Next Alice generates a random number (y) and calculates t, which is:

$$t = g^y$$

   Let's say that the value of v is 8. This gives:

$$t = 815,730,721$$

3. She now computes a hash value (c) which is created from g, y and t:

$$c = \text{Hash}(g + y + t)$$

   Let's say this gives us 12 (we would normally limit the range of the value produced).

4. Now she computes r of:

$$r = v - c \times t = -124$$

5. Now she sends out t and r to prove her identity:

$$[815730721, -124]$$

6. Everyone who now who wants to prove her identity will then compute (where c can be calculated as a hash of g, y and t):

$$t = g^r \times y^c$$

In this case the calculation gives 815,730,720, which is the same as the value of *t* that Alice sent, so they have proven her identity. Every time Alice generates a new random number and then she proves that she knows the value of *t* each time. An outline of the code used is:

```python
import random

p=59
g=13
x=11
v=9

def string2numeric_hash(text):
    import hashlib
    return int(hashlib.md5(text).hexdigest()[:8], 16)

if (len(sys.argv)>1):
        g=int(sys.argv[1])

if (len(sys.argv)>2):
         x=int(sys.argv[2])

v= random.randint(3, 8)

print 'g=',g
print 'x=',x, ' (the secret)'
print 'v=',v, ' (random)'
print '=====Alice computes========='

import hashlib

y= g**x
t= g**v

print 't=',t

print 'y=',y

c = string2numeric_hash(str(g)+str(y)+str(t))
c =c % p

print 'c=',c

r= v -c*x

print '=============='

print 'Alice sends (t,r)=(',str(t),',',(r),')'

t1 = (g**r)
t2= (y**c)
```

```
val=int(t1*t2)
print 'My calc for g∧r x y∧c=',val

if (val==t):
        print "Alice has proven her ID"

else:
    print "You are a fraud"
```

The following provides an example:

```
g= 13
x= 11   (the secret)
v= 7    (random)
==============
t= 62748517
y= 1792160394037
c= 13
==============
Alice sends (t,r)=( 62748517 , -136 )
My calc for g∧r x y∧c= 62748517
```

   📖 **Web link (ZPF):** http://asecuritysite.com/encryption/z3

11.7 A Fair Coin Flip

We can create a fair coin flip between Bob and Alice where there is no trusted third party. Bob and Alice thus want to play a game of coin tossing over the Internet. Bob says that Alice can trust him to flip the coin, but Alice doesn't trust him to call the coin correctly. Neither of them can find anyone that they both trust on the Internet, so how can they play the game?

To create a fair coin flip, we can get Bob and Alice to generate a random number, where we are only interested in the least significant bit. Next Bob and Alice each take a secure hash of their random value, including with salt value. Bob sends his salted hash of the random number to Alice, and Alice sends her hashed number to Bob. Once they have confirmed they have received their hashed values, Bob sends his random value to Alice, and Alice sends her random value to Bob. They both check the salted hash against the random value they have received. If they check-out, they then calculate the result of an exclusive-OR of least significant bit of Alice's value and Bob's value, and this gives the flip. If the result is zero, it is tails, and a one gives heads.

For example, Bob creates a random value and hash:

```
Bob random= 341
Bob hash (with salt)= baf9e421673c2fcde06d6883c813197472eaf06
cffa2fcbcee3f74fb5981e507:5ed82b81670b457995938cedecfc75bd
```

Alice creates a random value and hash:

```
Alice random= 386
Alice hash (with salt)= a1823a25ca0a0c6741ddc8944f24050a28169
350092b43ffab967495e1b6887f:45f3e4982f6348af89d28b76643f3b58
```

Bob receives Alice's hash, and Alice receives Bob's hash. Bob sends Alice his random value, and Alice sends Bob her random value. They check the hash, and mask off the least significant bit and Ex-OR them together.

So why did we use a salted hash? Well Eve could have replayed the hashed value at some time in the future, so the salted hash value will match to the salt and the random value which has been used.

An outline of the code is here:

```python
import uuid
import hashlib
import random

def hash_password(password):
    salt = uuid.uuid4().hex
    return hashlib.sha256(salt.encode() + password.encode()).hexdigest() + ':' + salt

def check_password(hashed_password, user_password):
    password, salt = hashed_password.split(':')
    return password == hashlib.sha256(salt.encode() + user_password.encode()).hexdigest()
bob=random.randint(1, 1000)
hash_bob = hash_password(str(bob))

alice=random.randint(1, 1000)
hash_alice = hash_password(str(alice))

print '\n===Bob random and hash=====\n'
print 'Bob random=',bob
print 'Bob hash=',hash_bob

print '\n===Alice random and hash=====\n'

print 'Alice random=',alice

print 'Alice hash=',hash_alice

coin=(bob & 0x1) ∧ (alice & 0x1)
```

```
if (coin==0):
      print 'Heads ',
else
      print 'Tails ',

print '\n====Checking the flips ====\n'
print 'Alice checks value with salt: ',check_password(hash_bob,str(bob))
print 'Bob checks value with salt: ',check_password(hash_alice,str(alice))

print '\n====10 random flips====\n'

for i n range(1,10):
      bob=random.randint(1, 1000)
      hash_bob = hash_password(str(bob))

      alice=random.randint(1, 1000)
      hash_alice = hash_password(str(alice))

      coin=(bob & 0x1) ∧ (alice & 0x1)
      if (coin==0):
            print 'Heads ',
      else:
            print 'Tails ',
```

A sample run is:

```
===Bob random and hash=====
Bob random= 627
Bob hash= b2bc6827ff07445563b5511d04c258f0a7da9804f156493827e
1aeb7b9ee4087:df09efc5fb214ba98e6d7ef478dc97dc

===Alice random and hash=====
Alice random= 49
Alice hash= aabf5f1bb6833553ff5f88a73aa112d994455773c7a5512c0
e514b6ab5aec762:f50476971ce34d8e9cbacdcb7eac80c2
Heads
====Checking the flips ====

Alice checks value with salt: True
Bob checks value with salt: True

====10 random flips====
Tails  Heads  Heads  Tails  Tails  Heads  Tails  Tails  Heads  Heads
```

📖 **Web link (ZPF):** http://asecuritysite.com/encryption/z4

11.8 ZKP: Paillier

We can use the Paillier homomorphic crypto system to add or multiply to cryptography values. In this way we can have an election where each person encrypts their vote. Let's look at an example where we have two votes, where vote1 is 100 votes, and vote2 is 200 votes. In the following example, we encrypt vote1 and then vote2, and then add the encrypted values together, and then return the result. The decrypted value will be the addition of the two votes:

```
from phe import paillier
import sys
vote1=100
vote2=200

public_key, private_key = paillier.generate_paillier_keypair()

keyring = paillier.PaillierPrivateKeyring()

keyring.add(private_key)

public_key1, private_key1 = paillier.generate_paillier_keypair(keyring)

print 'Votes 1=',vote1
print 'Votes 2=',vote2

encrypted1= public_key.encrypt(vote1)
print 'Encrypted1=',encrypted1

encrypted2= public_key.encrypt(vote2)

print 'Encrypted2=',encrypted2

print 'Result =',private_key.decrypt(encrypted1+encrypted2)
```

📖 **Web link (ZPF):** http://asecuritysite.com/encryption/votes

11.9 Oblivious Transfer (OT)

Oblivious transfer allows a sender to not know the information that a receiver has read. So, we are Bob the Investigator and investigating a serious crime, and we suspect that Eve is the person who is involved in the crime. We now need to approach her employer (Alice) and ask for some information on her. So how do we do this without Alice knowing that we suspect Eve? Well oblivious transfer (OT) performs this. Let's say that HackerZForU employ Eve and Trent, and we are only interested in getting information on Eve, and that Alice runs the company.

Now the method we will use is based on the Diffie-Hellman key exchange method, but is modified so that we generated two keys for Alice to pass the data. One will work and the other will be useless. Alice will then have no idea which of the keys will work, and the information that we can look at. In this case we'll ask for data for both Eve(M_1) and Trent(M_2), and Alice will not know which of them is the suspect.

First Alice and Bob generate random numbers (a and b). Alice then takes a value of g and raises it to the power of a:

$$A = g^a$$

She passes this to Bob. If Bob is interested in the first record ($c == 0$) he calculates g to the power b, else if it is the second record ($c == 1$), he calculates the value passed from Alice (A), and multiplies this value with g to the power of b. Bob then sends one of these back (Figure 11.4):

$$\text{if}(c == 0) : B = g^b$$

$$\text{if}(c == 1) : B = A \times g^b$$

Alice receives the value from Bob (B). She then calculates two keys:

$$K_0 = Hash\,(\,B^a\,)$$

$$K_1 = Hash\left(\frac{B^a}{A}\right)$$

which is the hash of B to the power of a, and the hash of B/A to the power of a. She then encrypts the two messages (M_0 and M_1) with each of the keys, and returns the ciphers to Bob:

$$e_0 = E_{k0}\,(M_0)$$

$$e_1 = E_{k1}\,(M_1)$$

Bob calculates the decryption key (which will only work for one of the them) as the hash of A to the power of b:

$$K_{Bob} = Hash\left(A^b\right)$$

Bob will then try to decrypt the two ciphers with K_{BOB} and only one will work (Figure 11.4).

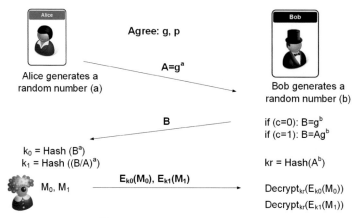

Figure 11.4   Oblivious Transfer.

Some sample code is:

```
from Crypto.Cipher import AES
import hashlib
import random
import sys

g=9
n=1001

a=random.randint(5, 10)

b=random.randint(10,15)

Alice=(g**a)  % n

c=1
if (len(sys.argv)>1):
        c=int(sys.argv[1])

print 'g: ',g,' n: ',n
print 'Alice value: ',Alice
print 'a (Alice random): ',a
print 'b (Bob random): ',b

# === Bob calculates ===

if (c==0):
        Bob=(g**b) % n
else:
        Bob=Alice*((g**b) % n)
```

```
# === Alice calculates ===

key0 = hashlib.sha256(str((Bob**a) %n)).digest()
key1 = hashlib.sha256(str(((Bob/Alice)**a) %n)).digest()

cipher1 = AES.new(key0, AES.MODE_ECB)
cipher2 = AES.new(key1, AES.MODE_ECB)

print '\nAlice calculates these keys'
print 'Key 0: ',key0
print 'Key 1: ',key1

en0=cipher1.encrypt('Bob did it       ')
en1=cipher2.encrypt('Alice did it     ')

## === Bob decrypts
print '\nBob calculates this key:'
Bob_key = hashlib.sha256(str((Alice**b) %n)).digest()
print 'Bob key: ',Bob_key

cipher1 = AES.new(Bob_key, AES.MODE_ECB)

message_0=cipher1.decrypt(en0)
message_1=cipher1.decrypt(en1)

print '\nBob decrypts the messages:'
print 'Message 0: ',message_0
print 'Message 1: ',message_1
```

A sample run gives:

```
g:  9  n:  1001
Alice value:  456
a (Alice random):  9
b (Bob random):  15

Alice calculates these keys
Key 0:  k†²sÿ4üá□k€NÿZ?WG-¤ê¢/IÀ-Rý ‡[K
Key 1:  _iëfÿÉo8ÚRxlmilyÂÛÂ9ýN'’g)×:'ûWé

Bob calculates this key:
Bob key:  k†²sÿ4üá□k€NÿZ?WG-¤ê¢/IÀ-Rý ‡[K

Bob decrypts the messages:
Message 0:  Bob did it
Message 1:  □Úzà|'x'UZ1óH
```

📖 **Web link (ZPF):** http://asecuritysite.com/encryption/ot

11.10 Scrambled Circuits

Let Bob and Alice agree a secure protocol, where they must either both agree to something (A AND B) or where just one of them has to agree (A OR B), or where just one agrees and the other does not agree (A $\oplus$ B). This simple protocol is thus a logic function.

Let's say they agree to both agree on something (with a YES or NO), so we have an AND function (A & B), and we now create a scrambled AND gate. For this we start with an AND function get:

A	B	Z
0	0	0
0	1	0
1	0	0
1	1	1

Now Bob creates four encryption keys K(a=0), K(a=1), K(b=0) and K(b=1). Next he will go ahead and encrypt the four possible outputs (in this case "0", "0", "0", and "1"), using the two encryption keys associated with the bits. For example, the output:

```
A=0, B=0, Z=0
```

will encrypt the output ("0"), with the keys of K(a=0) and K(b=0). The outline code is:

```
keyX_0 = Fernet.generate_key()
keyX_1 = Fernet.generate_key()
keyY_0 = Fernet.generate_key()
keyY_1 = Fernet.generate_key()

data =[]
for a in range(0,2):
    for b in range(0,2):
        data.append(str(eval(operator) & 0x01))

cipher_text00 = Fernet(keyY_0).encrypt(Fernet(keyX_0).encrypt(data[0]))
cipher_text01 = Fernet(keyY_0).encrypt(Fernet(keyX_1).encrypt(data[1]))
cipher_text10 = Fernet(keyY_1).encrypt(Fernet(keyX_0).encrypt(data[2]))
cipher_text11 = Fernet(keyY_1).encrypt(Fernet(keyX_1).encrypt(data[3]))
```

Bob then passes the cipher_text values (cipher_text00 ... cipher_text11) to Alice, and provides the key for his input. If he says YES, then he passes keyX_1, otherwise he will pass keyX_0.

Now Alice receives the four values, and Bob's key. Now she uses obviously transfer to gain the key for her answer. If she says YES, she obtains the key for keyY_1, without Bob actually knowing that she says YES. If she says NO, she gets keyY_0.

In the end she will have two keys and she tries all the ciphers:

```
try:
    print Fernet(keyB).decrypt(Fernet(keyA).decrypt(cipher_text00))
except:
    print ".",
try:
    print Fernet(keyB).decrypt(Fernet(keyA).decrypt(cipher_text01))
except:
    print ".",
try:
    print Fernet(keyB).decrypt(Fernet(keyA).decrypt(cipher_text10))
except:
    print ".",
try:
    print Fernet(keyB).decrypt(Fernet(keyA).decrypt(cipher_text11))
except:
    print ".",
```

and the only one she can open is the one that matches Bob and Alice's decision.

To illustrate, let's say that Bob and Alice say "NO", so their inputs to the AND function will be 0 and 0. Now Bob encrypts the four outputs with the four keys, but the only one which will open up the outputs will be the keys for a zero input for A and for B. Bob passes his key, without revealing his input as K(a=0), and Alice receives her key through an oblivious transfer (Figure 11.5).

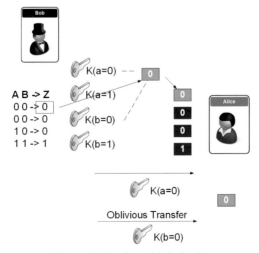

Figure 11.5 Scrambled circuit.

The outline code is:

```
from cryptography.fernet import Fernet
import sys
import binascii

operator = "a \& b"
x=0
y=0

operator=operator.replace('or','|')
operator=operator.replace('and','&')
operator=operator.replace('xor','^')
operator=operator.replace('not','~')

print "---Input parameters---"
print "Operation:",operator
print "Input:",x,y

keyX_0 = Fernet.generate_key()
keyX_1 = Fernet.generate_key()
keyY_0 = Fernet.generate_key()
keyY_1 = Fernet.generate_key()

data =[]
for a in range(0,2):
        for b in range(0,2):
                data.append(str(eval(operator) & 0x01))
print "Outputs of function:",data

print "\n---Keys generated---"

print "KeyX_0 (first 20 characters):"+binascii.hexlify(bytearray(keyX_0))[:20]
print "KeyX_1 (first 20 characters):"+binascii.hexlify(bytearray(keyX_1))[:20]
print "KeyY_0 (first 20 characters):"+binascii.hexlify(bytearray(keyY_0))[:20]
print "KeyY_1 (first 20 characters):"+binascii.hexlify(bytearray(keyY_1))[:20]

print "\n---Cipers send from Bob to Alice---"

cipher_text00 = Fernet(keyY_0).encrypt(Fernet(keyX_0).encrypt(data[0]))
cipher_text01 = Fernet(keyY_0).encrypt(Fernet(keyX_1).encrypt(data[1]))
cipher_text10 = Fernet(keyY_1).encrypt(Fernet(keyX_0).encrypt(data[2]))
cipher_text11 = Fernet(keyY_1).encrypt(Fernet(keyX_1).encrypt(data[3]))

print "Cipher (first 20 chars): "+binascii.hexlify(bytearray(cipher_text00))[:40]
print "Cipher (first 20 chars): "+binascii.hexlify(bytearray(cipher_text01))[:40]
print "Cipher (first 20 chars): "+binascii.hexlify(bytearray(cipher_text10))[:40]
print "Cipher (first 20 chars): "+binascii.hexlify(bytearray(cipher_text11))[:40]

if (x==0): keyB = keyX_0
if (x==1): keyB = keyX_1

if (y==0): keyA = keyY_0
if (y==1): keyA = keyY_1

print "\n---Bob and Alice's key---"
print "Bob's key: "+binascii.hexlify(bytearray(keyB))[:20]
print "Alice's key: "+binascii.hexlify(bytearray(keyA))[:20]
```

```
print "\n---Decrypt with keys (where '.' is an exception):"

try:
        print Fernet(keyB).decrypt(Fernet(keyA).decrypt(cipher_text00)),
except:
        print ".",
try:
        print Fernet(keyB).decrypt(Fernet(keyA).decrypt(cipher_text01)),
except:
        print ".",
try:
        print Fernet(keyB).decrypt(Fernet(keyA).decrypt(cipher_text10)),
except:
        print ".",
try:
        print Fernet(keyB).decrypt(Fernet(keyA).decrypt(cipher_text11)),
except:
        print ".",
```

A sample run is:

```
---Keys generated---
KeyX\_0 (first 20 characters):337853474c48344e4c72
KeyX\_1 (first 20 characters):555954354f6669706867
KeyY\_0 (first 20 characters):47525763366750782d55
KeyY\_1 (first 20 characters):30437934507343734e4a
---Ciphers send from Bob to Alice---
Cipher (first 20 chars): 674141414141425a5a6f6d58332d387944705844
Cipher (first 20 chars): 674141414141425a5a6f6d583458515843443149
Cipher (first 20 chars): 674141414141425a5a6f6d58456776384467777a
Cipher (first 20 chars): 674141414141425a5a6f6d584772525231326a38

---Bob and Alice's key---
Bob's key: 337853474c48344e4c72
Alice's key: 47525763366750782d55

---Decrypt with keys (where '.' is an exception):
0 . . .
```

📖 **Web link (ZPF):** http://asecuritysite.com/encryption/obf

11.11 Millionaire's Problem

So Bob and Alice have been doing rather well in computer security. After they were named as actors in computer security models, they have done well with licencing deals, and are now millionaires. But how can we tell who has more money, without them revealing how much each of them has? One method involves us using RSA encryption, so let's use the following for our RSA key selection:

$$e = 79$$

$$d = 1019$$

$$N = 3337$$

and select a prime number:

$$p = 631$$

We define I as Bob's millions, and J as Alice's millions:

$$J = 4$$

$$I = 5$$

Next we select a random number U:

$$U = \mathrm{randint}(0, 2000)$$

And compute the C value (which is the RSA encryption process):

$$C = U^e \bmod N$$

Now Alice calculates:

$$Aliceval = C - J + 1$$

She shares this with Bob, and Bob calculates 10 values:

$$Z_1 = (Aliceval + 1)^d \bmod N \bmod p$$

$$Z_2 = Aliceval + 2^d \bmod N \bmod p$$

$$\cdots$$

$$Z_{10} = (Aliceval + 10)^d \bmod N \bmod p$$

Bob then takes these values, and for the i-th value, and onwards, he will one he will add one onto these value. These are then sent back to Alice. Alice now computes:

$$G = U \bmod p$$

If the j-th value is the same as G, Alice has more money or the same, else Bob has more money. The following shows some sample code:

```
import sys
from random import randint

J = 4
I = 5

e=79
d=1019
N=3337

primes = [601,607,613,617,619,631,641,643,647,653,659,661,673,
677,683,691,701,709,719,727,733,739,743,751,757,761,769,773,
787,797,809,811,821,823,827,829,839,853,857,859,863,877,881,
883,887,907,911,919,929,937,941,947,953,967,971,977,983,991,997]
val=randint(0,len(primes))
p=primes[val]

U=randint(0,2000)

C=(U**e) % N

print 'Bob has',I,'millions'
print 'Alice has',J,'millions'
print '\ne=',e,'d=',d,'N=',N,'p=',p
print '\nRandom Value U is:\t',U
print 'C value is (U^e %N):\t',C

val_for_alice = C - J + 1
print "Alice shares this value (C-J-1):",val_for_alice

Z=[]

for x in range(0,10):
        val = (((val_for_alice+x)**d) % N) % p
        if (x>(I-1)):
                Z.append(val+1)
        else:
                Z.append(val)

G = U % p

print "\nG value is",G
print "Z values are:",
for x in range(0,10):
        print Z[x],

print '\n\nAlice checks U(',U,') against the ',J,'th value (',Z[J-1],')'
if (G==Z[J-1]): print "\nSame. Bob has more money or the same"
else: print "\nDiffer. Alice has more money"
```

A sample run is:

```
Bob has 5 millions
Alice has 4 millions

e= 79 d= 1019 N= 3337 p= 691
```

```
Random Value U is: 1903
C value is (U∧e %N): 644
Alice shares this value (C-J-1): 641

G value is 521
Z values are: 655 29 223 521 553 142 656 106 137 412

Alice checks U( 1903 ) against the  4 th value (521)

Same. Bob has more money or the same
```

📖 **Web link (ZPF):** http://asecuritysite.com/encryption/mill

11.12 RAPPOR

Before we look at RAPPOR, let's look at Bloom filters. For this we take a number of hash values of a string, and then set the corresponding bits for the Bloom filter. For example, if we take two bits to represent the hashed values at three things:

	01234567890123456789012345678901
Add fred:	00000000000000100000010000000000 fred [21,14]
Add bert:	00000000100000100000010000000100 bert [29,8]
Add greg:	00000000100100100000011000000100 greg [11,22]

If we now search for "fred" we will see that bits 21 and 14 are set, so "fred" may be in the Bloom filter. We now have bit position 8, 11, 14, 21, 22 and 29 set.

Now we can test for amy (which sets bits 12 and 16):
    amy is not there [16,12]
New we can test for greg (which sets bits 11 and 22):
    greg may be in there [11,22]

   Web link (ZPF): http://asecuritysite.com/encryption/bloom

RAPPOR uses hashes and adds noise. For this we have a k-bit Bloom filter. There are then three steps to the gathering of information.

1. The first part is the signal creation, and involves taking the message and hashing it with a number of hashing methods (h) and setting bits on a Bloom filter (B).

2. Next we have a Permanent Randomized Response (PRR), which will be memorized on the system. With this, for each of bits (0 to $k-1$), and we create a fake Bloom filter, which is set as:

B_fake[i] = 1, for a probability of $0.5 \times f$
B_fake[i] = 0, for a probability of $0.5 \times f$
B_fake[i] = B[i], for a probability of $1 - f$

So, if $f = 0.5$, we will set with a fake 0 for a 1-in-4 chance, and a fake 1 for 1-in-4 chance, and the actual bit in the Bloom filter (B) for 1-in-2. This is equivalent to having a coin flip, where if it is head, we will lie about the result, but if it is tails we will tell the truth. This will create a fake Bloom filter with noise (based on a coin flip), where half the time we will fake the bit, and the other time we will put the correct answer into the Bloom filter. This bit array is memorized for future reports. When required to read the data, we perform the Instantaneous Randomized Response (IRR), which takes B_fake and creates S using:

If the bit in B_fake is set to 1, the corresponding bit in S is set to 1 with probability q.
If the bit in B_fake is set to 0, the corresponding bit in S is set to 1 with probability p.

Figure 11.6 provides an example the process.

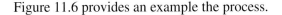

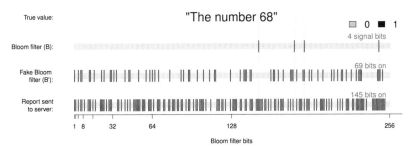

Figure 11.6 RAPPOR processing [1].

With this we have input data values of:

```
client,cohort,value
1,24,'hello'
1,24,'hello'
```

```
2,34,'goodbye'
1,43,'hello'
1,40,'help'
```

and it produces:

```
client,cohort,bloom,prr,irr
1,24,1000101000100000,0001001000110000,1011011011111100
1,24,1000101000100000,0001001000110000,1001001011001001
2,34,0000000001000111,0000000001011110,1000011100100111
1,43,0001000000100001,0101001110110011,1001001010101000
1,40,0010010000010000,0001110100011010,0101011111100010
```

We can see for the same client and cohort that we produce the same Bloom filter (1000101000100000), but the PRR has noise applied to it (0001001000110000). Finally the IRR value differs, even though we have the same input (1011011011111100 and 1001001011001001). In this way we cannot infer the input values.

If we use freq=1, all the bits will be random:

```
client,cohort,bloom,prr,irr
1,24,1000101000100000,0001011001111001,0011011010111000
1,24,1000101000100000,0001011001111001,0010100101110100
2,34,0000000001000111,1000100111011100,0000000011100011
1,43,0001000000100001,0101111111010111,0111000101100101
1,40,0010010000010000,0001100110101010,1000101111101011
```

If we look in more detail at the first entry, we see the bits that have changed (where an X is a change) and where we see that eight are changed and eight have not (so we see a random pattern):

```
1000101000100000
0001011001111001
X--XXX---X-XX--X
```

and with f=0.5, we have a 1-in-2 chance of a fake bit:

```
client,cohort,bloom,prr,irr
1,24,1000101000100000,0001001000110000,1001101011111100
1,24,1000101000100000,0001001000110000,0111101011111000
2,34,0000000001000111,0000000001011110,0011111111111110
```

```
1,43,0001000000100001,0101001110110011,0011010010101011
1,40,0010010000010000,0001110100011010,1011100101011001
```

This time we only see four bits changing:

```
1000101000100000
0001001000110000
X--XX------X----
```

and with freq=0 we will get all of the bits from B coming through, and where there will be no random bits in B_fake:

```
client,cohort,bloom,prr,irr
1,24,1000101000100000,1000101000100000,1010101100000111
1,24,1000101000100000,1000101000100000,0000010111111000
2,34,0000000001000111,0000000001000111,1101010010001011
1,43,0001000000100001,0001000000100001,0000001011010111
1,40,0010010000010000,0010010000010000,1010111011001001
```

If we take a simple example, let's say you have 10 friends on Facebook, and you want to know the split of male:female. So how can we do this without actually asking them to tell you their gender? Well, we get them to toss a coin. If it is heads, they should lie, else if it is tails they should tell the truth. So here we go:

Friend 1 (Male). Tosses coin: Tails. Tell: Male (Truth!)
Friend 2 (Male). Tosses coin: Heads. Tell: Female (Lie!)
Friend 3 (Male). Tosses coin: Heads. Tell: Female (Lie!)
Friend 4 (Female). Tosses coin: Tails. Tell: Female (Truth!)
Friend 5 (Female). Tosses coin: Heads. Tell: Female (Lie!)
Friend 6 (Female). Tosses coin: Heads. Tell: Male (Lie!)
Friend 7 (Male). Tosses coin: Tails. Tell: Male (Truth!)
Friend 8 (Male). Tosses coin: Tails. Tell: Male (Trust!)
Friend 9 (Male). Tosses coin: Tails. Tell: Male (Truth!)
Friend 10 (Female). Tosses coin: Heads. Tell: Male (Lie!)

Note: a lie has a 50/50 chance of being male or female.

So the number of heads are the same as tails (which is expected), and we have six males, and four females from the results, and this is the same as the cohort, but we can't tell who was actually telling the truth or not. This is the basis for methods of privacy within Big Data analysis, where we can lie (or add noise), and protect our data, but where statistics a still be gathered over a given population.

📖 **Web link (ZPF):** http://asecuritysite.com/encryption/rap

11.13 Secure Function Evaluation (SFE)

A Secure Function Evaluation (SFE) method can be used to verify a value, without releasing the original data. For example if we have a voting competition with Bob, Alice and Carol. Bob, Alice and Carol vote, and they want to keep their votes secret, but they need to calculate the overall total. Typically an independent person would tally up the votes, but what if they do not trust anyone? This is where SFE comes in, and where they can calculate the total with knowing the votes from the others. For example, let's say they generate some votes:

Bob:	Alice	Carol
2	7	7

Next Bob creates three random values which give a sum of (his vote+100) to give:

Bob 1	Bob 2	Bob 3
57	28	17

Next Alice creates three random values which give a sum of (her vote+100) to give:

Alice 1	Alice 2	Alice 3
8	89	10

Next Carol creates three random values which give a sum of (her vote+100) to give:

Carol 1	Carol 2	Carol 3
46	18	43

Bob gets Alice's second value, and Carol's second value, and adds it to his first value and calculates the sum as: **164**

Alice gets Bob's second value, and Carol's third value, and adds it to her first value and calculates the sum as: **79**

Carol gets Alice's third value, and Bob's third value, and adds it to her first value and calculates the sum as: **73**

Finally Bob, Alice and Carol announce their calculations, and it is added up to: **316**

which taken with modulo 100 to give: **16**

which should be the total of the votes. So ... Bob, Alice and Carol know the total, but not any of votes of the others.

  📖 **Web link (ZPF):** http://asecuritysite.com/encryption/sfe

11.14 Secure Remote Password (SRP) Protocol

We build systems which are often insecure and where we pass our passwords over channels which can contain sniffing agents, such as for man-in-the-middle ones, and which can discover our password. We also typically use HTTPs as a tunnel, and where we only authenticate one side to the other. The method often used to authenticate Bob the Server to Alice the User is with then a digital certificate. So how do we authenticate each side, and password the proof of the password, without actually storing the password?

One method to improve the process is SRP. In this protocol the server does not contain any password-related data, and involves the client providing a proof that it knows the password, without giving away what the password is.

So Alice has a password p, and wants to define it with Bob the server. Alice first selects some salt (s) and computes a hash of s and p:

$$x = Hash(s, p)$$

Next she then calculates v using a generator value (g):

$$v = g^x$$

Alice sends this (and s), and Bob the Server indexes (v, s) with a value of I. The password is now registered with Bob the Server. Now Alice will prove she knows the password. First she generates a random number (a) and sends I and A:

$$I, A = g^a$$

Bob then sends the salt (s) and B:

$$s, B = kv + g^b$$

Alice and Bob then compute:

$$u = \text{Hash}(A, B)$$

Alice computes a hash of s, I and p, and then calculate a shared session key of:

$$x = \text{Hash}(s, I, p)$$

$$S_c = \text{pow}(B - k \times \text{pow}(g, x, N), a + u \times x, N)$$

$$K_c = \text{Hash}(S_c)$$

Bob computes a session key of K_s:

$$K_S = \text{pow}(A \times \text{pow}(v, u, N), b, N)$$

$$S_c = \text{Hash}(S_S)$$

where pow(A, B, N) is A^B mod N. They should now have the same session key ($K_s == K_c$), and will send messages from Alice to Bob the Server:

Alice (cipher) = K_s("I am Alice")

and Bob the Server to Alice:

Bob(cipher) = K_c("I am Bob")

If the messages decode we authenticate Alice to Bob, and Bob to Alice. A sample run is:

```
#. H, N, g, and k are known beforehand to both client and server:

N = 0xc037c37588b4329887e61c2da3324b1ba4b81a63f9748fed2d8a410c2fc21b1232f0d3
bfa024276cfd88448197aae486a63bfca7b8bf7754dfb327c7201f6fd17fd7fd74158bd31ce7
72c9f5f8ab584548a99a759b5a2c0532162b7b6218e8f142bce2c30d7784689a483e095e7016
18437913a8c39c3dd0d4ca3c500b885fe3
g = 0x2
k = 0xb317ec553cb1a52201d79b7c12d4b665d0dc234fdbfd5a06894c1a194f818c4a
```

```
0. server stores (I, s, v) in its password database
I = fred
p = qwerty1234
s = 0x4aad95565d8b9e6
x = 0xd9e2d50f122a85483130284e2558ed22e2e83ddd5c57ff3e51844d0455b94e7f
v = 0x7e4ae36f11197bc44759835f4d18868ac18db173225afaf3d8ee97ce67794082db006d
4e702cf25c85cd1a49720447ad537d99b0dbe98cfa167b05dea379c403191011394e8f67fc21
414c1e6ad3f5fa38d3cdac113b7093060fc4c3f260656b1fc89153d67af0327991aef2d2834b
555d791c4280784ab9e2d903216b9fc03

1. client sends username I and public ephemeral value A
to the server
I = fred
A = 0x16c558041e1f5845a0b7b0d5011b4dcd3c284cc907e9521e81afaf581b0388e67852c36
0a4e650c79ed2cca26894394e84afe087f24b54f2b57cbb93dbb7844d4efe1e04dacd3c825ed3
de0947e5241d3eb17f21875b865b0d3d2e783137a24fd7d46ef4762ca1112f9695a01388464ff
4145172a6004e252472e6310b94b804

2. server sends user's salt s and public ephemeral value Bto client
s = 0x4aad95565d8b9e6
B = 0x9ac3d20882263a9d659b90bb3bdd72851f67223dcd1ce5b86c47c873f895e5ae82413f
e93d3347e54d565dd7509e56afa15ab0596b84870e511df39a4a5930af29d86835861e5ae4c0
bae21d87f149082e89d970e72259ecfcaa1ec51960d7ffc8994841c558105ab473218685dab6
263e5bb4319b4863e9499ae5229e0a6791

3. client and server calculate the random scrambling parameter
u = 0x6ff7fdb99f0032c19fcb1f9cb850fb6e6c518ed1e6bfb160234a1ca9335f7993

4. client computes session key
S_c = 0x70b2d0d02c7d6a3ec2820962159cb039cc3b6841b8c350e19ee330693589a1f613f19
6ce59c5d401750838c7839dff01178cd9797143d1b48bb6873a7e13c61fce6e77cbd2482db4f4
ce45794700cf54a50e36e0f9670cebbe7b05880a61f301dd9d5ecf3270d47f06acd3981e4e7b2
0984f70573d7bddc66639fe3107aea872
K_c = 0x521c877a97ad0b99e137893757cb49ad510bb73d878c89f602e707a2aa5560ab

5. server computes session key
S_s = 0x70b2d0d02c7d6a3ec2820962159cb039cc3b6841b8c350e19ee330693589a1f613f1
96ce59c5d401750838c7839dff01178cd9797143d1b48bb6873a7e13c61fce6e77cbd2482db4
f4ce45794700cf54a50e36e0f9670cebbe7b05880a61f301dd9d5ecf3270d47f06acd3981e4e
7b20984f70573d7bddc66639fe3107aea872
K_s = 0x521c877a97ad0b99e137893757cb49ad510bb73d878c89f602e707a2aa5560ab

6. client sends proof of session key to server
M_c = 0xa77a4c0b4586ef7bf80c512d03148cb5cf1f9e52ce8c99b37a6b06fc566c6622

7. server sends proof of session key to client
M_s = 0xdcf2d2f86f4a718ac0e04f6e671c97771a0ce5bde3001c8660d11455407ade83
```

📖 **Web link (SRP):** http://asecuritysite.com/encryption/srp

11.15 Lab/Tutorial

The lab and tutorial related to this chapter is available on-line at:

http://asecuritysite.com/crypto11

Reference

[1] Ú. Erlingsson, V. Pihur, and A. Korolova, "RAPPOR: Randomized Aggregatable Privacy-Preserving Ordinal Response," *Proc. 2014 ACM SIGSAC Conf. Comput. Commun. Secur. – CCS '14*, pp. 1054–1067, 2014.

12

Wireless Cryptography and Stream Ciphers

12.1 Introduction

We increasingly connect to networks using wi-fi connections, which can be at home, within a corporate environment, or within public spaces. As the communications channel is open for others to listen too, we must be careful about the security of the communications, and on the authentications methods used. In a home network we often connect to a single device – a wireless router – which is configured to support the required security levels. On most corporate networks, this policy is normally defined centrally and where we have light-weight access devices (AP – access points), which use a back-end authentication server to define the access policies.

Within Figure 12.1, we see that a client (a supplicant) connects to a wireless access point, and which can have local authentication. The user is then not allowed to connect to the wi-fi network unless they can provide the required access credentials. The encryption used either supports 40-bit RC4 (for WEP), 128-bit RC4 (for WPA – Wi-Fi Protected Access), or 128-bit/256-bit AES (for WPA-2). For a corporate network, the authentication request is sent to a remote authenticator, which then allows the organisation to centralise it authentication access method, and create user credentials which scale across multiple access points (Figure 12.2).

Before we analyse the latest standards for wi-fi encryption (WPA and WPA-2), it is important to understand the problems that have been caused by WEP, and which had many weaknesses, including having a global encryption key for a whole network, and in bit flipping. Overall, the method had many weaknesses which meant that any network key could be crack in just a few hours. Along with this, stream ciphers are often used within wireless networks with transmitters and receivers which can struggle to cope with the requirements for block ciphers. The chapter will thus cover two important ciphers: RC4 and ChaCha.

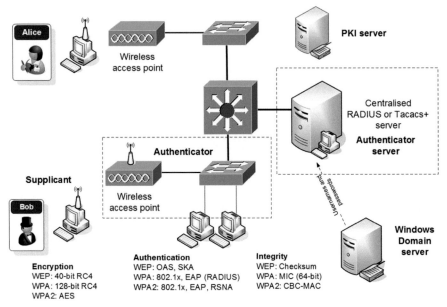

Encryption
WEP: 40-bit RC4
WPA: 128-bit RC4
WPA2: AES

Authentication
WEP: OAS, SKA
WPA: 802.1x, EAP (RADIUS)
WPA2: 802.1x, EAP, RSNA

Integrity
WEP: Checksum
WPA: MIC (64-bit)
WPA2: CBC-MAC

Figure 12.1   Wi-fi Overview.

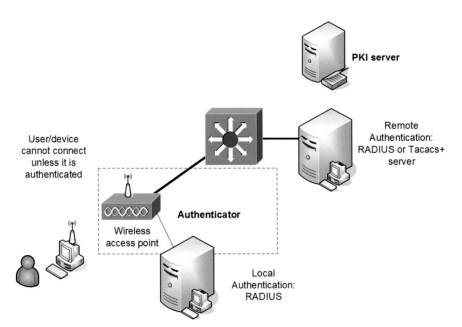

Figure 12.2   Wi-fi Authentication.

12.2 RC4

In the past networking adapters in wireless systems were fairly basic and often lacked the processing power required to implement a strong block cipher. These days, increasingly, too, we create sensor networks which can have limited space for memory and for processing. For these reasons wireless systems have often used stream ciphers rather than block ones, and where we create a key stream which is then EX-ORed with the plaintext stream. Within IEEE 802.11, wi-fi has evolved through the usage of the RC4 stream encryption method. Because of export restrictions, the size of the original key in RC4 was limited to 40 bits (for WEP) and was then increased to 128 bits (TKIP).

Within the RC4 encryption algorithm, we generate a key stream which is then XOR-ed with the plaintext. This makes the output processing of the cipher a fairly simple operation (and thus fast to process). There is then an 8×8 S-Box which gives a permutation of 8-bit input values, with values of 0 to 255, and where the permutation is a function of the variable length key. Overall we just need two counters (i and j) and which are set to zero at the start of the processing.

One of the core advantages of using RC4 is that the key length is variable, and can range from 1 to 256 bytes (from 8 bits to 2,048 bits). The key itself is used to initialise a 256-byte state table, and which is used to generate pseudo-random bytes for the key stream. Each element of the state table will be swapped at least one time. To generate the key stream we have [1]:

```
def rc4_key_setup(key):
    key_length = len(key)
    Sbox = range(256)
    j = 0
    for i in range(256):
        j = (j + Sbox[i] + key[i % key_length]) % 256
        Sbox[i], Sbox[j] = Sbox[j], Sbox[i]   # swap values in S-box
    return Sbox
def rc4_key_stream(Sbox):
    i = 0
    j = 0
    while True:
        i = (i + 1) % 256
        j = (j + Sbox[i]) % 256
        Sbox[i], Sbox[j] = Sbox[j], Sbox[i]   # swap values in S-box
        tmp = Sbox[(Sbox[i] + Sbox[j]) % 256]
        yield tmp # return infinite key
```

There are two main phases: key setup (RC4_setup) and key ciphering (RC4_key_stream). In the key setup phase we take an *n*-byte key and determine its length. This length value is then used to setup the key, along with swapping and modulo operations. The output of this will be an S-box of 256 values, based on the key length and the key contents. For example, if we use a key of "test" (length of four bytes), there will be 256 iterations to determine the S-box, and we get the following:

```
[116, 218, 23, 7, 64, 230, 30, 211, 24, 151, 56, 113, 92, 69, 245,
120, 43, 117, 127, 74, 50, 186, 87, 148, 252, 149, 243, 169, 89,
152, 76, 238, 115, 222, 161, 198, 162, 34, 131, 42, 250, 138, 96,
3, 95, 241, 146, 60, 90, 55, 224, 124, 99, 172, 160, 135, 47, 2, 44,
212, 9, 38, 157, 80, 62, 170, 37, 97, 206, 28, 233, 2, 36, 5, 72,
126, 125, 176, 12, 195, 94, 46, 110, 111, 0, 119, 66, 190, 147, 173,
227, 109, 221, 79, 133, 246, 184, 231, 182, 139, 163, 191, 188, 174,
137, 122, 118, 8, 103, 71, 25, 49, 17, 18, 1, 13, 58, 123, 21, 52,
213, 225, 235, 215, 179, 65, 158, 223, 247, 101, 177, 244, 193, 214,
81, 201, 189, 41, 255, 155, 100, 98, 167, 217, 181, 107, 175, 203,
67, 88, 220, 229, 39, 40, 6, 121, 45, 10, 207, 153, 15, 57, 32, 36,
234, 232, 242, 102, 29, 86, 200, 178, 145, 150, 205, 85, 105, 108,
154, 48, 165, 249, 129, 59, 136, 194, 164, 128, 180, 104, 75, 237,
130, 210, 61, 143, 216, 183, 197, 219, 202, 226, 91, 166, 20, 82,
26, 16, 63, 141, 199, 68, 156, 134, 251, 11, 78, 84, 253, 140, 106,
142, 132, 31, 73, 33, 83, 77, 54, 192, 51, 168, 240, 248, 204, 208,
228, 4, 70, 19, 254, 22, 171, 209, 27, 53, 14, 159, 35, 144, 112,
93, 114, 187, 196, 185, 239]
```

This has now scrambled the S-box based on the size of the key and key content. Note that a variation of either the key size and/or the contents of the key will change the S-box. After this there is then no need for the key again, as the S-box is passed into a scrambling process which takes two counters (*i* and *j*) and does modulus and S-box swapping operations (RC4 key-stream). To produce an pseudo-infinite key length, in Python, we use the **yield** statement to generate the required length of the key when it is used within the output X-OR stage.

The core weakness of RC4 is that if you have two ciphertext streams of the same plaintext and with the same key, it is possible to X-OR them together to recover the original plaintext. To overcome this, RC4 uses an IV (Initialisation Vector or nonce), and the method is then secure, unless the IV value repeats for the same key.

In many systems the IV value starts with a random value and is then incremented for every cipher value sent. This problem is overcome by making sure that a new key is generated before the IV value rolls-over. As we will see, within IEEE 802.11 networks, WEP (Wireless Equivalent Protocol), which uses RC4 and a 24-bit IV, could roll-around after a few hours. This method was replaced by TKIP (Temporal Key Integrity Protocol) and which had a 48-bit IV value, and where a session key was used as the seed for the S-box. With a session key, a new key is created within a given time period, and overcomes the problem of a roll-over of the IV value.

12.3 WEP

WEP was one of the first security methods that was used within IEEE 802.11b networks. As wireless devices often have limited processing capabilities, they can use the RC4 stream cipher. With this we have a 24-bit IV and a 40-bit key. RC4 then creates a pseudo-infinite key stream which is then X-ORed with the plaintext input. For each packet sent, the IV value is incremented, in order that the same plaintext value will not appear as the same ciphertext (Figure 12.3).

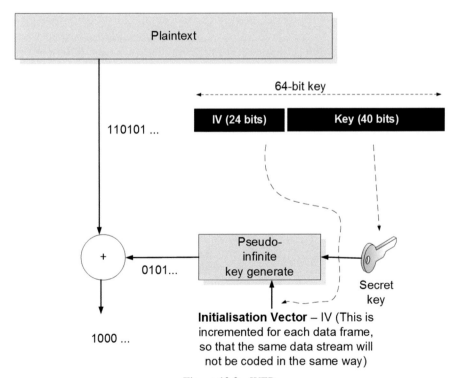

Figure 12.3   WEP.

In Figure 12.4 we see that the sender and the receiver share the same shared 40-bit key, and that the IV must be sent with the encrypted data packet. In this way the receiver can determine the IV value, and use it with the shared encryption key. This is similar to adding an IV to block ciphers and salt to hashing values, and where the IV and salt value must always be sent/stored with the cipher value, otherwise it will be difficult to decrypt without these. With WEP we had a global key for the whole network, which meant that anyone who gained the key, could sniff (and crack) the contents of the whole network.

One problem with WEP is that it has a simple checksum value, and which is calculated by taking four-byte values, and then creating either odd or even parity bit for each bit position (Figure 12.5). Eve can then flip bits within each of the columns and change the transmitted values, while also giving the correct checksum. For example, Eve could flip the bits in the IP packet header, and change the IP destination of a data packet (Figure 12.6), where Eve redirects the destination from alice.com to mal.com.

The major problem with WEP, though, is the small 24-bit IV value, and which can roll-over to the same value within a reasonable time space. Overall there can only be 2^{24} IV values (16,777,216), and if we use 1500 byte

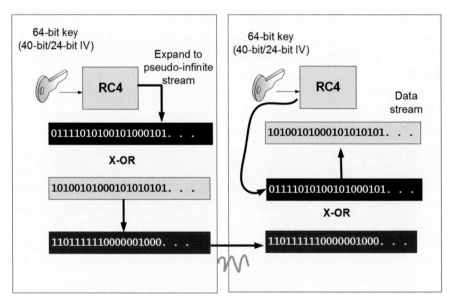

Figure 12.4 WEP transmitter and receiver.

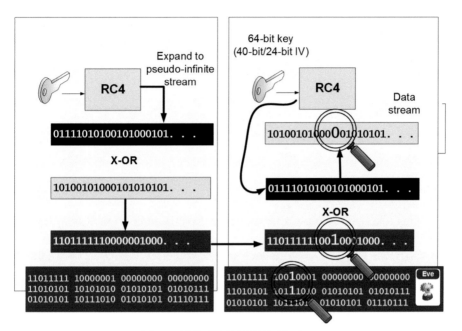

Figure 12.5 WEP RC4 bit flipping.

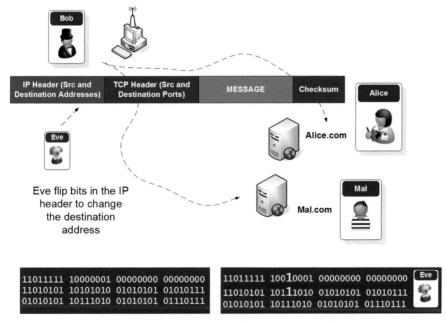

Figure 12.6 WEP RC4 bit flipping for IP header.

data frames, the minimum time to send each packet on a 10 Mbps link will be:

$$\text{Time to sent} = \frac{1500 \times 8}{10 \times 10^6} = 1.2 \text{ ms}$$

Thus, if the device is continually sending data packets, the same IV value will repeat after:

$$\text{Time to repeat} = 1.2 \times 10^{-3} \times 16,777,216 = 20,132.7 \text{ seconds}$$

which is around 5.6 hours. If we use a 100 Mbps link, the value will drop by a factor of 10.

Eve thus takes the two ciphertexts which have been encrypted with the same key and IV, and performs a statistical analysis on it. As shown in Figure 12.7, Eve just has to store each of the ciphertext values, and waits for the IV value to repeat and then simply XOR's the values together. For example if we encrypt for WEP for "hello" with the same IV value, we will get the same cipher value, and when we XOR them, we will get a result of zero. Using a frequency analysis method, we can quickly find the key.

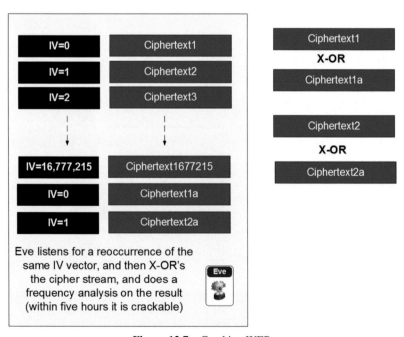

Figure 12.7   Cracking WEP.

If we EX-OR the two values together we get:

$$(MESS1 \hat{} [Passphase + IV]) \hat{} (MESS2 \hat{} [Passphase + IV])$$

When we EX-OR the same value together we nullify the EX-OR'ing, so we get:

$$(MESS1 \hat{} MESS2)$$

Thus the key (and IV value) has disappeared, and we do not have the key anymore. The intruder would then just run some frequency analysis, and reveal the original messages.

 📖 **Web link (Crack):** http://asecuritysite.com/encryption/rc4_wep
    Web link (RC4): http://asecuritysite.com/encryption/rc4

Overall WEP is weak from a number of viewpoints:

- Small value of IV. This meant that it repeated within a reasonable time, and the key could then be attacked.
- Construction of keys made it susceptible to the weak key attacks (FMS attack) [2].
- Lack of protection against message replay. There was no protection against cipher streams being played back over the network.
- Lack of message tampering identification. The method did not support the detection of message tampering.
- Directly used a master key. The method had no way of updating the keys.

12.4 Wi-fi Standards

Wi-fi systems have evolved through the development of IEEE standards. Figure 12.8 outlines the main wireless security standards used within IEEE 802.11, and where we have WEP, WPA or WPA-2 for encryption, and protocols such as EAPS, LEAP, EAP-TLS, and PEAP for authentication. Within IEEE 802.1x we have a mechanism which supports a range of authentication techniques and which can be used to access a network (Figure 12.9). This includes the usage of encryption keys, authentication and in defining a centralized policy.

Within a home type network, the authentication is likely to happen within a single device, but on an enterprise network, it is likely that users will migrate through a number of wireless access points, and where there is a

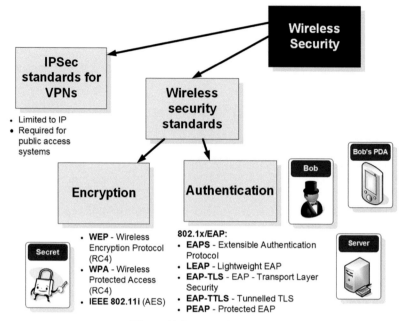

Figure 12.8   Wi-fi Standards.

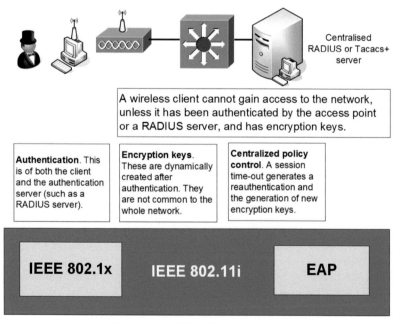

Figure 12.9   IEEE 802.1x and EAP.

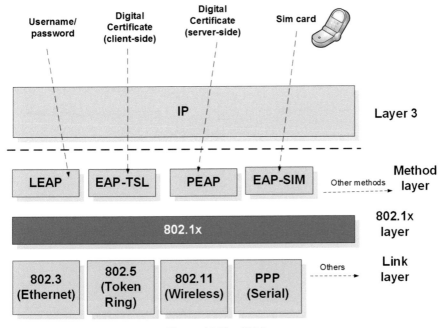

Figure 12.10 802.1x.

centralized authentication server (such as for RADIUS and Tacacs+). Within IEEE 802.1x we can thus support authentication mechanisms such as PEAP for server supplied digital certificates, and EAP-TLS for client provided digital certificates (Figure 12.10).

After WEP, there was a strong need to fix the problems, but to keep compatibility, thus WPA supported TKIP, and which increased the IV value to 48 bits (rather than 24 bits). Figure 12.11 outlines that WPA-2 enhances security by moving away from the RC4 stream encryption method towards AES. Overall, RC4 has been shown to suffer from several vulnerabilities, whereas AES is relatively free of vulnerabilities (apart from timing attacks).

12.5 WPA and WPA-2

WPA addressed the weaknesses of WEP, and without requiring significant hardware changes, and focused on two main methods: WPA-PSK and WPA Enterprise. WPA-PSK (Pre-Shared Key) is intended for a home environment and does not support an authentication server. The initial setup of the PSK is only defined within the initial AP (Access Point) session, and can be generated from an alphanumeric string. The key is then never passed over the

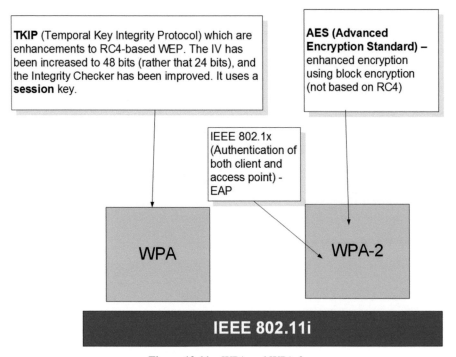

Figure 12.11   WPA and WPA-2.

wireless network, but the AP and client use mutual authentication to prove they know the key.

WPA Enterprise uses an authentication server with 802.1x and EAP, along with supporting TKIP (Temporal Key IntegrityProtocol) encryption, and which provides the session keys to be used. The 802.1x model uses a supplicant (the client), an authenticator (AP) and an authentication server (typically a RADIUS server).

With WPA Enterprise, as with WPA, no pre-shared key is used, and it also includes a MIC (Message Integrity Check). The MIC mainly guards against the bit flipping attacks identified within WEP [3]. To keep compatibility with WEP, TKIP still used the RC4 encryption method, but takes a hash of the IV and secret key in order to create the special key, and which is then fed into the RC4 process along with the IV value. Along with this the IV value was increased to 48-bits, in order to reduce the opportunity for the same IV values coming around within a reasonable time period. There is also a time-out for the usage of a key, so that the same key and IV value should never occur.

WPA-2 (IEEE 802.11i-2004) advanced the WPA standard, by keeping compatibility with WPA, but adding AES-CCMP (AES-Counter Mode CBC-MAC Protocol), which is a block encryption method. Again it supported two modes: Personal (with a pre-shared key) and Enterprise. Within PSK, the AP and client are preconfigured with a pass-phrase of up to 64 ASCII characters. Within Enterprise, it uses the 802.1x, EAP authentication framework, and secure key distribution. Within 802.11i there is improved support, though EAP-TLS, for a two-way authentication of the client (supplicant) and the AP (authenticator). The supplicant and authenticator then negotiate the cipher suite that will be used in the protection of the traffic (for pairwise unicast traffic); the group cipher suite (for multicast and broadcast traffic); and whether the clients wants PSK or 802.1x.

12.6 WPA-2 Handshaking

Within WPA-2 we aim to create an initial pairing between the client and the access point, and then to identify them without giving away the password which has been used. In the initial authentication, the client will either use pre-shared key (PSK), or use an EAP exchange through 802.1x (EAPOL).

The EAPOL exchange requires the usage of an authentication server. After this phase, a shared secret key is created and is known as the Pairwise Master Key (PMK). This uses PBKDF2-SHA1 as a hashing method, as the PBKDF2 part makes it difficult to crack the hash (as there are a number of rounds used to slow down the hashing process). Within PSK, the PSK is defined with the PMK, but within EAPOL, the PMK is derived from EAP parameters. Generally, EAPOL is more difficult to crack than using PSK. The PMK is generated from the PSK with:

```
PMK = PBKDF2(HMAC-SHA1, PSK, SSID, 4096, 256)
```

and where we use the SHA1 hashing function with HMAC as the message authentication code. In this case the PMK is generated from 4,096 iterations of the hashing method and creates a 256-bit PMK. A simple Python script to generate the PMK is:

```
from pbkdf2 import PBKDF2
ssid = 'home'
phrase = 'qwerty123'
print "SSID: "+ssid
print "Pass phrase: "+phrase
print "Pairwise Master Key: " + PBKDF2(phrase, ssid, 4096).
        read(32).encode("hex"))
```

and a sample run is:

```
SSID: home
Pass phrase: qwerty123
Pairwise Master Key: bbaf585c301dc4d4024523535f42baf04630f852e2
b01979ec0401edcdf0e9c8
```

Within WPA-2 we get the four-way handshake process, and which is illustrated in Figure 12.12. It is designed so that the access point and wireless client can prove that they know each other by showing that the know the PSK/PMK, without ever releasing the key. They must then encrypt messages to each other, and if they can decrypt them, they have successfully authenticated each other. In this way we can protect against a malicious spoof access point which is broadcasting a valid looking SSID.

Overall the PMK will last for the complete authentication of the devices and should be used sparingly. Thus, the four-way handshake uses a derived key known as the Pairwise Transient Key (PTK), and which is generated from: the PMK; a client nonce (ANonce); an access point nonce (SNonce); and the MAC addresses of the client and the access point (AP). These are then put into a pseudo-random function, and it generates a GTK (Group Temporal Key). The GTK is then used to decrypt multicast and broadcast traffic. The details of the handshake are thus:

- AP sends a nonce to the STA (ANonce). The client (STA) creates the PTK.
- Client sends a nonce (SNonce) to the AP and a Message Integrity Code (MIC), and which includes the authentication.
- The AP creates PTK and sends the GTK, along with a sequence number together and a MIC.
- The client sends a confirmation to the AP.

The Pairwise-Transient-Keys (PTK) process uses a combination of the PMK, AP MAC Address, Client MAC Address, AP Nonce, and the Client Nonce. This produces a 512-bit PTK, and comprises of five separate keys:

- Key Confirmation Key (KCK) – This used in the creation of the Message Integrity Code.
- Key Encryption Key (KEK) – This is used by the AP when using data encryption.
- Temporal Key (TK) – This is used for the encryption/decryption of unicast packets.

- MIC Authenticator Tx Key (MIC Tx) – This is used with TKIP setup for the unicast packets sent by APs.
- MIC Authenticator Rx Key (MIC Rx) – This is used with TKIP setup for the unicast packets sent by clients.

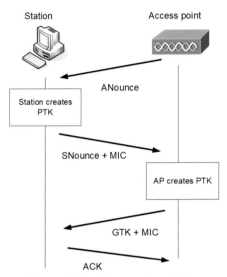

Figure 12.12   Four-way handshake.

12.7 Cracking WPA-2 PSK

On the second packet of the four-way handshake there is enough information for Eve to crack the Pairwise-Transient-Key (using a list of PSK passphrases). The MIC is calculated using HMAC-MD5, from the KCK Key within the PTK. We can then crack WPA-2 by capturing the four-way handshaking process, and cracking the hashed message, and matching it against a list of common passwords.

Only certain wireless chipsets support the capturing of the packets used in the four-way handshaking. We can test the chipset by running **airmon-ng**:

```
root@kali:~ airmon-ng
PHY Interface Driver  Chipset
null wlan0  ??????  Realtek Semiconductor Corp. RTL8188CUS 802.11n WLAN
Adapter
phy0 wlan1  ??????  Broadcom 43430
phy1 wlan2  rt2800usb Ralink Technology, Corp. RT2870/RT3070
```

In this case, the RT2870/RT3070 chipset supports the capture, so we can start the **airmon-ng** process to capture packets on wlan2:

```
root@kali:~ airmon-ng start wlan2
If airodump-ng, aireplay-ng or airtun-ng stops working after
a short period of time, you may want to run 'airmon-ng check kill'
  PID Name
  175 NetworkManager
  363 wpa_supplicant
  491 dhclient
  609 dhclient
PHY Interface Driver  Chipset
null wlan0  ??????  Realtek Semiconductor Corp. RTL8188CUS 802.11n WLAN
Adapter
phy0 wlan1  ??????  Broadcom 43430
phy1 wlan2  rt2800usb Ralink Technology, Corp. RT2870/RT3070

 (mac80211 monitor mode vif enabled for [phy1]wlan2 on [phy1]wlan2mon)
 (mac80211 station mode vif disabled for [phy1]wlan2)
```

We can see we are now monitoring on **wlan2mon**, and to test:

```
root@kali:~ airodump-ng wlan2mon
CH  5 ][ Elapsed: 1 min ][ 2017-02-19 12:10
 BSSID              PWR  Beacons     #Data, #/s  CH  MB    ENC  CIPHER AUTH ESSID

 XX:FC:AF:XX:XX:XX  -44      39        893   24   1  22e   WPA              ZZZZZ
 XX:A1:XX:XX:XX:XX  -49      34          0    0  11  54e   WPA2 CCMP   PSK  ZZZZZ
 XX:D3:XX:XX:XX:XX  -65      46          0    0   6  54e   WPA2 CCMP   PSK  ZZZZZ
 XX:21:XX:XX:XX:XX  -90       3          1    0  13  54e   WPA2 CCMP   PSK  ZZZZZ

 BSSID              STATION            PWR   Rate   Lost    Frames  Probe
 (not associated)   XX:XX:XX:XX:XX:XX  -44   0 - 1     0        10  ZZZZZ
 XX:XX:XX:XX:XX:XX  XX:XX:XX:XX:XX:XX   -1   0e- 0     0        46
 XX:XX:XX:XX:XX:XX  XX:XX:XX:2B:XX:XX  -20   0e- 0e    0       836
```

We can see a number that the first entry is using WPA (and transmitting on wireless channel 1), and the other three are using WPA-2 CCMP PSK. We can now grab the four-way handshake with:

```
airodump-ng -c 1 --bssid  XX:FC:AF:XX:XX:XX -w psk wlan2mon
```

This reads from the required BSSID on wireless channel 1, and will create a file with the prefix of "psk", and which has a .cap extension. The output when the four-way handshake is captured shows:

```
CH  1 ][ Elapsed: 18 s ][ 2017-02-19 21:38 ][ WPA handshake: XX:FC:AF:XX:XX:XX
BSSID              PWR RXQ  Beacons    #Data, #/s  CH  MB   ENC  CIPHER AUTH ESSID
XX:FC:AF:XX:XX:XX  -30  0     215       3077  90   1  54e  WPA2 CCMP   PSK  ZZZZZ
BSSID              STATION            PWR   Rate   Lost    Frames  Probe
XX:FC:AF:XX:XX:XX  XX:XX:XX:XX:XX:XX   3   -22    0e- 1e    0      2569
```

Next we create a list of passwords in password.lst, and analyse the cap files with:

```
aircrack-ng -w password.lst -b  XX:FC:AF:XX:XX:XX psk*.cap
```

This gives the results of (where some details have been removed):

```
                         Aircrack-ng 1.2 rc4
      [00:00:00] 2/1 keys tested (28.31 k/s)
      Time left: 0 seconds                              200.00%
                       KEY FOUND! [ ------- ]
      Master Key     : 5C ------------------ 0C
                       3A ------------------ 53
      Transient Key  : 6A ------------------ EB
                       4D ------------------ 72
                       7A ------------------ 87
                       80 ------------------ 21
      EAPOL HMAC     : C0 ------------------ 95
```

A full demo is given here:

📖**Web link (WPA-PSK Crack):** http://asecuritysite.com/subjects/chapter86

12.8 Stream Ciphers

While WEP contains many weaknesses, a properly defined stream cipher can be much faster than block ciphers, as they just have to create a key stream from an IV (also known as a nonce value) and a key. Google propose ChaCha20 – named as it has 20 rounds – as an alternative to AES, and actively use it within TLS connections. Currently it is three times faster than software-enabled AES, and is not sensitive to timing attacks. Overall it operates by creating a key stream which is then X-ORed with the plaintext, and has been standardised with RFC 7539.

Web link (ChaCha): http://asecuritysite.com/encryption/chacha

The encryption process involves taking 64 bytes as an input block, and then creates a key stream based on the key, a counter value, and a nonce value. This is then EXOR-ed with the current block:

```
chacha20_encrypt(key, counter, nonce, plaintext):
  for j = 0 upto floor(len(plaintext)/64)-1
     key_stream = chacha20_block(key, counter+j, nonce)
     block = plaintext[(j*64)..(j*64+63)]
     encrypted_message += block ^ key_stream
  end
  if ((len(plaintext) % 64) != 0)
     j = floor(len(plaintext)/64)
     key_stream = chacha20_block(key, counter+j, nonce)
     block = plaintext[(j*64)..len(plaintext)-1]
     encrypted_message += (block^key_stream)[0..len(plaintext)%64]
  end
  return encrypted_message
end
```

Overall the key stream generation has 20 rounds, and 80 quarter rounds. To generate the key stream, a basic quarter round involves an operation on four 32-bit values (a, b, c and d) with:

1. a += b; d ^= a; d $<<<$= 16;
2. c += d; b ^= c; b $<<<$= 12;
3. a += b; d ^= a; d $<<<$= 8;
4. c += d; b ^= c; b $<<<$= 7;

In this case "+" is an integer addition modulo 2^{32}, "^" defines a bitwise XOR operation, and "$<<<$ n" defines an *n*-bit left rotation. This is defined in JavaScript as:

```
function chacha20_round(x, a, b, c, d) {
    x[a] += x[b]; x[d] = rotl32(x[d] ^ x[a], 16);
    x[c] += x[d]; x[b] = rotl32(x[b] ^ x[c], 12);
    x[a] += x[b]; x[d] = rotl32(x[d] ^ x[a], 8);
    x[c] += x[d]; x[b] = rotl32(x[b] ^ x[c], 7);
}
```

The basic key stream generation is then achieved in JavaScript with:

```
function chacha20_keystream(ctx, dst, src, len) {
    var x = new Array(16);
    var buf = new Array(64);
    var i = 0, dpos = 0, spos = 0;
```

```
        while (len > 0) {
            for (i = 16; i--;) x[i] = ctx.input[i];
            for (i = 20; i > 0; i -= 2) {
                chacha20_round(x, 0, 4, 8, 12);
                chacha20_round(x, 1, 5, 9, 13);
                chacha20_round(x, 2, 6, 10, 14);
                chacha20_round(x, 3, 7, 11, 15);
                chacha20_round(x, 0, 5, 10, 15);
                chacha20_round(x, 1, 6, 11, 12);
                chacha20_round(x, 2, 7, 8, 13);
                chacha20_round(x, 3, 4, 9, 14);
            }
            for (i = 16; i--;) x[i] += ctx.input[i];
            for (i = 16; i--;) store32(buf, 4 * i, x[i]);

            ctx.input[12] += 1;
            if (!ctx.input[12]) {
                ctx.input[13] += 1;
            }
            if (len <= 64) {
                for (i = len; i--;) {
                    dst[i + dpos] = src[i + spos] ^ buf[i];
                }
                return;
            }
            for (i = 64; i--;) {
                dst[i + dpos] = src[i + spos] ^ buf[i];
            }
            len -= 64;
            spos += 64;
            dpos += 64;
        }
    }
```

Within RFC 7905, Google defines a number of cipher suites that can be used in TLS which support ChaCha stream cipher methods:

```
TLS_ECDHE_RSA_WITH_CHACHA20_POLY1305_SHA256   = {0xCC, 0xA8}
TLS_ECDHE_ECDSA_WITH_CHACHA20_POLY1305_SHA256 = {0xCC, 0xA9}
TLS_DHE_RSA_WITH_CHACHA20_POLY1305_SHA256     = {0xCC, 0xAA}

TLS_PSK_WITH_CHACHA20_POLY1305_SHA256         = {0xCC, 0xAB}
TLS_ECDHE_PSK_WITH_CHACHA20_POLY1305_SHA256   = {0xCC, 0xAC}
TLS_DHE_PSK_WITH_CHACHA20_POLY1305_SHA256     = {0xCC, 0xAD}
TLS_RSA_PSK_WITH_CHACHA20_POLY1305_SHA256     = {0xCC, 0xAE}
```

12.9 A5 Ciphers

The mobile phone network typically uses the A5/1 or A5/2 stream encryption method, but almost on its first day of operation it has been a target for crackers, and the source code to crack A5/2 was released within one month of being made public. While blocked ciphers like AES are robust, the A5/1 stream cipher has been shown to be weak against certain types of attack [4]. Now with the increase in cracking with GPUs (such as on NVIDIA cards), it is becoming an easy target.

For GSM cellular networks, there were two main choices for countries with their encryption: A5/1 and A5/2. A5/2 is intentionally weak, so that nation states can easy crack the cipher, but was cracked generally within a month of it being publicly released. A5/1 was meant to be stronger, but, as it has a relatively short key, it can be cracked with powerful computers.

With A5/1 we use three shift registers with feedback (Figure 12.13). With a stream cipher, we often generate an infinitely long key stream which is then EX-OR'ed with the data stream.

In the first shift register the bits at positions 18, 17, 16 and 13 are X-OR'ed together to produce a new bit which is put at the end. This pushes all the bits to move one position to the left. The last bit (the one at Position 18) will then

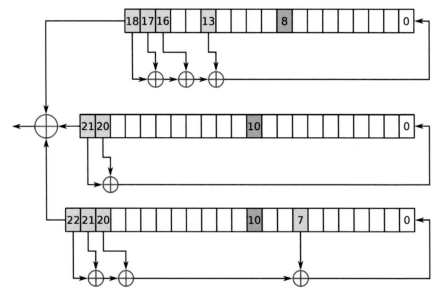

Figure 12.13   https://en.wikipedia.org/wiki/A5/1

be pushed off and X-OR with the output from the other shift registers. The shift registers are 19, 22 and 23 bits long, thus the key used is 64-bits long (19+22+23). In the diagram the bits 8, 10 and 10 are highlighted, and are the clocking bits. These are examined each cycle. In these are will either be more 1's than 0's or more 0's than 1's. The registers with the most popular bit value will advance their bits, and the other will not advance.

The algorithm itself was help secret while it was installed in 100 million mobile phones, and was part of the GSM (Global System and Mobile Communications) standard, and has since become a standard with 3G (even though it is seen as weak). The US and Europe adopted the strong A5/1 algorithm, but many selected the weaker one. It was finally made public in August 1999, and within a month the A5/2 method had been cracked. For A5/1 it is thought that the NSA can crack it (as they have the computing power to crack 64-bit keys).

The attack typically use known plaintext attacks, but new ones now allow the cipher stream to be decrypted in real-time. When first proposed, in 1982, it is thought that that the A5/1 key would be 128-bits long, but it finalised ended up with a 64-bit key (which can be cracked on expensive hardware using brute force). It is likely that government pressure forced the key to be much shorter. In fact it is thought that the UK wanted just 48 bits, while the West German government of the time pushed for larger key sizes (as they were worried that the East German government would be able to break their ciphers).

📖 **Web link (A5/1):** http://asecuritysite.com/encryption/a5

A stream cipher allows for automatic synchronisation, and where each bit can be decoded as it is received.

12.9.1 Practical Systems

SRLabs (Security Research Labs) focus on the capture of data from the mobile phone network and the cracking of A5/1 encryption. They take two encrypted known plaintext messages from the communications, and then use their Kraken utility to find the secret key for a 90% success rate with within seconds, using a set of rainbow tables (40 tables of 2TB in total size).

Gendrullis et al [5] cracked A5/1 within hours using a parallel processing system, but these days NVIDIA GPUs running in the Cloud could achieve the same result, but significantly less expensive.

12.9.2 A5/3

The A5/3 encryption system – known as KASUMI – the Japanese word for "mist" – is the upgrade to A5/1 and uses a block cipher. A5/1 is designed to be used for the GSM network, whereas A5/3 is for 3GPP, and is based on the MISTY1 cipher created (and patented) by Mitsubishi, but was modified to reduce processing restrictions on mobile devices.

In 2010, researchers (Orr Dunkelman, Nathan Keller and Adi Shamir) showed that it could be cracked with a related key attack [4]. If the standards organisation has stuck with the original MISTY1 specification, the attack would not have been possible.

A newer attack – a sandwich attack – on A5/3 has shown that it is possible to crack A5/3 using an "unoptimised PC", with 96 key bits recovered in a few minutes, and the rest of the 128-bit key within two hours.

☰ **Web link (A5/3):** http://asecuritysite.com/encryption/kasumi

12.10 Lab/Tutorial

The lab and tutorial related to this chapter is available on-line at:

http://asecuritysite.com/crypto12

References

[1] B. Zhu, "RC4 cipher." [Online]. Available: https://asecuritysite.com/encryption/rc4. [Accessed: 18-Jul-2017].

[2] S. Fluhrer, I. Mantin, and A. Shamir, "Weaknesses in the Key Scheduling Algorithm of RC4," *Sel. Areas Cryptogr.*, vol. 2259, pp. 1–24, 2001.

[3] N. Borisov, I. Goldberg, and D. Wagner, "Intercepting mobile communications: the insecurity of 802.11," *Proc. 7th Annu. Int. Conf. Mob. Comput. Netw.*, pp. 180–189, 2001.

[4] O. Dunkelman, N. Keller, A. Shamir (2010-01-10). "A Practical-Time Attack on the A5/3 Cryptosystem Used in Third Generation GSM Telephony".

[5] T. Gendrullis, M. Novotný, and A. Rupp. "A real-world attack breaking A5/1 within hours." Cryptographic Hardware and Embedded Systems– CHES 2008. Springer Berlin Heidelberg, 266–282, 2008.

Appendix A

Table A.1 ASCII characters

Char	Decimal	Binary	Hex	Oct	Html
null	0	00000000	0	0	&#0;
start of header	1	00000001	1	1	&#1;
start of text	2	00000010	2	2	&#2;
end of text	3	00000011	3	3	&#3;
end of transmission	4	00000100	4	4	&#4;
enquiry	5	00000101	5	5	&#5;
acknowledge	6	00000110	6	6	&#6;
bell	7	00000111	7	7	&#7;
backspace	8	00001000	8	10	&#8;
horizontal tab	9	00001001	9	11		

line feed	10	00001010	a	12	

vertical tab	11	00001011	b	13	&#11;
form feed	12	00001100	c	14	
enter/carriage return	13	00001101	d	15	
shift out	14	00001110	e	16	&#14;
shift in	15	00001111	f	17	&#15;
data link escape	16	00010000	10	20	&#16;
device control 1	17	00010001	11	21	&#17;
device control 2	18	00010010	12	22	&#18;
device control 3	19	00010011	13	23	&#19;
device control 4	20	00010100	14	24	&#20;
negative acknowledge	21	00010101	15	25	&#21;
synchronize	22	00010110	16	26	&#22;
end of trans. block	23	00010111	17	27	&#23;
cancel	24	00011000	18	30	&#24;
end of medium	25	00011001	19	31	&#25;
substitute	26	00011010	1a	32	&#26;
escape	27	00011011	1b	33	&#27;
file separator	28	00011100	1c	34	&#28;
group separator	29	00011101	1d	35	&#29;
record separator	30	00011110	1e	36	&#30;

(*Continued*)

Table A.1 Continued

Char	Decimal	Binary	Hex	Oct	Html
unit separator	31	00011111	1f	37	&#31;
space	32	00100000	20	40	&#32;
!	33	00100001	21	41	&#33;
"	34	00100010	22	42	&#34;
#	35	00100011	23	43	&#35;
$	36	00100100	24	44	&#36;
%	37	00100101	25	45	&#37;
&	38	00100110	26	46	&#38;
'	39	00100111	27	47	&#39;
(	40	00101000	28	50	&#40;
)	41	00101001	29	51	&#41;
*	42	00101010	2a	52	&#42;
+	43	00101011	2b	53	&#43;
,	44	00101100	2c	54	&#44;
-	45	00101101	2d	55	&#45;
.	46	00101110	2e	56	&#46;
/	47	00101111	2f	57	&#47;
0	48	00110000	30	60	&#48;
1	49	00110001	31	61	&#49;
2	50	00110010	32	62	&#50;
3	51	00110011	33	63	&#51;
4	52	00110100	34	64	&#52;
5	53	00110101	35	65	&#53;
6	54	00110110	36	66	&#54;
7	55	00110111	37	67	&#55;
8	56	00111000	38	70	&#56;
9	57	00111001	39	71	&#57;
:	58	00111010	3a	72	&#58;
;	59	00111011	3b	73	&#59;
<	60	00111100	3c	74	&#60;
=	61	00111101	3d	75	&#61;
>	62	00111110	3e	76	&#62;
?	63	00111111	3f	77	&#63;
@	64	01000000	40	100	&#64;
A	65	01000001	41	101	&#65;
B	66	01000010	42	102	&#66;
C	67	01000011	43	103	&#67;
D	68	01000100	44	104	&#68;
E	69	01000101	45	105	&#69;
F	70	01000110	46	106	&#70;
G	71	01000111	47	107	&#71;
H	72	01001000	48	110	&#72;

I	73	01001001	49	111	&#73;
J	74	01001010	4a	112	&#74;
K	75	01001011	4b	113	&#75;
L	76	01001100	4c	114	&#76;
M	77	01001101	4d	115	&#77;
N	78	01001110	4e	116	&#78;
O	79	01001111	4f	117	&#79;
P	80	01010000	50	120	&#80;
Q	81	01010001	51	121	&#81;
R	82	01010010	52	122	&#82;
S	83	01010011	53	123	&#83;
T	84	01010100	54	124	&#84;
U	85	01010101	55	125	&#85;
V	86	01010110	56	126	&#86;
W	87	01010111	57	127	&#87;
X	88	01011000	58	130	&#88;
Y	89	01011001	59	131	&#89;
Z	90	01011010	5a	132	&#90;
[	91	01011011	5b	133	&#91;
\	92	01011100	5c	134	&#92;
]	93	01011101	5d	135	&#93;
^	94	01011110	5e	136	&#94;
_	95	01011111	5f	137	&#95;
`	96	01100000	60	140	&#96;
a	97	01100001	61	141	&#97;
b	98	01100010	62	142	&#98;
c	99	01100011	63	143	&#99;
d	100	01100100	64	144	&#100;
e	101	01100101	65	145	&#101;
f	102	01100110	66	146	&#102;
g	103	01100111	67	147	&#103;
h	104	01101000	68	150	&#104;
i	105	01101001	69	151	&#105;
j	106	01101010	6a	152	&#106;
k	107	01101011	6b	153	&#107;
l	108	01101100	6c	154	&#108;
m	109	01101101	6d	155	&#109;
n	110	01101110	6e	156	&#110;
o	111	01101111	6f	157	&#111;
p	112	01110000	70	160	&#112;
q	113	01110001	71	161	&#113;
r	114	01110010	72	162	&#114;
s	115	01110011	73	163	&#115;

(Continued)

Table A.1   Continued

Char	Decimal	Binary	Hex	Oct	Html
t	116	01110100	74	164	t
u	117	01110101	75	165	u
v	118	01110110	76	166	v
w	119	01110111	77	167	w
x	120	01111000	78	170	x
y	121	01111001	79	171	y
z	122	01111010	7a	172	z
{	123	01111011	7b	173	{
|	124	01111100	7c	174	|
}	125	01111101	7d	175	}
~	126	01111110	7e	176	~
delete	127	01111111	7f	177	&#127;

Table A.2   Base-64 conversion

Char.	Dec.	Hex.	Char.	Dec.	Hex.	Char.	Dec.	Hex.
A	0	00	W	22	16	s	44	2C
B	1	01	X	23	17	t	45	2D
C	2	02	Y	24	18	u	46	2E
D	3	03	Z	25	19	v	47	2F
E	4	04	a	26	1A	w	48	30
F	5	05	b	27	1B	x	49	31
G	6	06	c	28	1C	y	50	32
H	7	07	d	29	1D	z	51	33
I	8	08	e	30	1E	0	52	34
J	9	09	f	31	1F	1	53	35
K	10	0A	g	32	20	2	54	36
L	11	0B	h	33	21	3	55	37
M	12	0C	i	34	22	4	56	38
N	13	0D	j	35	23	5	57	39
O	14	0E	k	36	24	6	58	3A
P	15	0F	l	37	25	7	59	3B
Q	16	10	m	38	26	8	60	3C
R	17	11	n	39	27	9	61	3D
S	18	12	o	40	28	+	62	3E
T	19	13	p	41	29	/	63	3F
U	20	14	q	42	2A			
V	21	15	r	43	2B	=	(pad)	(pad)

Index

About the Author

Bill Buchanan OBE is a Professor in the School of Computing at Edinburgh Napier University, and a Fellow of the BCS and the IET. He was appointed an Officer of the Order of the British Empire (OBE) in the 2017 Birthday Honours for services to cyber security.

Currently he leads the Centre for Distributed Computing, Networks, and Security at Edinburgh Napier University and The Cyber Academy (http://thecyberacademy.org). His main research focus is around information sharing, IoT, e-Health, threat analysis, cryptography, and triage within digital forensics. This has led to several World-wide patents, and in three highly successful spin-out companies: Zonefox; Symphonic Software; and Cyan Forensics.

Bill regularly appears on TV and radio related to computer security, and has given expert evidence to both the Scottish and UK Parliaments. He has been named as one of the Top 100 people for Technology in Scotland for in every year from 2012 onwards. Bill was also included in the FutureScot "Top 50 Scottish Tech People Who Are Changing The World". Recently his work on Secret Shares received "Innovation of the Year" at the Scottish Knowledge Exchange Awards, for a research project which involves splitting data into secret shares, and can then be distributed across a public Cloud-based infrastructure. He was included in the JISC Top 50 Higher Education Social Media Influencers.

He has published over 250 research papers and 27 academic books. His current work focuses on the secret share methods, and especially in how documents can be stored securely in public cloud based systems. The current cryptography work around secret shares has won several awards, and is the basis for new funded work. This is further enhanced with work around sticky policies and identity based encryption, and which aims to integrate access rights on document within public cloud systems. This includes the storage of high risk documents, such as health care records and financial information. Along with this he has new research work which integrates machine learning into insider threat detection, and within side channel analysis on embedded systems.

Bill's work around information sharing models for trust and governance, which was funded through two EPSRC grants, is now showing significant impact with the integration of over 7,000 health and social care entities within London, and which will be able to share information in a secure and trusted way.